**W9-CSP-115**

ACS SYMPOSIUM SERIES **654**

# Environmental Biomonitoring

## Exposure Assessment and Specimen Banking

**K. S. Subramanian,** EDITOR
*Health Canada*

**G. V. Iyengar,** EDITOR
*Biomineral Sciences International Inc.*

Developed from a symposium sponsored by the
International Chemical Congress of Pacific Basin Societies
and the ACS Division of Environmental Chemistry, Inc., at the
1995 International Chemical Congress of Pacific Basin Societies

American Chemical Society, Washington, DC

**Library of Congress Cataloging-in-Publication Data**

International Chemical Congress of Pacific Basin Societies (1995:
Honolulu, Hawaii)

Environmental biomonitoring: exposure assessment and specimen
banking / K. S. Subramanian, G. V. Iyengar, editors.

p.    cm.—(ACS symposium series, ISSN 0097–6156; 654)

"Developed from a symposium sponsored by the International
Chemical Congress of Pacific Basin Societies and the ACS Division of
Environmental Chemistry, Inc., at the 1995 International Chemical
Congress of Pacific Basin Societies, Honolulu, Hawaii, December 17–22,
1995."

Includes bibliographical references and indexes.

ISBN 0–8412–3477–9

1. Indicators (Biology)—Congresses.   2. Environmental
monitoring—Congresses.

I. Subramanian, K. S., 1944–  . II. Iyengar, G. V.  (Govindaraja V.)
III. Title. IV. Series.

QH541.15.I5I525   1995
363.73'63—dc21                                              96–48151
                                                                 CIP

This book is printed on acid-free, recycled paper.

PRINTED IN THE UNITED STATES OF AMERICA

# Foreword

THE ACS SYMPOSIUM SERIES was first published in 1974 to provide a mechanism for publishing symposia quickly in book form. The purpose of this series is to publish comprehensive books developed from symposia, which are usually "snapshots in time" of the current research being done on a topic, plus some review material on the topic. For this reason, it is necessary that the papers be published as quickly as possible.

Before a symposium-based book is put under contract, the proposed table of contents is reviewed for appropriateness to the topic and for comprehensiveness of the collection. Some papers are excluded at this point, and others are added to round out the scope of the volume. In addition, a draft of each paper is peer-reviewed prior to final acceptance or rejection. This anonymous review process is supervised by the organizer(s) of the symposium, who become the editor(s) of the book. The authors then revise their papers according to the recommendations of both the reviewers and the editors, prepare camera-ready copy, and submit the final papers to the editors, who check that all necessary revisions have been made.

As a rule, only original research papers and original review papers are included in the volumes. Verbatim reproductions of previously published papers are not accepted.

ACS BOOKS DEPARTMENT

# Contents

Preface.................................................................................................... ix

1. Environmental Biomonitoring and Specimen Banking:
   Bioanalytical Perspectives...................................................... 1
   G. V. Iyengar and K. S. Subramanian

MONITORING

2. General Aspects of Heavy Metal Monitoring by Plants
   and Animals.......................................................................... 19
   Bernd Markert, Jörg Oehlmann, and Mechthild Roth

3. Secondary Ion Mass Spectroscopy in the Analysis of the Trace
   Metal Distribution in the Annual Growth Rings of Trees............. 30
   R. R. Martin, E. Furimsky, Jinesh Jain, and W. M. Skinner

4. Human Hair and Lichen: Activities Involving the Use
   of Biomonitors in International Atomic Energy Agency
   Programs on Health-Related Environmental Studies..................... 42
   Susan F. Stone

5. The Silver Content of the Ascidian *Pyura stolonifera* as an
   Indicator of Sewage Pollution of the Metropolitan Beaches
   of Sydney, Australia .............................................................. 49
   T. M. Florence, J. L. Stauber, L. S. Dale, and L. Jones

6. The Moose (*Alces alces* L.), A Fast and Sensitive Monitor
   of Environmental Changes ..................................................... 57
   A. Frank and V. Galgan

7. Use of Manganese Concentration in Bivalves as an Indicator
   of Water Pollution in Japanese Brackish Lakes ......................... 65
   Yoshiko Yano

8. Mutagenesis and Acute Toxicity Studies on Saliva-Leached
   Components of Chewing Tobacco and Simulated Urine Using
   Bioluminescent Bacteria.......................................................... 77
   Shane S. Que Hee

9.  Distribution of Polychlorinated Biphenyl Congeners in Bear
    Lake Sediment.................................................................... 83
    Min Qi, S. Bierenga, and J. Carson

EXPOSURE ASSESSMENT

10. Challenges to Health from the Major Environmental Chemical
    Contaminants in the Saint Lawrence River.............................. 96
    S. Raman and K. S. Subramanian

11. Body Burden Concentrations in Humans in Response to Low
    Environmental Exposure to Trace Elements ........................... 105
    Jane L. Valentine

12. Placenta: Elemental Composition, and Potential as a Biomarker
    for Monitoring Pollutants to Assess Environmental
    Exposure............................................................................ 115
    K. S. Subramanian and G. V. Iyengar

13. Cadmium, Lead, Mercury, Nickel, and Cesium-137
    Concentrations in Blood, Urine, or Placenta from Mothers
    and Newborns Living in Arctic Areas of Russia and Norway ....... 135
    J. O. Odland, N. Romanova, G. Sand, Y. Thomassen,
    B. Salbu, Eiliv Lund, and E. Nieboer

14. Total Mercury and Methylmercury Levels in Scalp Hair
    and Blood of Pregnant Women Residents of Fishing Villages
    in the Eighth Region of Chile .............................................. 151
    Carlos G. Bruhn, Aldo A. Rodriguez, Carlos A. Barrios,
    Victor H. Jaramillo, José Becerra, Nuri T. Gras,
    Ernesto Nuñez, and Olga C. Reyes

15. Personal Exposure to Indoor Nitrogen Dioxide ....................... 178
    Toshihiro Kawamoto, Koji Matsuno, Keiichi Arashidani,
    and Yasushi Kodama

BIOMARKERS

16. Biomarkers of Inherited and Acquired Susceptibility to Toxic
    Substances ....................................................................... 184
    E. Nieboer

17. Biological Monitoring and Genetic Polymorphism..................... 190
    Toshihiro Kawamoto and Yasushi Kodama

18. Biomarkers Used for Assessment of Cancer Risk in Populations Living in Areas of High Potential Carcinogenic Hazard (Near Communal and Chemical Waste Dumping Sites) ............... 195
J. A. Indulski and W. Lutz

19. Molecular Epidemiology in Cancer Risk Assessment ..................... 206
J. A. Indulski and W. Lutz

SPECIMEN BANKING

20. Environmental Specimen Banking and Analytical Chemistry: An Overview ..................................................................................... 220
G. V. Iyengar and K. S. Subramanian

21. Environmental Specimen Banking: Contributions to Quality Management of Environmental Measurements ............................. 246
Rolf Zeisler

22. Ethical and Legal Aspects of Human Tissue Banking ................... 254
E.-H. W. Kluge and K. S. Subramanian

23. Establishing Baseline Levels of Elements in Marine Mammals Through Analysis of Banked Liver Tissues .................................... 261
Paul R. Becker, Elizabeth A. Mackey, Rabia Demiralp, Barbara J. Koster, and Stephen A. Wise

24. A Pilot Biological Environmental Specimen Bank in China ......... 271
P. Q. Zhang, C. F. Chai, X. L. Lu, Q. F. Qian, and W. Y. Feng

INDEXES

Author Index ............................................................................................... 279

Affiliation Index ........................................................................................ 279

Subject Index ............................................................................................. 280

# Preface

ENVIRONMENTAL BIOMONITORING has become an important tool for identifying and instituting remedies to the threat of contamination of the biosphere. The available information suggests that current pollutant levels have become stressful to a large number of sensitive flora and fauna, as well as to human beings. The deleterious impacts, both real and perceived, of natural and man-made pollutants have created the need to monitor their release and subsequent movement through the physical and biological components of the environment. It has become necessary to institute environmental surveillance systems to monitor noxious substances for early identification of, and response to, health and environmental problems caused by xenobiotic toxicants. In addition, increasing attention is being focused on the monitoring and assessment of human exposure to environmental contaminants throughout the world.

Environmental biomonitoring involves several aspects of environmental health, including the determination and identification of low levels of pollutants and their metabolites, exposure assessment, and measurement of biological markers of exposure. Another important facet of environmental biomonitoring is the emerging field of environmental specimen banking. A specimen bank acts as a bridge connecting real-time monitoring with future trend-monitoring activities.

This book is based on the symposium entitled "Environmental Biomonitoring and Specimen Banking", held at the 1995 International Chemical Congress of Pacific Basin Societies (Pacifichem '95) in Honolulu, Hawaii, on December 17–22, 1995. The purpose of the book is to assimilate current developments in the diverse field of environmental and human health. In particular, we have highlighted recent developments in monitoring strategies, exposure assessment, bioindicators (or biomarkers), and specimen banking.

The term "environmental biomonitoring", as used in this book, refers to the measurement and identification of pollutants and their metabolites in environmental and biological media. The term "media" refers to the environments in which the pollutants are present. The term "biological monitoring" is used when the concentration of a pollutant or its metabolite is monitored in a human indicator medium such as blood, urine, hair, saliva, or the placenta. Although the emphasis of the book is on monitoring the biosphere in the context of public health, monitoring efforts related to occupational exposure are also touched upon because of their methodological relevance. A group of internationally renowned research-ers was assembled to address these issues and to facilitate an assessment of the state-of-the-art of the field of environmental biomonitoring and specimen banking.

The book is organized into four sections: monitoring, exposure assessment, bioindicators, and specimen banking. Chapter 1 provides a comprehensive overview of the subject of environmental biomonitoring to set the tone of the book. Other chapters in the book have been developed from only 23 of the 60 papers presented at the symposium in order to avoid an incoherent collection of papers and to synchronize the flow of information from one chapter to the next. The mix of review papers, original research manuscripts, and reports of new work presented here best reflect the current state of the subject. This book represents a modest beginning of our efforts to achieve a comprehensive perspective of the field of environmental biomonitoring.

The contents of the book should appeal to scientists in the fields of analytical chemistry, clinical chemistry, ecology, medicine, and environ-mental science and health. We hope that the readers will find this book interesting and useful in their many and varied quests into the realm of environmental biomonitoring and specimen banking.

K. S. SUBRAMANIAN
Environmental Health Directorate
Postal Locator 0800B3
Health Canada
Tunney's Pasture, Ottawa
Ontario K1A 0L2, Canada

G. V. IYENGAR
Biomineral Sciences International Inc.
6202 Maiden Lane
Bethesda, MD 20817

August 17, 1996

# Chapter 1

# Environmental Biomonitoring and Specimen Banking
## Bioanalytical Perspectives

G. V. Iyengar[1] and K. S. Subramanian[2]

[1]Biomineral Sciences International Inc., 6202 Maiden Lane, Bethesda, MD 20817
[2]Environmental Health Directorate, Postal Locator 0800B3, Health Canada, Tunney's Pasture, Ottawa, Ontario K1A 0L2, Canada

The health impact of toxic substances in the environment is a widely studied subject. The primary goal here is to develop reliable basis for risk assessment. In this context, recent advances in analytical chemistry have made it practical to measure very small quantities of chemicals in tissues, body-fluids, food and air. Consequently, it has been possible to quantify contaminants such as pesticides, other organic and inorganic constituents, metals, and solvent and drug residues in our environment. However, much emphasis during this developmental phase is placed on instrumentation, while factors affecting the "total quality" of a bioenvironmental investigation tend to be overlooked. Unless other tools for risk assessment such as sound biomonitoring programs (real-time and long-term) based on environmental epidemiological concepts are properly used in generating data, interpreting the significance of these findings will continue to be a challenge. Therefore, dedicated efforts are needed for (i) consolidating the biologic basis for selection of specimens for environmental surveillance, (ii) developing strategies for long-term preservation of sampled materials, (iii) accomplishing harmonization of analytical measurements, and (iv) recognizing the multidisciplinary expertise, for understanding the pollution trends and to provide a reliable basis for health risk assessment in exposure situations.

The concern about toxic substances in the biosphere and the need for developing strategies to eliminate or minimize the health hazards caused by these pollutants is well recognized. This is particularly relevant in the context of the estimated 60,000 or so chemicals of industrial consequence, of which only a handful are currently being examined for their environmental impact *(1)*. For example, pesticides (chlorinated hydrocarbons and their degradation products) are ubiquitous in ecosystems. There is evidence in Finland that although the use of DDT and its derivatives ended in the early 1970s, these compounds are still encountered in trace amounts in fish *(2)*. Thus, contaminants such as pesticides, polycyclic aromatic hydrocarbons, halogenated hydrocarbons, heavy metals and other trace elements, enter the food chain through various sources (e.g. anthropogenic) in the biosphere. The organic pesticides are generally highly stable, lipophilic substances that tend to accumulate in adipose tissues over time, culminating in environmental toxicity and mutagenic changes.

The above mentioned events call for stringent measures for environmental surveillance, but the scenario is not easy to comprehend; there are too many chemicals and practically endless routes of exposure requiring comprehensive schemes for monitoring. Moreover, the state-of-the-art of analytical competence is not the same for all classes of pollutants. These problems warrant skillfully conceived multidisciplinary approaches including identification of appropriate specimens to assess human exposures to xenobiotic toxicants in the environment. However, any monitoring program should be designed to meet both real time pursuit (tracking short-term trends) and provisions for retrospective research (identifying long-term trends) as analytical capabilities improve. The ability to reevaluate data retrospectively extends the scope of data accumulation, and provides a sound basis for evaluation of the biological effects of various pollutants.

Therefore, dedicated efforts are needed for (i) consolidating the biologic basis for selection of appropriate specimens for environmental surveillance, (ii) developing strategies for long-term preservation of sampled materials, (iii) accomplishing harmonization of analytical measurements, and (iv) recognizing the multidiscilpinary expertise *(3)* to evaluate pollution trends. The benefits are: improvements in problem oriented analytical approaches, establishment of reliable baseline values for numerous chemical constituents in selected environmental media, and a proven experimental tool for assessment of the environmental health criteria. These aspects will be addressed in the following sections.

## Chemicals in the Environment

Inorganic constituents: From the environmental pollution point of view, various anthropogenic activities, especially the burning of fossil fuels required in various industrial operations, are a   major source of several toxic trace elements, including Se *(4)*. Among these, As, Cd, Hg, and Pb deserve special mention since they have profound effects on both domestic animals and human beings. Froslie and co-workers *(5)* have shown that heavy-metal contamination of natural surface soils from atmospheric deposition occurs even at very long distances from the major point source.

Several examples of human exposure to As, Cd, Hg, and Pb have been recorded in the scientific literature. It is known, for instance, that sections of human populations from the south and far east Asia such as Japan, Taiwan and the Philippines have high levels of As in their blood, milk or hair *(6)*. The source of As is linked to the consumption of fish which is a significant component of an average diet. The As content of the soil also plays a role.

Among the innumerable sources of industrial pollution, gasoline is the major source of environmental Pb in those countries where unleaded petrol is not mandatory. In many urban populations blood-Pb levels have been shown to be well over 200 ng/mL *(6)*. This figure may be compared with the moderate or low levels of 90 ng/mL in Japan and Sweden, or even lower levels of 30 ng/mL observed in remote parts of Nepal and the interior regions of Venezuela *(6)*. Similarly, urban mothers have been shown to secrete more Pb into their milk than their corresponding rural controls *(7)*. A common problem is faced by young children who ingest several mg of soil (along with paint scrapings and house dust) per day resulting in significant extra-dietary exposure to not only Pb, but also Al, Si, Ti and V. This is a serious bioenvironmental problem faced in pediatric fields, since it is not easy to precisely estimate a child's daily intake of soil and other non-food sources. The ever widening understanding of Pb toxicity indicates that Pb intake by human populations has increased about one hundred-fold above the natural level based on analysis of samples of prehistoric human skeletons *(8)*.

It has also been reported that subjects residing in the vicinity of Cu smelters near Cluj-Napooa in Rumania excreted higher levels of As in their urine and hair than did controls. A five-fold increase in urinary As (6.4 vs 31 ng/mL) and an increase of up to 32 times in hair content (0.25 vs 8 ng/mL) was demonstrated *(9)*. Analysis of the body burden of As clearly indicated excessive exposure to this element resulting from Cu-ore smelting activities. Environmental air particulate analyses have shown a correlation between As and Cu near smelters.

Cadmium is a toxic industrial environmental pollutant. Tobacco smoking is a significant contributor to the high levels of Cd found in smokers. Blood levels of Cd in smokers are higher than in those in nonsmokers by a factor of 3 to 12 *(6)*. Excess exposure to Cd can cause renal tubular damage and obstructive lung disease. The amount of Cd transferred from soil to plant to animal or human can be of concern, especially where sewage sludge is used to fertilize the soils *(10,11)*. The degree of risk depends on several factors: type of food affected (e.g. commonly consumed foods and foods consumed in large quantities) by the application of the sludge, soil pH and the amount of Cd entering the soil.

Sea-food is a major source of Hg intake. For example, it has been shown that Alaskan Eskimo mothers have rather high concentrations of this element in placenta, hair, blood and milk *(12)*. It is also known that these concentrations of Hg are directly proportional to the quantity of seal meat consumed. Subjects who consumed seal meat daily had the highest levels of Hg *(12)*. This brings into focus considerations of infant nutrition since breast-fed babies of these mothers are potentially subjected to toxic doses of Hg.

Mineral oil contains about 0.2 $\mu$g/g of Se whereas coal contains as much as 3 $\mu$g/g, and in some exceptional cases levels can be much higher *(4,13,14)*.

Coal ash (especially the fine fraction of fly ash), which contains up to several hund.ed $\mu$g/g of Se is carried over long distances and distributed over soil surfaces. The availability of this Se to plants depends upon soil characteristics such as degree of alkalinity. A combination of soil and environmental pollution factors has been reported to be the cause of Se toxicity in the neighboring counties of selenium deficient areas in Northeastern China (15). A mortality rate of about 50 percent was recorded in some villages. The source of this environmental Se was found to be a stony coal that contained an average of 300 $\mu$g/g of Se. The Se entered the soil and was readily taken up by plants (vegetables and maize) because of the traditional use of lime as fertilizer in that region, leading to an outbreak of human selenosis.

**Organic Pollutants:** The determination of halogenated aromatic and aliphatic compounds, aromatic amine compounds and other miscellaneous class of chemicals (e.g. benzene) and specific metallo-organic compounds such as methyl mercury and organotins in environmental, biological and food related matrices have been the focal point of several biomonitoring investigations. This is particularly relevant to marine monitoring programs because of increasing concerns about the effects of marine pollution on the health of marine mammals.

### Geochemical: The High Altitude Context

Many interesting associations of I and Se with environment, high altitude and human and animal health are beginning to unfold (16). Similarly, it may be postulated that if quantities of Pb in high-altitude environments are appreciable, or if Pb levels in foods consumed by these populations are high, Pb accumulation becomes a potential danger since more than 95 per cent of the absorbed fraction of Pb resides in blood, namely in erythrocytes (17). In view of the elevated haematocrit in high altitude populations, the potential dangers are obvious. The erythrocyte-turnover cycle is 120 days, which means that accumulated Pb will be partially released and redeposited in bone in which lead has a long biological half-life. Therefore, for high altitude populations, it is imperative that environmental Pb levels be kept low.

There are some examples of altitude-related differences in Se levels from the animal world too. It has recently been shown in the USA that in rock squirrels captured at different elevations, ranging from grass-land to pinyon juniper ecosystems (elevations of 1600 to 2400 meters), the Se content of kidney and liver was highest in grass land animals and decreased with increasing elevation gradient (18). As grass-land soil is alkaline, it would appear that the mobilization and uptake of Se by plants, and thereby also in animals, was the key contributory factor.

### Bioanalytical Approaches

**Harmonization of Environmental Measurements:** The present day analytical techniques are capable of detecting extremely small quantities of chemical substances in the biosphere and have eached the potential to serve as routine tools for ultra-trace measurements. Undoubtedly, the technological progress has greatly contributed to the advances in metrology (the science of measurements) of trace analysis of bioenvironmental systems. Similarly, analytical quality assurance brought about by the development and use of a variety of Standard Reference Materials (SRM) have further enhanced the metrological excellence. However, this capability is somewhat neutralized because of failure to observe proper procedures, analyzing inadequately prepared samples and generalizing the findings. Hence, in spite of the technological supremacy, inconsistencies in the quantitative data of environmental measurements are still prevailing. This is an indication that high detection capability and sensitivity of analytical techniques alone is not the solution to the problem of reliable data generation in bioenvironmental systems. Thus, there is much concern in the minds of the life sciences researchers that improved capability for quantification is not harmonized with appropriate analytical and biological perceptions. First of all, there is a lack of appreciation for multidisciplinary approaches, leading to poor planning of bioenvironmental studies.

"Bio"sources of analytical errors: Two major sources of errors need to be considered in this context: *conceptual* (arising from limited understanding of the "bio" dimension of the specimens such as biological variations that are identifiable but not always quantifiable, wrong statistical approach to data processing, wrong basis for expressing the data, and  incompatible data interpretation); and *analytical* (stemming from sampling, sample preparation, calibration, matrix effects, procedural/instrumental, etc.). Of particular concern is the fact that a significant portion of the existing analytical information is derived from analyses of uncontrolled and often "spot" samples (e.g. hair, urine, soil and foods) with deficient sampling plans and inadequate AQA procedures. Yet, major decisions of public health concern are being made that are dependent on the analytical information obtained by monitoring and other biomedical studies. Apart from the analytical reliability expected of such studies, measures must be taken to ensure that the experimental design for obtaining such information is conceptually relevant *(19)*.

## Analytical Quality Assurance

Bioenvironmental investigations require provisions for a "total" Analytical Quality Assurance (AQA) to yield meaningful results. These are measures that encompass all stages of an investigation such as experimental design, collection of analytically and biologically valid specimens, analytical measurement processes and proper evaluation of analytical data, including data interpretation. Further, one of the critical decisions that the analyst needs to make is the degree of quality (quality standard), or tolerance limits, required for the purpose of the investigation for which the analyses are being made. If the tolerance limits are set narrower than the investigation really requires or not feasible under practical laboratory conditions, it can cause unnecessary expense and loss of time.

In initiating an AQA program use of two or more independent analytical methods to verify the accuracy of an analytical finding is a crucial requirement. This is reflected through the development and certification of a wide variety of reference materials for many inorganic constituents *(20)*. Therefore, reasonably well-founded baseline data for trace elements in biomaterials have been generated by few selected laboratories around the world.

On the other hand, in the area of organic analysis procedures for AQA of many constituents are still evolving. Although there is ample awareness among the researchers concerning the role of AQA in generating reliable results, this perception has not translated into action other than the recognition of inconsistencies stemming from intercomparison trials. This is not swiftly followed up by identification of the sources of discrepancy, initiation of the remedy, and subsequent AQA exercises. One obvious reason for this slow progress is the paucity of funding for basic research, thus shifting a major fraction of the developmental work to a few institutions such as those dealing with Reference Materials Programs. In addition, the problem of lack of certified reference materials (CRM) for organic constituents has been existing for a long time. This is of course a reflection of genuine methodological and technical hurdles; by far the most challenging step being preparation of a natural matrix and extended preservation of the compositional integrity of the sampled material.

**CRMs for Organic Constituents:** NIST has developed several CRMs ranging from simple solutions for calibration of analytical instruments to complex natural matrix material, some fortified, for suitable for validation of methods adopted for biomonitoring and related programs. For example, the mussel tissue (Mytilus edulis; NIST SRM 1974) is certified for anthracene, benzo(b)fluoranthene, benzo(ghi)perylene, benzo(a)pyrene, fluoranthene, pyrene, perylene and phenanthrene. It also contains information values for a number of other PAHs, PCBs and chlorinated pesticides. Similarly, CRMs of organics in marine sediments (NIST SRM 1941) and cod liver oil (NIST SRM 1588), and PCBs in human serum (NIST SRM 1589) are available. Recently, frozen whale blubber (NIST SRM 1945) has been developed at the NIST for use as control material *(21)* for organic and inorganic contaminants including methyl mercury. This SRM has been analyzed for 30 polychlorinated biphenyl congeners and 16 chlorinated pesticides. Similarly, IAEA has released the shrimp homogenate (MA-A-3/OC) and lyophilized fish tissue (MA-B-3/OC) certified for chlorinated hydrocarbons. The Bureau of Community Reference Materials (BCR) has CRMs for pesticides in milk powder (BCR-150, BCR-51, BCR152), and for aflatoxins in milk powder (BCR-282, BCR-283, BCR-284, BCR-285). Concerning organo-metallic contaminants, besides NIST-SRM-1945 (whale blubber) certified for methyl mercury, fish tissue from the National Institute of Environmental Studies (NIES-11) has been certified for total Sn, tributyl and triphenyl Sn. The National Research Council of Canada has issued dogfish muscle (NRCC-DORM-1) and lobster hepatopancreas (NRCC-TORT-1) certifed for methyl mercury. Information on additional matrices can be found elsewhere *(22)*.

## Database for Reference Concentrations

Baseline data in a well defined group of individuals reflect reference values. Obviously, factors such as age, sex, living environment and diet, among others, influence the concentration levels of certain trace elements, but some of these parameters can be well defined. In some cases, even the habits such as smoking tobacco (e.g. elevation of blood Cd) and consuming alcohol (e.g. elevation of blood Pb), should be considered and several reference sources are now available *(23-26)*.

   In the inorganic area, over the last decade, there has been  considerable progress in understanding the laboratory related analytical problems in improving the quality of trace element data generated *(26)*. With increased understanding of the sources of variations in elemental concentrations arising from physiological changes, pathological influences, and occupational and environmental exposures *(23,27)*, efforts to generate reliable reference data bases for elemental composition of human tissues and body fluids are showing signs of success.

   In the organic area, the increasing emphasis on developing natural matrix reference materials has been an important step *(21,22)*. Data compilations developed by the U.S. EPA *(28)*, NOAA *(29)* and other sources *(22)* provide reference information on many organic substances in various media.

## Bioenvironmental Surveillance

The role of bio-environmental monitoring (BEM) or surveillance is that of an early warning system to identify factors responsible for adverse health effects and measures to prevent them. However, BEM per se has different meanings and these should be understood properly. In the context of human health, the expression biological monitoring (BM) of health is commonly used and is aimed at the detection of biological effects arising from exposure to chemicals in the environment (e.g., exposure to metals resulting in proteinurea or perturbation of enzyme levels or other metabolites). BM is performed by measuring the concentrations of toxic agent and its metabolites on representative biological samples from the exposed organism. If appropriate indicator specimens, e.g. blood, urine, expired air and hair are used, the body burden of certain pollutants absorbed or retained in the organism during a specific time interval can be assessed, dose-response relationships can be established, and the data can be used for risk-assessment purposes.

**Biostatistical considerations:** The primary consideration for a viable Biomonitoring Program concerns the problem of sample size (number of samples needed for a meaningful study) and defining the target population. They should be randomly selected  to preserve representativeness and to avoid selection bias. The sample size is related to the ability to detect differences between groups, and to some extent, it is also dictated by the nature of the problem studied. If very small differences between groups are to be differentiated, the burden rests on both the capability of the analytical methodology and a large enough sample size.

Therefore, the statistical plan has to be formulated based on the purpose of the investigation, reliability and detection capability of the analytical method, and the cost factor. Also, the question of analysis of single samples versus pooled samples has to be considered. If sampling expenses far exceed the rest of the analytical costs, it is not desirable to pool the samples. Conversely, analysis cost alone cannot be a deciding factor if pooling of samples is statistically not permissible *(30)*.

The size of the sample is directly related to the ability to detect important differences among subgroups in the target population using statistical tests of significance. The process of measurement depends on the nature of the variable and the level of measurement. These may be nominal, ordinal, interval, or ratio. Instrumental sophistication alone is not adequate to dictate the level of measurement since on many occasions cost and efficiency considerations dictate the choice of the level of measurement. While a single measurement, however precise, may not permit discrimination among significant risk groups, the method of discriminant analysis strategically exploits the correlation structure among the variable to obtain optimum discrimination with efficiency and cost reduction.

**Exposure Assessments and Biomarkers**

Exposure of humans and animals to some contaminant or the other present in the environment occurs constantly. An assessment of such exposure to environmental chemicals requires that both external and internal doses be assessed to understand the adverse effects. The duration of the exposure, and the concentration levels at which a toxic substance (i.e. external exposure) is present are important attributes in health risk assessment. However, the toxicity potential of a given chemical can vary widely and therefore, the amount of toxicant absorbed (i.e. internal dose) by a biological system should be known to determine the severity of the damage. The objective here is to establish a dose-response relationship, and identify the threshold points. The choice of methods for exposure assessment depends on intended use of the exposure data, economic and logistic constraints and the availability and interpretability of measures of exposure *(31)*. In assessing human exposure to toxic metals, several factors influence the interpretation of biomonitoring results as critically reviewed by Christensen in a recent review *(32)*.

Biological markers are defined as cellular, biochemical or molecular alterations that is measurable in biochemical media such as human tissues, cells or fluids *(33)*. Some sources act as target tissues thus enhancing the sensitivity for quantification of specific properties of biomarkers. Several examples can be cited: methyl-Hg in hair and organic pollutants in lipid-rich organs. As practical clinical specimens, blood and urine have been extensively investigated for toxicants or for one of their metabolites. Pb in blood is perhaps the most common example that has served as a sensitive tool in detecting even low level exposures to the central nervous system. With the help of the pharmacokinetics of a given toxicant in biological systems such as blood and urine, very important conclusions related to biomonitoring can be reached. For example, fluoride is rapidly eliminated whereas As lingers longer in the blood *(34)*. Similarly, volatile organic compounds are efficiently removed, while chlorinated pesticides reside for extended periods. Therefore, information on the clearance rate is of great value to determine whether or not a certain exposure is of recent origin. Equally important in this context is the knowledge of reliable baseline values, as discussed in an earlier section.

Bioenvironmental surveillance of pollutants is a complex task which requires careful attention to several aspects: an understanding of the multidisciplinary perspectives involved; an objective evaluation of the suitability of bioenvironmental specimens; proven methods for collection of valid samples and their processing; availability of appropriate analytical methodologies to meet the desired accuracy and precision criteria; expertise required for data processing and meaningful interpretation of results; effective mechanism for dissemination of the acquired information; and provisions for a technical facility for long-term storage of samples for retrospective analysis (see chapter by Iyengar and Subramanian on Specimen Banking).

Sample selection and collection are critical components of a biomonitoring system and any compromise at this stage would vitiate the purpose. Special attention should be paid to minimize the impact of presampling factors to safeguard the biological validity of the sample *(35)*. For a comprehensive bioenvironmental monitoring program, the basic planning should take into consideration the requirements of both inorganic and organic pollutants. In developing analytical schemes it is prudent to plan for a broad range of analyte coverage since the aim is environmental surveillance and the need is to establish baseline values for as many parameters as possible. If the focal point is organic pollutants, then retention of the biochemical integrity of the specimen (e.g., by cryogenic preservation) is of highest consideration during pre-, and post-sampling stages. On the other hand, sampling tools (e.g., tools made of Ti) and ambient conditions play a crucial role if inorganic pollutants are of primary concern *(36)*.

## Real-time and Long-term Monitoring of Pollutants

Real Time Monitoring (RTM) is a means of frequently checking short-term changes of pollutant profiles using samples such as blood, milk, saliva and urine. The samples collected for RTM are analyzed as soon as laboratory facilities permit. RTM differs from ESB in terms of time interval between sample collection and analysis. The usefulness of RTM as a tool using bovine milk as suitable specimen has been demonstrated in a recent Spanish study that successfully monitored residue levels of the organochlorine pesticides (e.g. lindane, heptachlor-epoxide, aldrin, endrin, dieldrin, and DDT and related components *(37)* in bovine milk. On the other hand, samples for Long-Term Monitoring (LTM) should be reliable indicators of long-term body burden of the chemicals identified in them. Examples of specimens suitable for LTM are hair (with some limitations), adipose tissue (for organic pollutants), and liver (for both organic and inorganic pollutants).

The Canadian National Specimen Bank *(38)*, in collaboration with other Canadian Agencies, has initiated several LTM projects covering the Atlantic Coast (Seabirds), Eastern Forest (Woodcock Wings), Prairie Provinces (Prairie falcons), Pacific Coast- Fraser Estuary (Great Blue Heron) and Great Lakes (Herring Gulls). Since 1970, several studies have been carried out to study organic pollutant trends in the Great Lakes area specimens of birds, eggs, mammals, fish, reptiles, amphibians and Herring Gulls. Herring Gull eggs collected between the years 1971 and 1982 have been examined for chlorinated organics such as PCBs and DDE to establish contaminant levels. A substantial decline in both PCB and DDE levels has been demonstrated over time. The remedial measures introduced in Canada and the United States sine the 1970s, limiting the use and disposal of PCBs are believed to have contributed to the observed decline in their levels.

The German monitoring program *(39)* for organic compounds is extensive and several classes of compounds such as halogenated hydrocarbons and polychlorinated biphenyls (PCB), polycyclic aromatic hydrocarbons (PAH), aromatic amines, and phenolic compounds have been determined. Pesticide monitoring of whole blood, blood serum, and urine is carried out routinely as part of the RTM. Over a period of 5 years between 1982 and 1987, a decrease in the level of pentachlorophenol (PCP) in urine and whole blood has been demonstrated. This trend has been linked to the 1979 German legislation restricting the use of PCP as a wood protectant medium. The monitoring for inorganics (mainly metals) is primarily aimed at Al, As, Be, Cd, Hg (including methyl-Hg), Pb, Sb, Se, Sn, and Tl. A few other elements are determined periodically to establish reference levels. Both Cd and Pb are determined in whole blood on a routine basis as a part of RTM.

The United States program *(40)* is operated from the National Institute of Standards and Technology in Gaithersburg. The NIST facility is set up for human liver, marine specimens and food composites. Presently, major efforts are directed towards marine mammal specimens and the human liver program. The liver specimens have been preserved for over 12 years now. Since the inception of the program over 550 human liver samples have been collected and stored under cryogenic conditions. The organic contaminants that have been determined in human livers include: hexachlorobenzene, the beta isomer of hexachlorocyclohexane, heptachlor epoxide, trans-nonachlor, dieldrin, p,p'-DDE, and p,p' DDT. The results showed that the most abundant pesticide residue was p,p'-DDE, the metabolic derivative of p,p'-DDT. In the inorganic analysis program, up to 30 elements are being determined in livers to establish baseline concentrations. It was notewothy that the concentration range values observed were narrower and lower for many trace elements (e.g. Al, As, Sn and Pb) than the previously reported values.

### Human Specimens for Pollutant Biomonitoring

**Organic contaminants:** Adipose tissue and liver have been recognized as accumulators of lipophilic constituents, particularly organochlorine pesticides;

similarly, they are also proven accumulators of many other organic contaminants. However, the collection of these specimens involves invasive techniques and the logistics including medico-legal aspects have to be in place for implementation. In an investigation reported recently, concentrations of organochlorine insecticide residues such as DDT, DDE, and HCH were determined in human adipose tissue samples obtained during autopsies from urban and suburban areas of Vercruz city, Mexico. The results for the mean levels of total DDT, which constitutes the accumulated pesticide, showed a decrease over the years from 1988 to 1991 *(41)*. The highest levels were observed among persons over 51 years old. The adipose tissue proved to be a very useful specimen.

Alternatively, one may also consider specimens that are meaningful and easy to obtain under practical conditions (e.g., whole blood) or those collected by non-invasive procedures (e.g. breast milk, urine, hair, placenta). The suitability of whole blood and milk for RTM of organic pollutants, especially the former as indicator of aromatic hydrocarbons *(42)*, offers a good starting point. Several organic chemicals are excreted in the breast milk. Reasonable quantities of samples, especially for RTM purposes, can be obtained.

**Organohalogens in breast milk:** Several studies have been conducted around the world to investigate the presence of organohalogens in human milk as documented in a monograph *(43)*. The following organohalogens have been determined: DDT and its metabolites; Cyclodiene pesticides such as dieldrin, aldrin, endrin, heptachlor and its epoxide, chloradane, oxychloradane and trans-nonachlor; and other organohalogens such as hexachlorocyclohexanes, hexachlorobenzene, polychlorinated biphenyls, polychlorinated terphenyls, polybrominated biphenyls, polychlorinated dibenzo-p-dioxins, polychlorinated dibenzofurans, chlorobenzenes, pentachlorophenol, mirex (dechlorane), toxaphene, chloroethers, and polychlorinated napthalenes. Investigations carried out in several countries have demonstrated a link between human milk fat for p,p'-DDE and PCBs, p,p'-DDT and PCBs, among several other compounds *(43)*.

Extensive investigations have been carried out on human milk to detect numerous organic pollutants *(43)*. DDT and its metabolites are still the most frequently determined compounds, especially in samples from developing countries. For example, in a recent study from Venezuela, levels of DDT residues in human milk from various rural populations were monitored. All mothers showed measurable levels of DDT residues in their milk ranging from 5.1 to 68.3 ng/mL and the levels were found to be significantly increased with age *(44)*. Other chlorinated pesticides determined are dieldrin, heptachlor epoxide, hexachlorocyclohexanes and hexachloro benzene. Among PCBs, polychlorinated dibenzo-p-dioxins (PCDDs), and polychlorinated dibenzofurans (PCDFs) have been reported in samples from developed countries *(43)*.

In the case of breast milk, the fat content of individual donors fluctuates during feeding, within the day and between days. It has been shown that hindmilk samples having higher fat content than foremilk samples *(45)*. This is an important source of error when lipophilic pollutants are assayed. Hence different fractions from a single donor during the day should be collected and mixed. The results of breast milk analysis should be expressed both on whole milk (for calculating intake of pollutants by the infants) and milk fat basis (for risk assessment purposes). For lipophilic contaminants, expressing the results on milk fat basis eliminates errors from sampling heterogeneity and also permits a better comparison with adipose tissue results.

Among other easy to obtain specimens, urine but not hair has been shown *(1)* to be a good specimen for RTM of pentachlorphenol (PCP). Placenta as a potential specimen for biomonitoring organic pollutants does not appear to have been explored and offers possibilities to carry out some research work (see chapter on placenta by Subramanian and Iyengar).

**Inorganic Contaminants:** Several clinical specimens such as whole blood (for Cd and Pb, and their metabolites), hair (for As and methyl-Hg), and urine (for both organic and inorganic pollutants) are good specimens for RTM *(46)*. Determination of urinary As is a good example of RTM for environmental exposure to As *(9)*. Breast milk has also been shown to be useful for monitoring the concentration levels of As, Cd, F, Hg and Mn *(47)*, and Pb *(7)*. Placenta has been used to identify elevated levels of Cd *(48)* and Hg *(49)* under different exposure conditions. Under normal environmental situations, the information obtained from placental analysis is an indicator of a time aggregate. However, the data interpretation should include the effect of placental age on its weight. Liver (many elements) is also a good specimen for RTM.

**Environmental Specimen Banking**

Environmental Specimen Banking (ESB) is an emerging discipline that has made impressive progress during the past ten years. It is an intrinsic part of a comprehensive analytical process. ESB enables a systematic collection and careful preservation and storage of an array of environmental samples for deferred chemical characterization and evaluation. Specimen banking enables tracing newly recognized pollutants (retrospective evaluation) and permits reevaluation of an environmental finding when new and improved analytical techniques become available.

Presently, specimen banking activities are taking place in several countries of which two facilities in Germany and one in the United States of America have made substantial progress in setting up guidelines for collection and long-term preservation of both human and non-human specimens. Further, systematic approaches for sample selection, collection, preservation, processing and measurement of the analytical signal through data handling have been developed for the characterization of organic and inorganic constituents. The U.S. program has successfully completed the pilot phase, has established the potential usefulness of specimen banking that has led to the establishment of a National Biomonitoring Specimen Bank *(40, 50)*, while in Germany a national facility is already in operation *(39)*.

The ESB has entered the application phase as witnessed by several examples. The Canadian investigations with Herring Gull eggs *(38, 51, 52)* through retrospective determination of biphenyls (PCBs) using improved methodology have dispelled earlier high values. The NIST has conducted several investigations in collaboration with other governmental agencies and representative specimens of several ecosystems have been preserved. For example, 550 human livers have been collected, and about 20 % of these livers have been analyzed to establish baseline values. As a result of this effort, significant findings such as continued decline in liver lead concentrations have emerged *(40)*. The German investigations focusing on both ESB and real-time monitoring (RTM), have observed a shift in the distribution of pentachlorophenol (PCP) towards the lower end of the scale in recent years. Another notable example in this context is the establishment of anthropogenic sources to account for the existence of chlorinated dioxins in human tissues thus disproving the trace chemistry theory of the origin of dioxins as naturally occurring in woods not treated with chemicals *(53)*. Other examples include monitoring activities related to grain and soil *(54)*, seal oil and sediment *(55)*, and fish *(56)*.

## Conclusion

Human biomonitoring is a direct source of exposure information. Not all metabolites are presently measurable and therefore research efforts are required. In the nutrition area, too, biomonitoring is of interest for diagnostic purposes. Examples assembled by Subramanian and Iyengar *(57)* on measurable parameters in trace metal deficiencies point out the problems and prospects in these types of investigations.

It is of vital importance in an environmental surveillance effort to establish baseline data for as many constituents as possible. It should also be recognized that concentrations established at a given time may not necessarily be "valid" baseline values in cases where a chemical has already assimilated with various environmental pathways.

The realization of reliable chemical measurements depends on the state of the analytical techniques at a given time. The tasks to be accomplished are: (a) accurate measurement of total element concentrations at natural concentration levels; (b) reliable assessment of the desired organic constituents; (c) ability to answer questions of chemical speciation problems in natural biological matrices; and finally (d) meeting the requirements of "total" analytical quality control (biological-, and measurement- related standardization, including data treatment and interpretation). Concerning (a) and (d) it may be said that there is no lack of methodology but lack of proper practice and applications *(58)*. On the other hand, concerning (b) and (c), there are still some very significant measurement problems to be resolved. Hence the value of preserving the sample in its natural state through specimen banking for resolving the problem when technology is available, cannot be over-emphasized.

## Literature Cited

1. Kemper F.H. *Umweltprobenbank fuer Human-Organproben* (Report), University of Muenster, Muenster, Germany, 1992.
2. Tulonen J.; Vuorinen P.J. *Sci. Total Environ.* **1996**, *187*, 11-18.
3. Iyengar G.V. *Sci. Total Environ.* **1981**, *100*, 1-15.
4. Bowen H.J.M. *Environmental Chemistry of Elements;* Academic Press: London, 1979.
5. Froslie A.; Norheim G.; Ramback J.P.; Steinnes E. *Bull. Environ. Contam.* **1985**, *34*, 75-182.
6. Iyengar G.V. *Biol. Trace Ele. Res.* **1987**, *12*, 263-295.
7. Ghelberg N.W.; Ruckert J.; Strauss H. *Igiena (Bucharest).* **1972**, *21*, 17-22.
8. *Historical Monitoring.* MARC Report 31, University of London: London, 1985.
9. Ghelberg N.W.; Bodor E. *Management and Control of Heavy Metals in the Environment*; CEP Consultants: Edinburgh, 1979, 163-166.
10. Berrow M.L.; Burridge J.C. *Inorganic Pollution and Agriculture*; MAFF-Book-326 (H.M.S.O.), London 1979.
11. Kitagishi K.; Yamane I. *Heavy Metal in Soils of Japan*; Japan Scientific Society Publishers: Tokyo, 1981.
12. Galster W.A. *Environ. Health Perspect.* **1976**, *15*, 135-45.
13. *Applied Environmental Geochemistry*; Thornton I, Ed.; Academic Press: London, 1983.
14. Adriano D.C. *Trace Elements in the Terrestrial Environment;* Springer: New York, 1986.
15. Yang G,; Wang.S; Zhou R.; Sun S. *Amer. J. Clin. Nutr.* **1983**, *37*, 872-81.
16. Iyengar G.V.; Gopal-Ayengar A.R. *Ambio.* **1988**, *17*, 31-35.
17. Iyengar G.V.; Kollmer W.E. *Trace Ele. Med.* **1986**, *3*, 25-33.
18. Rail D.; Kidd D.E. *Ecotox. Environ. Safety.* **1982**, *6*, 2-8.
19. Iyengar G.V. *Anal. Sci.* **1992**, *8*, 399-404.
20. *Survey of Reference Materials.* Int. Atomic Energy Agency: Vienna. IAEA-TEC-DOC-854, Vol. 1, 1995; IAEA-TEC-DOC-880, Vol. 2, 1996.
21. Wise S.A.; Schantz M.M.; Koster B.J.; Damiralp R.; Mackey E.A.; Greenberg R.R.; Burow M.; Ostapczuk P.; Lillestolen T.I. *Fres. J. Anal. Chem.* **1993**, *345*, 270-277.
22. *Standard and Reference Materials for Marine Science;* NOAA Technical Memorandum NOS ORCA 94, NOAA: Rockville, MD, 1994.
23. Iyengar G.V.; Woittiez J.R.W. *Clin. Chem.* **1988**, *34*, 474-481.
24. Hamilton E.I. *The Chemical Elements and Man;* Charles Thomas: Spring Fields, IL, 1979.
25. Heydorn K. *Neutron Activation Analysis in Clinical Trace* Element Research, CRC Press: Boca Raton, FL, 1984.
26. Versieck J. *CRC Critical Rev. Clin. Lab. Sci.* **1985**, *22(2)*.
27. Iyengar G.V. *J. Path.* **1981**, *134*, 173-180.
28. *Environmental Protection Agency*; Report. EPA-560/5-86-036, Washington, DC, 1986.

29. *NOAA Technical Memorandum NOS OMA 44;* NOAA: Rockville, MD, 1988.
30. *Environmental Epidimiology. Effects of Environmental Chemicals on Human Health;* Draper W.M. Ed.; Advances in Chemistry Series 241, ACS, Washington, DC, 1994.
31. Needham L.L. In *Environmental Epidimiology. Effects of Environmental Chemicals on Human Health;* Draper W.M. Ed.; Advances in Chemistry Series 241, ACS, Washington, DC, 1994, Chapter 10, pp. 121-135.
32. Christensen J.M. *Sci. Total Environ.* **1995,** *166,* 89-135.
33. Hulka B.B. In *Biological Markers in Epidemiology;* Hulka B.S.; Wilcosky T.C.; Griffiths J.D. Eds.; Oxford University: New York, 1990, Chapter 1, pp. 3-15.
34. Iyengar G.V.; Iyengar V. In *Determination of Trace Elements;* Alfassi Z.B. Ed.; Verlag Chemie: Weinheim, Chapter 13, pp. 544-94.
35. Iyengar G.V. *Anal. Chem.* **1982,** *54,* 554A-558A.
36. Iyengar G.V. Manual of Methods of Analysis; O'Neill I.; Schuller P; Fishbein L. Eds.; International Agency for Cancer: Geneva, 1986, 141-158.
37. Losada A.; Fernandez N.; Diez M.J.; Teran M.T. Garcia J.J.; Sierra M. *Sci. Total Environ.* **1996,** *181,* 131-135.
38. Elliott J.E.; Norstrom R.J.; Kennedy S.W.; Fox G. *NBS Special Publ. 740;* NBS: Gaithersburg, MD, 1988, 131-142.
39. *Umweltprobenbank;* Springer: Heidelberg, 1988.
40. Wise S.A.; Koster B.J.; Parris. R.M.; Schantz. M.M.; Stone S.F.; R. Zeisler, *Int. J. Environ. Anal. Chem.* **1989,** *37,* 91-106.
41. Waliszewski, S.M.; Pardio V.T.S.; Chantiri J.N.P.; Infanzon R.M.R.; Rivera J. *Sci. Total Environ.* **1996,** *181,* 125-131.
42. *Monitoring Human Tissues for Toxic Substances;* National Research Council: Washington, DC, 1991.
43. *Chemical Contaminants in Human Milk;* Jensen A.S.; Slorach S.A., Eds.; CRC Press: Boca Raton, FL, 1991.
44. Brunetto R.; Leon A.; Burguera J.L.; Burguera M. *Sci. Total Environ.* **1996,** *186,* 203-207.
45. Iyengar G.V. *The Elemental Composition of Human and Animal Milk;* IAEA-TECDOC-269, Int. Nat. Atomic Energy Agency: Vienna, 1982.
46. Iyengar G.V. *Elemental Analysis of Biological Systems;* CRC Boca Raton, FL, 1989.
47. Parr R.M.; DeMaeyer E.M.; Iyengar G. V.; Byrne A.R.; Kirkbright G.F.; Schoech G; Ninisto L.; Pineda O.; Vis H.L.; Hofvander Y.; Omololu A. *Biol. Trace Ele. Res.* **1991,** *29,* 51-75.
48. Thuerauf J; Schaller K.H.; Engelhardt E.; Gossler K. *Int. Arch. Occup. Environ. Health.* **1975,** *36,* 19-27.
49. Thieme R.; Schramel P.; Kurz E. *Geburthilfe u. Frauenheilkd.* **1977,** *37,* 9-17.
50. *International Review of Environmental Specimen Banking.* Wise S.; Zeisler R. Eds.; NBS Special Publication 706, NBS: Gaitehrsburg, MD,1985.
51. Turle R.; Norstorm J.; Won H.T. *NBS Special Publ. 740.* NBS: Gaithersburg, MD, 1988, 171-183.

52. Elliot J.E.; Wise S.; Zeisler R. Eds. *NBS Special Publication 706*, NBS, Gaithersburg, MD, 1985, 4-12.
53. Schecter A.; Gross M.; Dekin A. *Abstracts of Dioxin '86*. 6th Int. Symp. on Chlorinated Dioxins and Related Compounds, Fukuoka, Japan, 1986.
54. King N. *Monitoring Environmental Materials and Specimen Banking;* Luepke N.-P., Ed.; Nijhoff Publishers: The Hague 1979, 74-85.
55. Beeton A.M. In *Environmental Specimen Banking and Monitoring as Related to Banking;* Lewis R.A.; Stein N.; Lewis C.W., Eds.; Nijhoff Publishers: The Hague, 1984, 143-152.
56. Lewis R.A. *Report No. 12;* U.S. Department of Interior, 1987.
57. Subramanian, K.S.; Iyengar G.V. In *Handbook of Metal Ligand Interactions in Biological fluids;* Berthon, G., Ed.; Chapter 5, section D, 1995, pp. 543-548.
58. *Biological Trace Element Research: Multidisciplinary Perspectives;* Subramanian K.S.; Iyengar G.V.; Okamoto K., Eds.; ACS Symposium Series 445, ACS, Washington D.C., 1991.

# MONITORING

# Chapter 2

# General Aspects of Heavy Metal Monitoring by Plants and Animals

Bernd Markert[1], Jörg Oehlmann[1], and Mechthild Roth[2]

[1]International Graduate School Zittau (IHI), Markt 23,
02763 Zittau, Germany
[2]Technical University of Dresden, Institute of Forest Botany
and Forest Zoology, P.O. Box 10, 01737 Tharandt, Germany

The idea of using organisms or communities of organisms to register and evaluate certain characteristics of the environment is based on the ecological theorem of congruence between the prominence of environmental factors and the requirements of the species (1) and can be traced back to the 16th century (2). Certain forms of plant cover were already known to indicate the presence of ores in the ground, and the composition of the vegetation was used to judge the fertility of the soil (3). With the beginning of the industrial era and the resulting increase in emissions it became plain that organisms are not only capable of indicating the "natural" characteristics of a location but also provide qualitative and quantitative information on changes in the environment brought about by man. As far back as 1860, Nylander (4) drew conclusions on air pollution from the species composition of the lichens occurring naturally in Luxembourg. Since then an immense volume of literature has been published on bacteria, fungi, plants and animals from aquatic and terrestrial biotopes that provide information on the abiotic condition of their environment (5, 6, 7, 8).

## Definitions

Almost as great as the variety of organisms said to have bioindicative properties is the variety of definitions of the technical terms in use. In the last few years, especially, the rapid development of this field of biological science has resulted in an extremely polysemous use of the nomenclature. Many terms are still not unequivocally or uniformly defined and properly distinguishable from each other; usage differs from one country to another, and also within the international literature (5, 7, 8, 9, 10). Consider, for example, the meaning of the term "bioindicator" and how it can be distinguished from the term "biomonitor". Considerable differences in definitions have also emerged, depending on the field of research (e.g. microbiology, mycology, botany, zoology, physiology, morphology) in which the "systems" with bioindicative properties originated (5, 8). On the basis of our present knowledge of the ability of organisms to indicate pollution , and especially in respect of the bioindication of

**Table I. Definitions of technical terms**

| | |
|---|---|
| **Bioindicators** | Organisms or communities of organisms whose content of certain elements or compounds and/or whose morphological, histological or cellular structure, metabolic-biochemical processes, behaviour or population structure(s), including changes in these parameters, supply information on the quality of the environment or the nature of environmental changes. |
| **Biomonitors** | Organisms or communities of organisms whose content of certain elements or compounds and/or whose morphological, histological or cellular structure, metabolic-biochemical processes, behaviour or population structure(s), including changes in these parameters, supply information on the quantitative aspects of the quality of the environment or changes in the environment. |

**Classification of organisms (or communities of these) according to their "mode of reaction"**

| | |
|---|---|
| **Accumulation indicators / monitors** | Organisms that accumulate one or more elements and/or compounds from their environment. |
| **Effect or impact indicators / monitors** | Organisms that demonstrate specific or unspecific effects in response to exposure to a certain element or compound or a number of substances. Such effects may include changes in their morphological, histological or cellular structure, their metabolic-biochemical processes, their behaviour or their population structure. |

**Classification of organisms (or communities of these) according to their "origin"**

| | |
|---|---|
| **Active bioindicators / biomonitors** | Organisms usually bred in laboratories that are examined for accumulation of elements or compounds and specific or unspecific effects after exposure for a defined period in the area studied. |
| **Passive bioindicators / biomonitors** | Organisms that are taken from their natural biotope and analysed for accumulation of elements or compounds and specific or unspecific effects. |

metals, the definitions (9) may contribute to reaching agreement on standard nomenclature at the interdisciplinary and international level. (**Table I.**)

Thus, both bioindicators and biomonitors may be considered as organisms or communities of organisms that react to changes in environmental conditions with a change in their "signs of activity". These changes may manifest themselves:

* in element concentrations or levels of certain compounds;
* in morphological structures from the cellular to the organological or habitus level;
* in the course taken by intracellular and intercellular biochemical processes;
* in the pattern of innate or acquired behaviour; and
* in the structure or dynamics of the population or community of organisms.

The difference between bioindicators and biomonitors lies in the information they supply. While bioindicators only provide information on the quality of the environment or its changes, the "reactions" of biomonitors make it possible to determine the quantity of the environment or its changes (*11, 12*). Especially where the bioindication of metals is concerned, the literature often makes a distinction between "accumulation indicators" and "effect indicators" in respect of the reaction of the indicator/monitor to changes in environmental conditions (*5, 8*). Here we should bear in mind that this differentiation does not imply a pair of opposites; it merely reflects two aspects of analysis. As the accumulation of a substance by an organism already constitutes a reaction to exposure of this substance which - at least in the case of high accumulation factors - is measurably reflected in at least one of the parameters used in defining the term "effect indicator/monitor" (e.g. morphological changes at the cellular level: formation of metal-containing intracellular granules in many invertebrates after metal accumulation (*13*)), we should discuss whether it is worthwhile distinguishing between accumulation and effect indicators or whether both terms fall under the more general expression **reaction indicator**. Often, too, it is not until a substance has been accumulated in organisms that intercellular or intracellular concentrations are attained that produce effects which are then analysed in the context of effect and impact monitoring (Figure 1).

When studying accumulation processes it would seem useful to distinguish between the paths by which organisms take up elements/compounds. Various mechanisms contribute to overall accumulation (**bioaccumulation**), depending on the species-related interactions between the indicators/monitors and their biotic and abiotic environment. **Biomagnification** is the term used for absorption of the substances from nutrients via the epithelia of the intestines. It is therefore limited to heterotrophic organisms and is the most significant contamination pathway for many land animals except in the case of metals that form highly volatile compounds (e.g. Hg, As) and are taken up through the respiratory organs (e.g. trachea, lungs) (*14, 15*). **Bioconcentration** means the direct uptake of the substances concerned from the surrounding media, i.e. the physical environment, through tissues or organs (including the respiratory organs). Besides plants, that can only take up substances in this way (mainly through roots or leaves), bioconcentration plays a major role in aquatic animals. The same may also apply to soil invertebrates with a low degree of sclerotization when they come into contact with the water in the soil (*14, 15*).

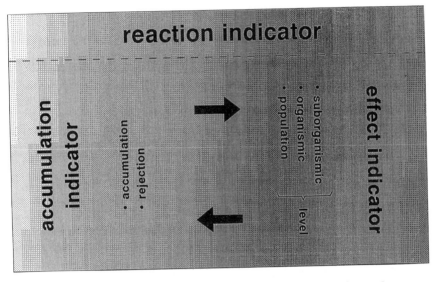

Figure 1. Illustration of the terms reaction, accumulation and effect/impact indicator

## Instruments of Bioindication

Bioindication and biomonitoring can be carried out at various levels of organismic life (macromolecule, organelle, cell, tissue, organ, organism, population or biocoenosis). As bioindication/biomonitoring is more complex at the higher organizational levels in biological systems and the detectable reactions, both accumulation and biological effects, become less and less specific, there has been a definite trend towards the use of lower degrees of organization in recent years in particular. The "boom" in biomarkers and biosensors is partly explained by the possibilities they offer of conducting more substance-specific biomonitoring, but their drawback is that the events have less relevance for the ecosystem. Many well-established biomarkers such as the induction of metallothiocines into animal organisms or phytochelatines into plant organisms after exposure to heavy metals, or the induction of cytochrome P-450 after exposure to hydrocarbons, are only signs of existing contamination (indicators of exposure). At the population or biocoenosis level they must rather be interpreted as signs that the stress caused by exposure is being dealt with, and not as a risk to the population or its fitness or even as a hazard to the biocoenosis.

Besides the classic floristic, faunal and biocoenotic investigations that primarily record rather unspecific reactions to pollutant exposure at higher organizational levels of the biological system, various newer methods have been introduced as instruments of bioindication. The most important of these are:

**Biomarkers**: measurable biological parameters at the suborganismic (genetic, enzymatic, physiological, morphological) level in which structural or functional changes indicate environmental influences in general and the action of pollutants in particular in qualitative and sometimes also in quantitative terms. **Examples**: enzyme or substrate induction of cytochrome P-450 and other Phase I enzymes by various halogenated hydrocarbons; the incidence of forms of industrial melanism as markers for air pollution; tanning of the human skin caused by UV radiation; changes in the morphological, histological or ultra-structure of organisms or monitor organs (e.g. liver, thymus, testicles) following exposure to pollutants.

**Biosensors**: measuring equipment that produces a signal in proportion to the concentration of a defined group of substances through a suitable combination of a selective biological system, e.g. enzyme, antibody, membrane, organelle, cell or tissue, and a physical transmission device (e.g. potentiometric or amperometric electrode, optical or optoelectronic receiver). **Examples:** Toxiguard bacterial toximeter; EuCyano bacterial electrode.

**Biotest (bioassay)**: routine toxicological-pharmacological procedure for testing the effects of agents (environmental chemicals, pharmaceuticals) on organisms, usually in the laboratory but occasionally in the field, under standardized conditions (with respect to biotic and abiotic factors). In the broader sense this definition covers cell and tissue cultures when used for testing purposes, enzyme tests and tests using micro-organisms, plants and animals in the form of single-species or multi-species procedures in model ecological systems (e.g. microcosms and mesocosms). In the narrower sense the term only covers single-species and model system tests, while the other procedures may be called suborganismic tests. Bioassays use certain biomarkers or - less often - specific biosensors and can be used in bioindication and biomonitoring.

## Quality Assurance

While the importance of quality assurance and quality control was recognized years ago in the field of chemical analysis, there is much less awareness of the quality problem when it comes to monitoring biological effects. This is all the more surprising since co-ordinated international programmes involving several laboratories in different countries have repeatedly revealed instances of results that could not be compared or were even obviously wrong; examples are the investigations into EROD (ethoxyresorufin deethylase) activity in the context of the North Sea Task Force Monitoring Master Plans (NSTF MMP) of 1992. More recently, however, efforts are being made to establish standard operating procedures (SOPs) accompanied by analytical quality control (AQC). As in chemical analysis there is now an increasing trend towards periodic checks for proper calibration within the individual laboratories, including the running of blank tests. In biomarker analysis, especially, performance checks are made on an interlaboratory basis, identical samples being periodically analyzed by several participants. In the case of some techniques, for example in imposex analysis (see below), we may in future expect to see the use of standard reference materials produced, certificated and distributed in the same manner as has been done in chemical analysis for several years. In the routine use of bioindicators/biomonitors the quality of the data acquired largely depends on a number of criteria that can be formulated parallel to the quality criteria for chemical analyses:

**Accuracy** is a measure of the closeness with which it agrees with the true levels of an element, compound or mixture of these and the structural, physiological, enzymatic, genetic or ethological changes in organisms or communities of organisms.

**Reproducibility** describes the extent to which identical results are obtained when an analysis is repeated. The **sensitivity** of a biomonitor is a measure of its readiness to react. The lower the threshold concentration at which a reaction is found to occur, the greater is the biomonitor's sensitivity. A biomonitor is said to possess **specificity** if an effect occurs after the action of a particular environmental variable (pollutant, radiant flux density etc.).

## Possibilities and Limitations

The advantages of using bioindicators/biomonitors rather than direct chemico-physical measuring techniques for environmental control have often been described (5,8,9). Quite apart from their cost, which is generally lower than that of direct immission measurements, bioindicators/monitors can be used for measuring and evaluating (metal) emissions throughout large areas provided that they are common enough and widely distributed. This criterion is met by a large number of lower and higher organisms - mosses, lichens, Lumbricidae, to mention but a few. The balance of the substances occurring in mosses is relatively independent of the substrate on which the plants grow. This means that their element concentrations - unlike those of many other organisms - can provide direct information on the wet and dry deposition of metals. Since the pioneer work of Rühling and Tyler in the late 1960s (16), mosses have been used regularly and successfully to monitor the emission, atmospheric transportation and deposition of metals over wide areas, for example across national boundaries (17, 18, 19, 20). What is more, some biomonitor species permit an insight into the man-made pollution of past times if material from herbariums or museums is analyzed (9). An interesting aspect in this connection is the analysis of core samples from raised bogs; when supported by radiocarbon dating, these can be used for large-scale retrospective analyses of the development of heavy metal pollution over the course of time (21).

A further interesting point is the integrative function of biomonitors. Organotin compounds are an example of how the time-integrating effect can be a great advantage in monitoring environmental pollution. Because of their biocidal properties these compounds are used in antifouling paints. Within this class of substances the tributyltin compounds (TBTs) are characterized by a particularly high degree of toxicity. TBT concentrations in seawater are subject to seasonal fluctuations, with maximum pollution during the summer months and minimum values in winter (22). When freshly painted pleasure craft are launched in the spring, the TBT levels rise; when most of these pleasure craft are taken ashore again in the autumn, the aquatic concentrations fall. Whereas physico-chemical methods only provide information on the pollution of the sea with TBT compounds at a particular time, accumulation indicators even out these fluctuations by acting as integrators. The xenobiotics are accumulated fairly gradually during peak pollution times, and only slowly eliminated from the organisms during times of low pollution because of their biological half-lives of more than 50 days in marine molluscs (23). The pollutant concentrations of bioindicators are also subject to seasonal fluctuations caused by biotic processes in the organisms themselves. Tests carried out on mosses show that the seasonal fluctuations in chromium concentrations (lowest levels in spring) are due to the dilution effect that occurs during peak biomass production at the beginning of the vegetation period (24).

The integrative effect of bioindicators does not only apply to the time axis. Bioindicators occupy defined ecological niches in ecosystems and are interrelated with their biotic environment in many ways (e.g. through symbiosis, mutualism, parasitism, predation or competition). By occupying niches in ecosystems, bioindicators make it possible to integrate analytical results in an overall biological system. This permits ecologically relevant statements not only concerning the bioindicators themselves but also on the community of organisms (or at least parts of it) through biotic interrelationships; this is scarcely possible with direct physico-chemical methods. The connection becomes very obvious when metal concentrations in food webs are analyzed.

Nevertheless, there are limits to the ability of organisms to indicate exposure. It is precisely the many interactions between organisms and their abiotic environment that so often make a causal analysis of the measurements difficult. Multifunctional and multistructural associations between bioindicators and other abiotic or biotic compartments of ecosystems often obscure individual mechanisms. Moreover, it is often impossible to determine synergistic or antagonistic effects of pollutants on organisms qualitatively and quantitatively under field conditions, or to distinguish them from each other.

The limits to the indication of exposure by organisms become especially obvious when attempting to quantify environmental qualities, i.e. in biomonitoring in the stricter sense of the term. While the number of potential bioindicators is virtually growing by the hour, it is difficult to find organisms (in nature) that meet the criteria of an active or passive biomonitor. For instance, the analysis of individual accumulation

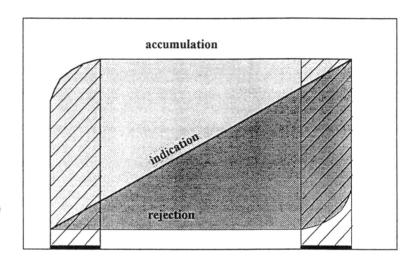

increasing environmental pollutant concentration

**Figure 2. Correlation between environmental concentration of the pollutant to be monitored and the concentration in the organism (explanations in the text).**

indicators for body burdens of certain substances does not necessarily permit conclusions on concentrations in the environment. Many plants and animals display high accumulation factors for certain substances at low environmental concentrations, but the accumulation factors decrease sharply at higher environmental levels. The result is more or less a plateau in the curve for environmental concentration/body burdens (Figure 2).

On the other hand, many organisms succeed in keeping their uptake of toxic substances very low over a wide range of concentrations in the environment. Not until acutely toxic levels in the environment are exceeded do the regulatory mechanisms break down, resulting in a high degree of accumulation (Figure 2). Exceptions are, of course, substances that are not taken up actively but enter the body by way of diffusion processes - doubtless rare in the case of inorganic metal compounds.

This means that the relationship between the bioindicator/monitor and its environment in respect of the concentration of the compound to be accumulated is often not linear but logarithmic. Even when linearity of the logarithmic function is achieved by mathematical conversion, the linear relationship between the two measurements is restricted to a small range. But organisms can only provide unequivocal information on their environment if a linear relationship exists (comparable with the calibration line on measuring instruments).

Metals, especially those that have essential functions, are subject to physiological regulation mechanisms that restrict the possible uses of organisms as bioindicators. Tests on numerous invertebrates have shown that many of these are able to keep their levels of nutritional elements constant over a wide range of concentrations in the surrounding medium, depending on their "physiological state" (e.g. stage of development, age, sex) (25, 26, 27). According to tests carried out on sapro(phyto)phagous species (e.g. Lumbricidae, Enchytraeidae, Diplopoda, Collembola, Tipulidae, Sciaridae, Cecidomyiidae) and predatory epigean arthropods, this applies to metals such as Zn, Cu and Fe, that may contribute to regional pollution of the environment at high levels, as well as the macronutrients (Na, K, Mg and Ca) (28, 29). Such homeostasis of the nutrient element level, which is probably controlled physiologically, affects the behaviour of elements during their transportation through food chains and thus the suitability of the organisms as bioindicators/biomonitors. In many invertebrates the accumulation of essential elements is determined by the animals' physiological requirements and the availability of the nutrients in the substrate.

Consequently it has so far been difficult to find organisms that allow a genuine quantification of environmental pollution in respect of space and time under field conditions. In recent years, some species of molluscs have shown promise in connection with the monitoring of tributyltin compounds. Their high degree of sensitivity to tributyltin compounds makes this group of animals good potential biomonitors in two respects. Firstly, they accumulate TBT in their tissues until a steady state, consisting of degradation or excretion on the one hand and uptake on the other, is reached between their body and ambient water; secondly, they demonstrate measurable morphological and physiological changes of an intensity that correlates with the TBT concentration in their habitat (23). As tests on Prosobranchia, the predatory dog whelk *Nucella lapillus* and the netted whelk *Hinia reticulata* show, the bio*accumulation* factors (BCFs) largely depend on the pollution of the seawater and amount to $9 \times 10^4$ for *H. reticulata* and $1.5 \times 10^5$ for *N. lapillus* at 2 ng TBT-Sn/l in the water (*23*).

By far the most sensitive bioindicator system for TBT is based on the pseudohermaphroditism or imposex phenomenon (*31 and 32 respectively*) in Prosobranchiata. In these heteroecious species the females develop parts of the male reproductive system, usually a penis and/or vas deferens. Imposex has now been described in over 120 marine species from all parts of the world (*30*). In all the species so far investigated, the development of imposex can be divided into the following stages:

- Stage 0 is a normal female without male parts.
- From Stage 1 to Stage 4 the size of the penis and/or the extent of the vas deferens steadily increase, but without closing the vaginal orifice and thereby impairing the animal's fertility.
- In Stage 5 the vagina either fails to develop (Stage 5a) or the existing vaginal orifice (vulva) is subsequently closed by hyperplastic vas deferens tissue (Stage 5b). Both alternatives prevent the depositing of egg capsules and result in sterility.
- Due to the oviduct blockade the egg capsules become abortive, causing the death of the female by a distension and finally a rupture of the capsule gland (*33, 34*).

Using this classification it is possible to calculate the intensities of the imposex phenomenon as a vas deferens sequence index (VDS: mean of the imposex stages in a population) (*32, 33*). In all the species examined in detail, imposex has shown itself to be a promising biomonitoring system. In the vicinity of TBT emission sources in harbours or marinas there is a sharp increase in imposex intensities, measured as a VDS index, and in the above species a significant correlation ($p << 0.001$) was found between body burdens or aquatic TBT concentrations and the VDS index. By simultaneously analyzing TBT levels in the water or in the tissues of the species and determining the imposex intensities it is possible to calibrate TBT biomonitors, provided that a sufficiently large number of stations and populations with different degrees of pollution are examined. Conversely, when a calibrated biomonitor is used, it is possible to draw conclusions about the TBT pollution in the habitat of a population from the imposex intensity in this population.

**Literature Cited**

1. Begon, M.; Harper, J.L.; Townsend, C.R. *Ecology - Individuals, Populations, Communities*. Blackwell Scientific Publications, Oxford, Great Britain, 1990, 1024 pp.
2. Thalius, J. *Sylvia hercynia, sive catalogus plantarum sponte nascentium in montibus*. Frankfurt, Germany, 1588.
3. Ernst, W. *Schwermetallvegetation der Erde*. Gustav Fischer, Stuttgart, Germany, 1974.
4. Nylander, W. *Bull. Soc. Bot. France*, **1866**, *13*, 364-371.
5. Arndt, U.; Nobel, W.; Schweizer, B. *Bioindikatoren: Möglichkeiten, Grenzen und neue Erkenntnisse*. Ulmer Stuttgart, Germany, 1987, 388 pp.
6. Jeffrey, D.W.; Madden, B. *Bioindicators and Environmental Management*. Academic Press, London, Great Britain, 1991, 458 pp.

7. Markert, B.: *Plants as Biomonitors*. VCH Weinheim, Germany, 1993, 644 pp.
8. Schubert, R.: *Bioindikation in terrestrischen Ökosystemen*. Gustav Fischer, Jena, Germany, 1991, 338 pp.
9. Markert, B. *UWSF-Z. Umweltchem. Ökotox* 1994, *6 (3)*, pp145 - 149.
10. Wittig, R. in: *Plants as Biomonitors* (Markert, B., Ed), VCH Weinheim, 1993, pp 3 - 27.
11. Markert, B. *Vegetatio* 1991, *95*, 127 - 135.
12. Martin, M.H.; Coughtrey, P.J. *Biological Monitoring of Heavy Metal Pollution*. Applied Science Publishers, London, Great Britain, 1982, 475 pp.
13. Hopkin, S.P. *Ecophysiology of Metals in Terrestrial Invertebrates*. Elsevier Applied Science Publishers Ltd., London, Great Britain, 1989, 366 pp.
14. Fleming, J.R.; Richards, K.S. *Pedobiol.* 1982, *23*, pp 415 - 418.
15. Roth, M. *Pedobiol.* 1994, *37*, pp 270 - 279.
16. Rühling, A.; Tyler, G. *Bot. Notiser.* 1968, *121*, pp 321 - 342.
17. Markert, B.; Weckert, V. *Toxicol. Environ. Chem.* 1993, *40*, pp 43 - 56.
18. Wolterbeek, H.Th.; Kuik, P.; Verburg, T.G.; Herpin, U.; Markert, B.; Thöni, L.: *Environ. Monitoring and Assessment*, 1995, *35*, pp 263 - 286.
19. Bruns, I.; Siebert, A.; Baumbach, R.; Miersch, J.; Günther, D.; Markert, B.; Krauss, G.-J.: *Fresenius' Z. Anal. Chem.* 1995, *353*, pp 101 - 104.
20. Rühling, A. *Atmospheric Heavy Metal Deposition in Europe - Estimation Based on Moss Analysis*. Nordic Council of Ministers (NORD 1884:9). 1994.
21. Ernst, W.H.O.; Mathys, W.; Salaske, J.; Janiesch, P. *Abh. Landesamt f. Naturkunde (Münster)*, 1974, *2*, pp 3 - 31.
22. Oehlmann, J.; Stroben, E.; Fioroni, P. *Cah. Biol. Mar.* 1993, *34*, pp 343 - 362.
23. Oehlmann, J.; Stroben, E.; Schulte- Oehlmann, U.; Bauer, B.; Bettini, C.; Fioroni, P.; Markert, B. *Münchner Beiträge zur Abwasser-, Fischerei und Flussbiologie*, 1995 (in press).
24. Markert, B.; Weckert, V. *Sci.Total Environ.* 1994, *158*, pp 93 - 96.
25. White, S.L.; Rainbow, P.S. *Mar. Ecol. Prog. Ser.*, 1984, *16*, pp 135 - 147.
26. Ireland, M.P.*Environ. Pollut.*, 1979, *19*, pp 201 - 206.
27. Coughtrey, P.J.; Martin, M.H. *Oecologia*, 1976, *27*, pp 65 - 74.
28. Roth, M. in: *Biogeochemistry of Trace Metals* (Adriano, D.C., Ed.), Lewis Publishers, Boca Raton, 1992, pp 299 - 328.
29. Roth-Holzapfel, M.; Funke, W. *Biol. Fertil. Soils* 1990, *9*, pp 192 - 198.
30. Jenner, M.G. *Science*, 1979, *205*, pp 1407 - 1409.
31. Smith, B.S. *Veliger*, 1981, *23*, pp 212 - 216.
32. Fioroni, P.; Oehlmann, J.; Stroben, E. *Zool. Anz.* 1991, *226*, pp 1 - 26.
33. Gibbs, P.E.; Bryan, G.W.; Pascoe, P.L.; Burt, G.R *J. Mar. Biol. Ass. U.K.* 1987, *67*, pp 507 - 523.
34. Oehlmann, J.; Stroben, E.; Fioroni, P.*J. Moll. Stud.*, 1991, *57*, pp 375 - 390.

Chapter 3

# Secondary Ion Mass Spectroscopy in the Analysis of the Trace Metal Distribution in the Annual Growth Rings of Trees

R. R. Martin[1], E. Furimsky[2], Jinesh Jain[3], and W. M. Skinner[4]

[1]Department of Chemistry, University of Western Ontario,
London, Ontario N6A 5B7, Canada
[2]Natural Resources, Canada, 555 Booth Street,
Ottawa, Ontario K1A 0G1, Canada
[3]Department of Geological Sciences, University of Saskatchewan,
Saskatoon, Saskatchewan S7N 0W0, Canada
[4]Ian Wark Research Institute, South Australia Surface
Technology Centre, University of South Australia, The Levels,
South Australia 5095, Australia

Secondary Ion Mass Spectroscopy (SIMS) was used to measure the distribution of trace metals in the annual growth rings of Eastern white pine (Pinus strobus L.) at selected sites throughout Southern Ontario, Canada. Previous results show systematic variation in individual metals between trees at different sites in rings corresponding to the same year, and between different years at the same site. In this work SIMS imaging combined with Scanning Electron Microscopy (SEM) has shown that trace metals are often confined to regions a few microns in size throughout the ring structure. In addition, there is evidence for marked seasonal variation in the metal content within individual rings. Finally, Inductively Coupled Plasma Mass Spectroscopy (ICP/MS) has been used to calibrate SIMS secondary ion yields so that rough quantisation of SIMS data is possible.

Considerable effort has been expended on using tree rings as evidence for such events as climate variation, solar flares, and forest fires*(1)*. The elemental composition in tree rings has also been used to monitor environmental events, thus neutron activation analysis (NAA) has been used to follow the long-term relationship between molybdenum and sulphur in red cedar*(2)*. Inductively coupled plasma atomic emission spectroscopy (ICP/AES) has been used to assess the impact of atmospheric deposition on forests*(3)* and the effects of altered soil chemistry have been examined using proton induced X-ray emission (PIXE)*(4)*. In addition, inductively coupled plasma mass spectroscopy (ICP/MS) has been used for multielement analysis at exceptionally low concentrations in tree rings*(5)*. ICP/MS has been used in conjunction with laser ablation to investigate the chemistry of tree rings downstream from a mine site(6). These studies and others have shown that dendrochemistry is now a well established means of providing a chronology of environmental events(7-10), though it is an approach that should be used with caution(11).

Secondary ion mass spectroscopy has shown promise as a tool for tree ring studies(6,7), adding to and complementing other methods. In SIMS a (focused) beam of energetic primary ions is used to bombard a solid sample surface causing ions characteristic of the sample surface composition to be ejected. These secondary ions are subjected to conventional mass spectrometric analysis. The process is capable of detecting virtually all elements and their isotopes in the low parts per million, and for some elements in the parts per billion range. Since the primary ion beam can be used to ablate sample material it can be used both to remove surface contamination and to probe possible changes in concentration as a function of analysis depth. In addition, the yield of specific secondary ions can be measured as a function of position while the primary ion beam is moved across the sample (steps scan). If the primary beam is rastered over the sample surface while the secondary ion yield is monitored images may be formed of the surface distribution of specific elements ( ion images ). Either approach effectively maps the surface distribution of elements of interest.

Three problems are commonly associated with SIMS analysis. Surface charging, which can be reduced by gold coating of the sample surface is difficult to control and results in variable ion yields as well as distortion in images obtained when using the imaging mode of the instrument. Molecular secondary ions, especially from organic samples, can make identification of ions of the same nominal mass difficult. Finally, since the secondary ion yield is dependent on such factors as bond energy, ionization energy, surface charging and secondary electron yield SIMS ion intensities cannot generally be used to calculate absolute elemental concentrations unless accurate standard materials are available to calibrate the instrument. Relative ion yields are usually established by ratioing individual ion yields to a second ion in the sample matrix. Our initial results have been reported elsewhere(12) and show that SIMS is capable of detecting significant differences between years in the metal content of trees within a given stand. There is also a significant difference between the metal content of rings representing specific years in different stands, and the differences can be correlated with known site influences such as road salt, and pollution from landfills and/or industrial activity. No significant differences were detected between individual trees or between individual rings in the same tree due to the high variability in secondary ion yield

Information relating secondary ion yield to wood structure can be obtained by using scanning electron microscopy with energy dispersive X-ray spectroscopy(13) (SEM/EDXS). In this work SEM/EDXS has been employed to examine the ion pits excavated by the SIMS instrument during analysis so that SIMS ion yields may be correlated with identifiable microstructures within the individual tree rings. In addition, SEM images could be used to provide better spatial resolution of individual wood features than could be achieved by SIMS.

A second complementary technique, ICP/MS, has been used with matching samples to correlate SIMS secondary ion yields with metal concentrations within the bulk of individual tree rings. This approach provides a means of calibrating the SIMS ion yields with bulk concentrations as measured by ICP/MS.This work, then, has the following objectives :

First, since we[12] and others[6], have noticed large variations in metal content between analytical areas of the order of 100 microns$^2$ in size, we would like to establish whether SIMS imaging can be combined with SEM to identify unique features within the wood structure that are enriched in specific metals.

Secondly, to learn if SIMS analysis can be used to establish seasonal variations in the metal content of the wood in the annual rings of eastern white pine.

Finally, to determine if ICP/MS can be used in conjunction with matching samples used in SIMS analysis to calibrate the SIMS instrument, thus making it an effective tool for quantitative analysis.

**Experimental**

Eastern white pine, Pinus strobus L., was chosen as the target species because it is widely distributed in the study area, specimens are of sufficient age to reflect long-term changes in metal availability, foliar uptake of metals is enhanced in conifers, and samples can be collected with relative ease.

Samples were collected using HAGLOF A 558 increment borers supplied by Canadian Forestry Equipment Limited, producing samples 4mm in diameter with a maximum length of 40 cm taken from trees at breast height (1.5 meters) and stored in plastic drinking straws. The samples were sanded to produce a uniform flat surface using 400 mesh sandpaper with doubly distilled conductance water as lubricant on a specially constructed sanding table. The individual rings were then dated using a low power optical microscope. Individual dated rings were subsequently removed for analysis. Following air drying for 24 hours the samples were sputter coated with a thin layer of gold which effectively reduced surface charging. No further pretreatment, such as embedding the samples in resin, in order to minimize the opportunities for outside contamination.

The SIMS used was a Cameca IMS/3f instrument at Surface Science Western. When collecting mass spectra a 500nA $^{16}O^-$ primary beam with a net energy of 17.4 keV/ion was rastered over a square 150 microns on edge while collecting positive secondary ions from a central spot 60 microns in diameter. The mass range from 1 to 250 amu was scanned. A 100 V offset voltage was applied to the sample holder to suppress molecular secondary ions. The mass spectra obtained were used to determine optimal analysis conditions and to identify the more abundant species present in the wood.

When the instrument was run in the depth profile mode the primary beam current was reduced to 250 nA rastered over a square 150 microns on edge while secondary ions were collected from a central spot 60 microns in diameter. In this mode of operation the SIMS primary ion beam was allowed to dwell on a single site on the sample surface while secondary ions of specified masses were collected sequentially for ten second intervals. This sequence was carried out repeatedly for twenty minutes. During this period the sample was ablated to greater depth and a new surface exposed for SIMS analysis. This mode of operation (Depth profiling) allowed a large number of ions to be collected, thus reducing statistical errors while providing information with depth in the wood samples. In this work masses corresponding carbon, oxygen, sodium, potassium, calcium, chromium, copper and zinc were included in most depth profiles because they were the only elements detected in all samples while collecting mass spectra. When studying seasonal variations in metal content potassium and zinc were omitted to save time.

The depth represented by this analysis was determined by bringing the top edge of the ion pits excavated during analysis into focus on a Zeiss optical microscope, then using a calibrated focusing system to bring the crater bottom into focus as well.

A set of samples was prepared by taking matching core samples from individual trees so that paired samples of individual rings were obtained, one member of the pair was subjected to SIMS the second to ICP/MS analysis. In addition, SIMS analysis was carried out in the portion of the ring representing growth early in the year and later growth in the same year (early wood vs late wood). Samples used in conjunction with ICP/MS and to examine seasonal variation within rings were analyzed in the depth profile mode as described above.

Ion images were collected for the ions of highest yield i.e. $C^+$, $Ca^+$, $K^+$, and $Na^+$. The conditions used during imaging were often optimized during the run since imaging of insulating materials is complicated by surface charging. In general the conditions were the same as those used during the depth profiles except that the raster area was 250 microns on edge while images were collected from a central spot 125 microns in diameter.

## Scanning Electron Microscopy ( SEM )

The SEM used was an ISI-DS130 Dual Stage Scanning Electron Microscope with energy dispersive X-ray analysis. The same samples were used for both SIMS and SEM.

Tree ring samples (100 mg) for ICP/MS were weighed in precleaned Teflon digestion cups. 2 mL of doubly distilled concentrated nitric acid and 2 mL of ultra-pure water were added and the apparatus was capped and subjected to microwave digestion for 20 minutes. The resulting solutions were transferred to polypropylene bottles and diluted with ultra-pure water to an acid concentration of 0.5 M.

Elemental analysis of the digested samples was carried out using a Perkin-Elmer-Sciex model 5000 ICP/MS. A nickel sampler with a 1.1 mm aperture and a platinum skimmer with a 0.89 mm aperture was used . The instrument was operated at a power setting of 1000 W, with a nebulizer flow rate of 0.8 mL/min , B ,P, E1, and S2 lens settings of 45, 45, 25, and 35 respectively, and an argon power supply of 55 psi. The argon flow rate was 15 L/min with an auxiliary flow rate of 0.8 L/min. Liquid analyte flow rate was controlled with a peristaltic pump. The samples were supplied to the apparatus using a Gilson 212b Liquid Sampler autosampler. All analyses used an external calibration procedure. Indium was used as an internal standard for drift and matrix correction.

## Results and Discussion

Figure 1 is a typical SIMS spectrum collected from within a tree ring. Repeated spectra taken at different sites in other rings showed wide variations in all ion yields. Iron, calcium, potassium, chromium, copper and zinc were the only metals found to be common to all sites interrogated and this suite of metals was used in subsequent depth profiles. Iron was omitted from this portion of the study because of possible confusion between $^{56}Fe$ and the molecular ion $CaO^+$.

Table I presents the secondary ion yield with time in a representative depth profile for carbon, oxygen, sodium, and potassium. All ion yields are referenced to oxygen since the primary ion beam was O⁻ insuring a constant oxygen concentration at the sample surface. Results are collected from approximately 400 seconds onward. The initial period is used to allow the sample to establish equilibrium with the primary ion beam, to ablate possible contamination from the surface and to reach a depth within the sample at which soluble ions like sodium do not show depletion as a result of exposure to water during sample preparation. The results show large variation in ion yield between sites and with depth. Measurements on ion pit depth suggest an ablation rate of about 0.25 microns per minute. These results are consistent with our previous observation that SIMS data can only be used to compare the metal content between rings if a sufficiently a large sample set is obtained to achieve statistically meaningful results. While some of the variation in the secondary ion yield is the result of such factors as surface charging it is thought to be more likely that metal concentrations occuring on a micron scale and seasonal variations , ie the location of the ion beam within a growing season in a ring, are the most important sources of differences in secondary ion yield. Certainly seasonal variations have been observed in other species(14,15) and models such as the cation exchange model for ion transport proposed by Bondietti et al(9) suggest that metals are localized in specific structures in wood. It should also be noted that during a typical depth profile the secondary ion yields are referenced to O⁺ as an internal standard with its yield corrected to 1000 counts. Under these conditions the yield for sodium rises till around 400 seconds into the profile reflecting a surface depletion that results from the use of water as lubricant during sample preparation, while the carbon yield remains essentially constant.

Figure 2 shows SIMS secondary ion images obtained from a circular region 125 microns in diameter for carbon, sodium, potassium and calcium. The metals are clearly confined to regions of the order of 5 to 10 microns. Surface charging limits the resolution of the instrument to about 5 microns.

Figure 3 is an SEM micrograph obtained at the bottom of the ion pit created by the SIMS during a depth profile. The crystalline feature displayed an X-ray fluorescence spectrum characteristic of a mixed crystal of sodium and potassium chloride. This crystal was probably formed as a result of the drying the sample in the high vacuum in the SIMS instrument..

Figure 4 shows a feature rich in copper and zinc. The object has been tentatively identified as a resin mass by Dr. Richard Greyson of the Department of Plant Sciences at the University of Western Ontario. If this assignment is correct it suggests that toxic metals may be sequestered in resins in white pine.

The SIMS images and SEM micrographs confirm the deposition of metals in discrete regions within individual tree rings. Clearly these techniques can be used in tandem to probe the physiology of metal transport in the tree ring system.

Table II shows the average secondary ion yields for sodium, calcium, chromium and copper obtained during a series of depth profiles obtained within regions corresponding to early wood and late wood respectively within individual rings. A

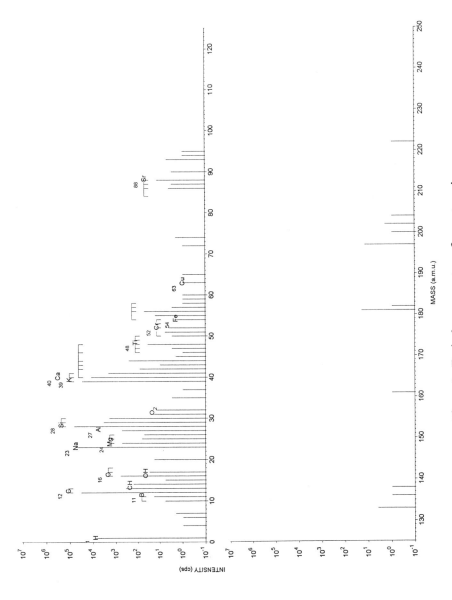

Figure 1. Typical mass spectrum from a tree ring.

**Table I. SIMS depth profiles for two different tree rings, referenced to oxygen**

| SITE 1 | | | | | | | |
|---|---|---|---|---|---|---|---|
| Carbon Ion Counts | Time/ Sec | Oxygen Ion Counts | Time/ Sec | Sodium Ion Counts | Time/ Sec | Potassium Ion Counts | Time/ Sec |
| 28474.9 | 424 | 998 | 436 | 7486.39 | 448 | 44793 | 461 |
| 29715.2 | 545 | 998 | 558 | 7236.21 | 570 | 44964.1 | 582 |
| 27664 | 647 | 998 | 659 | 7143.56 | 671 | 45062.7 | 684 |
| 26933 | 748 | 998 | 761 | 6963.33 | 773 | 43949.4 | 786 |
| 26915.7 | 850 | 998 | 862 | 6959.58 | 875 | 44327.1 | 887 |
| 26778.6 | 972 | 998 | 984 | 6952.75 | 996 | 44371.8 | 1009 |
| 26646.6 | 1073 | 998 | 1086 | 6904.16 | 1098 | 44569.4 | 1111 |
| 26301.9 | 1175 | 998 | 1188 | 6827.59 | 1200 | 44322 | 1213 |
| Average 27428.74 | | Average 998 | | Average 7059.196 | | Average 44619.26 | |
| SITE 2 | | | | | | | |
| 21815.9 | 427 | 998 | 440 | 1192.93 | 452 | 22449.8 | 465 |
| 22980.7 | 528 | 998 | 541 | 1066.06 | 573 | 24508 | 586 |
| 26289.8 | 649 | 998 | 662 | 1787.43 | 674 | 29593.7 | 687 |
| 30426.5 | 750 | 998 | 763 | 1922.8 | 775 | 32695.4 | 788 |
| 30862.7 | 852 | 998 | 865 | 1941.6 | 877 | 33339.5 | 890 |
| 34310.7 | 953 | 998 | 966 | 1444.96 | 978 | 33844.5 | 991 |
| 33526.9 | 1054 | 998 | 1066 | 1483.26 | 1079 | 33018.9 | 1091 |
| 33066.1 | 1175 | 998 | 1188 | 1680.2 | 1200 | 34336.6 | 1213 |
| Average 29159.91 | | Average 998 | | Average 1564.905 | | Average 30473.3 | |

Note the variability between rings and between individual measurements.

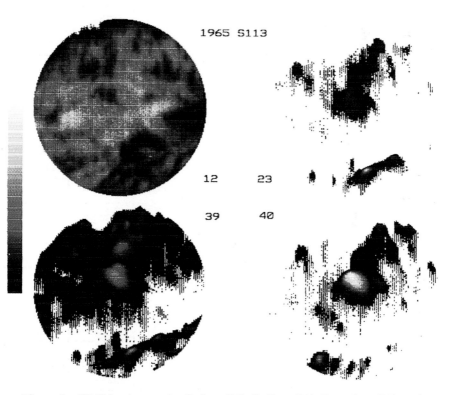

Figure 2. SIMS ion images for Carbon (12), Sodium (23), Potassium (39), and Calcium (40) in wood. Image diameter is 125 microns.

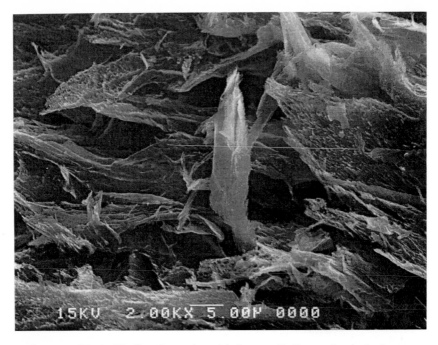

Figure 3.  SEM of Sodium/Potassium rich feature.  Reference bar is 5 microns.

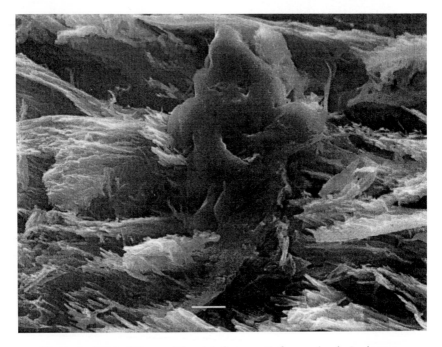

Figure 4.  SEM of Copper/Zinc rich feature.  Reference bar is 5 microns.

Table II. Student's t-test, SIMS ion counts in early wood vs. late wood
for selected elements.  n = 22

| Element | Average Counts in Early Wood | Variance Early Wood | Average Counts Late Wood | Variance Late Wood | t | t-critical |
|---|---|---|---|---|---|---|
| Sodium | 490 | 76562 | 15486 | 71086487 | -8.3 | 2.1 |
| Calcium | 11549 | 1443039 | 362420 | 5.496 E9 | -22.2 | 2.1 |
| Chromium | 4.2 | 29.2 | 451.6 | 13213 | -18.2 | 2.1 |
| Copper | 5.4 | 52.7 | 29 | 449 | -4.9 | 2.1 |

student's t-test performed on the data , also recorded in Table II, shows significant differences in ion yield at the 95% confidence level for the metals reported. Sodium, rather than potassium, was used because there is no reasonable molecular secondary ion having mass 23 that might interfere with interpretation of the results, in addition its behaviour should reflect that of potassium. Calcium was chosen because it is the most abundant metal in pine wood. Chromium was associated with regions suspected of having high metal burdens in our previous work as were copper and zinc. Copper was used as a surrogate for the copper/zinc pair.

While it is possible that the differences we observd are due to seasonal changes in ion transport in the trees studied, it is also possible that the different ion yields are due to differences in the wood density between early and late wood. Differencial charging is unlikely since $^{16}O$ from the primary ion beam is used as an internal standard. It is clear that SIMS secondary ion yields will show less variability if the interrogation sites are selected to represent either early or late wood. More significantly this approach provides a means of comparing differences in metal content between years in trees in a given stand and between trees in different stands that has a time resolution of the order of months.

Table III presents the results obtained when repeated SIMS analysis was carried out on  the paired samples which had been subjected to ICP/MS analysis. ICP/MS provides an average concentration for the entire sample while SIMS provides local concentrations (as we have shown). Accordingly the ICP/MS data have been averaged from five rings and the SIMS data have been averaged from three sites in each ring. The resulting SIMS average counts have been divided by the ICP/MS average yield, the primary ion current, the sampling time for the element in question and the area interrogated by the primary ion beam. The results are thus reported as ion counts $ppm^{-1}$ $amp^{-1}$ $micron^{-2}$ $second^{-1}$. This provides a means of quantitating the SIMS yield . The results reported here should not be regarded as having an accuracy much better than an order of magnitude, however better attention to the observed seasonality of the results and better control of sample charging will improve the method considerably.

**Table III.  Estimated SIMS yield for selected elements in wood
(based on average yields)**

| Element | ICP/IMS Concentration (ppm) | SIMS Ion Counts | SIMS Yield Ion $ppm^{-1}AMP^{-1}\mu m^{-2}s^{-1}$ |
|---------|-----------------------------|-----------------|---------------------------------------------------|
| Sodium | 38.8 | 4679 | $16.9 \times 10^{-6}$ |
| Calcium | 616 | 137155 | $31.3 \times 10^{-6}$ |
| Chromium | 0.3 | 228 | $107.3 \times 10^{-6}$ |
| Copper | 1.9 | 40 | $2.9 \times 10^{-6}$ |
| Zinc | 11.2 | 19.4 | $0.24 \times 10^{-6}$ |

## Conclusions

Clearly SIMS provides a useful tool to probe the metal content of individual tree rings and to compare metal uptake between years in a specific stand as well as between different stands of trees. Interpretation of the data is complicated by the fact that the metal content varies dramatically within regions having spatial dimensions in the micron range.

SIMS imaging can be combined with other techniques such as SEM to examine the metal content of wood on a micron scale and can thus contribute to an understanding of tree physiology.

There is a marked difference in secondary ion yield between the early wood and late wood within individual tree rings, while it is not clear if this represents different metal uptake and/or wood density, it does provide a means of reducing the scatter of SIMS results by confining analysis to either region of the tree rings. In addition these differences mean that the time scales over which trees may be compared can be reduced to months rather than years.

Finally, ICP/MS can be combined with SIMS analysis so that the secondary ion yields might be used for quantitative analysis.

## Acknowledgments:

Tree Ring Research Laboratory, University of Arizona, Tucson. Surface Science Western, Unversity of Western Ontario. Professors Richard Greyson and Anwar Maun, Department of Plant Sciences, University of Western Ontario. Professor R. Protz, Department of Land Resource Science, University of Guelph, Guelph, Ontario.

## References

(1)  Amato, I. *Anal. Chem.* **1988**, *60*, 1103-1107.
(2)  Guyette,R.P.; Cutter,B.E.; Henderson,G.S. *J. Environ. Qual.* **1989**, *18*, 385-389.
(3)  Bondietti,E.A.; Baes III, C.F.; McLaughlin, S.B. *Can. J. Forest. Res.* **1989** *19*, 586-594.
(4)  McClenahen, J.R., Vimmerstedt, J.P.; Scherzer, A.J. *Can. J. Forest. Res.* **1989**, *19*, 880-888.
(5)  Hall,G.S.; Yamaguchi, D.K.; Retterg, T.M.; *J. Radioanal. Nucl. Chem. Letters* **1990**, 146, 255-265.
(6)  Gough, L.P.; Yanoski, T.M.; Liche, F.E.; Balistrieri, L.S. *The Chemistry Of Tree Rings Downstream From The Summitville Mine.* Colorado Geological Survey, Special Publication 38, Proceedings: Summitville Forum '95. **1995** 236-243.
(7)  Vroblesky, D.A. and Yanosky, T.M. **1990**, 28, 677-684.
(8)  Hupp, C.R.; Woodside, M.D.; Yanosky, T.M. ?? **1993**, *13*, 95-104.
(9)  Momoishima, N. and Bondietti, E.A. *Can. J. Forest Res.*, **1990**, *20*, 1840-1849.
(10) McClenahen, J.R.; Vimmerstedt, J.P.; Scherzer, A.J. *Can. J. Forest Res.* **1989**, *19*, 880-888.
(11) Hagenmeyer, J. *Markert, B.: Plants As Biomonitors*, VCH-Publisher, Weinheim, New York, **1993**, 549-563.
(12) Martin R.R.; Zanin, J.P.; Bensette, M.; Lee, M.; Furimski, E. *Can. J. of Forest Res.*, in press. ??
(13) Selin, E.; Standzenieks, P.; Boman, J.; Teeyasoontranont *X-ray Spectrometry* **1993**, *22*, 281-285.
(14) Hagenmeyer, J.; Schafer, H. *Sci. Total Environ.* (**1995**), *166*, 77-87.
(15) Marhert, B.; Weckert, V. *Toxicological and Environmental Chemistry* **1993** vol. 40, 43-46.

# Chapter 4

# Human Hair and Lichen

## Activities Involving the Use of Biomonitors in International Atomic Energy Agency Programs on Health-Related Environmental Studies

Susan F. Stone[1]

International Atomic Energy Agency, Division of Human Health, P.O. Box 100, A–1400 Vienna, Austria

Environmental studies have often involved the collection and analysis of non-biological (abiotic) samples such as soil, air and water, where trace element and other analyte concentrations are often at very low levels and where it may be difficult to distinguish or to interpret the results of the analyses. The collection and analysis of biomonitor samples has many advantages over abiotic samples, including bio-concentration of pollutants, simpler sampling techniques, more relevance of the information obtained on the biological significance of concentrations of specific pollutants, and providing a measure of integrated exposure over time. The International Atomic Energy Agency (IAEA) is including the study of biomonitors for evaluation of environmental status in its programmes on health-related environmental studies. Biomonitoring activities from two co-ordinated research programmes are presented; the collection and analysis of human hair and that of lichens. In addition, quality assurance materials were developed and characterized in support of these activities. These materials include IAEA-336, Lichen, and two human hair materials, IAEA-085 and IAEA-086, the latter two with elevated and low levels of methylmercury, respectively.

As the financial and manpower resources in many Member States are increasingly coming under pressure, the design of programmes for environmental studies is becoming more focused, especially in the number and types of samples that are taken. The data obtained from the collection and analysis of biomonitor samples is often less complicated and costly than other types of abiotic samples, and has the additional advantage of indicating which pollutants in the abiotic environment have a direct impact on the individual and populations of organisms, human and otherwise. The Nutritional and Health-Related Environmental Studies Section of the IAEA has included the study of biomonitors in several of its co-ordinated research programmes (CRPs); the

[1]Current Address: 7038 Ridge Road, Frederick, MD 21702

monitoring of populations for methylmercury using human hair as a biomonitor, and the monitoring of air particulate matter and total deposition by collecting and analyzing lichen.

## Methylmercury and Human Hair

The toxicity of high levels of methylmercury (MeHg) on the human nervous system has been well-established, through investigations of the poisonings in Minamata Bay in Japan and in Iraq. There are still significant inputs of mercury into the environment in many parts of the world, either through natural sources in mercury-rich environments, or via human activities, including gold mining areas (where large amounts of mercury are used in the extraction process), and areas with chloralkali plants and other industries. In aquatic systems, the elemental mercury that is discharged into the environment, either by natural sources or anthropogenically, is transformed by various mechanisms to methylmercury. There is a rapid accumulation of methylmercury through the food chain, and it is concentrated in predatory species of fish and consumers of fish (*e.g.* mammals). In humans, the MeHg ingested through food items, such as fish, is almost completely absorbed into the bloodstream. Recently, there has been renewed concern over MeHg, due to the recognition that prenatal exposure at previously considered low levels of this mercury species can affect the developing nervous system (*1,2*). Although the general population is not at risk, there is special concern about population groups with a high consumption of fish in their diets or live in areas with high mercury levels in the surroundings.

Hair analysis has been established as an effective monitoring method for MeHg and is frequently employed in studies of mercury exposure (*3,4*). Compared with either blood or urine, the use of hair as a biomonitor is more widespread due to bio-concentration, ease of sample collection and the possibility of an integrated history of exposure, from months to even years. The determination of Hg and MeHg in hair has been the core of an IAEA research programme, "Assessment of Environmental Exposure to Mercury in Selected Human Populations as Studied by Nuclear and Other Techniques". The programme was aimed at identifying population groups which might be at risk from elevated methylmercury intake from diet, and method development and validation were specifically stressed. There have been indications that hair mercury levels as low as 6 mg/kg in pregnant women are correlated with effects on the developing fetal nervous system (*5,6*). In view of these indications, the emphasis of the studies in the CRP has been on pregnant women and their babies. Table I gives a list of the participants of the co-ordinated research programme, their respective institutes, and the population areas being monitored.

The methods for MeHg determination are not simple and straightforward, and until recently, have not been readily available in many countries. However, especially in biological samples, it is important to determine methylmercury, compared to total mercury alone, as a constant ratio of MeHg to total Hg cannot be assumed, and because MeHg is the more toxic form. The methods utilized by the participants for MeHg determination varied due to availability of instrumentation, and range from volatilization followed by detection by NAA (7) to ion exchange-CVAAS (8) to distillation coupled with gas chromatography (9).

The results from the individual studies showed that, in most of the groups, the mercury results from the hair samples in the study populations were lower than 6 mg/kg, which has been proposed as a guideline level (2,10). This level has been extrapolated from the Provisional Tolerable Weekly Intake (PTWI) of 0.3 mg mercury (or <0.2 mg MeHg), recommended by an expert group of the FAO/WHO (11). One example, however, where individual mercury levels were elevated, was in an area near Marano, Italy, where the Isonzo river crosses the mercury-rich area of the Idria in Slovenia. Although the median value of hair mercury values in the study group did not exceed the proposed guideline level, (5.1 mg/kg vs. 6), individual values were determined up to 30 mg/kg (12).

**Table I. IAEA Co-ordinated Research Programme on "Assessment of Environmental Exposure to Mercury in Selected Human Populations as Studied by Nuclear and Other Techniques"**

| Participant | Country | Population Area Being Monitored |
|---|---|---|
| M.B.A.Vasconcellos | Brazil | Amazon, Billings Dam, São Paulo |
| C.G. Bruhn | Chile | Eighth Region (Southern Coast) |
| Chai Chifang | China | Northeast China; Beijing |
| K. Kratzer | Czech Republic | Prague |
| S. Gangadharan | India | Western Coast |
| G. Ingrao | Italy | Bagnara Calabra; Fuimicino; Ravenna; Marano |
| S.B. Sarmani | Malaysia | Kuala Juri; Chendering |
| A.R.Byrne | Slovenia | (Reference Laboratory) |
| Tac Ahn Nguyen | Viet Nam | Dalat; Ho Chi Minh City; Nha Trang |
| (M. Horvat) | (IAEA-MEL) | (Reference Laboratory) |

Another significant exception was one of the population groups studied by the counterpart in Brazil. This group consisted of Indian tribes in the Amazon region of Brazil, and the hair mercury levels did show significantly elevated mercury levels compared with a control group from São Paulo, who had a very low consumption of fish. At the present time, hair samples have been collected from subjects in ten different tribes living in this area, and the results of the analyses have shown significantly elevated levels of Hg and MeHg compared to the control population, and significantly higher than the proposed guideline level. The major form of Hg was found to be MeHg, as shown in Figure 1, although the ratios of MeHg to Hg varied, between 0.5 to 1.0. This variation in the MeHg to total Hg ratio emphasizes again the importance of determining MeHg, as well as total Hg, in the hair samples. It is not clear at this time if these levels are significant with regard to health effects, since no neurological symptoms have been observed. The cause of these elevated levels is also unclear. However, a more detailed study of levels of Hg and MeHg in diet items, especially fish, is planned, and a careful

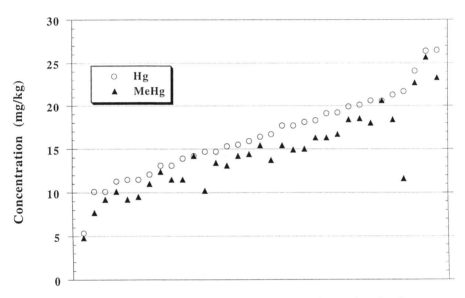

Figure 1. Mercury and methylmercury concentrations in hair samples of Indians from the Amazon region in Brazil (ranked in order of the value for total mercury).

epidemiological study is also required. Another study area from Brazil was located near Billings Dam, which is in one of the most heavily industrialized parts of the country. Although it was expected that this group would show elevated mercury levels from consuming fish from the Dam, no statistical difference was found between the levels from this group compared with the control group from São Paulo (*13*).

The combined data obtained from the CRP indicated that although most of the individuals studied did not have levels greater than the proposed guideline level of 6 mg/kg, there were significant differences in the hair MeHg values in fish-consuming populations compared with associated control populations. However, it must again be emphasized that there are no confirmed health effects in those groups who do have levels over 6 mg/kg. Since the dose-response relationship of MeHg at low levels of exposure has not yet been extensively studied, these populations would be ideal for such epidemiological studies. In addition, a Joint FAO/WHO committee has noted that more studies at the low end of the exposure range are needed to set a guideline value for pregnant and nursing women (*14*). The other important outcome of this CRP was that methods for MeHg determination were established in the participating groups, many for the first time. The comparable methodologies that were employed in the CRP have resulted in data that can easily be used as a reference because of the minimization of bias.


**Trace Elements in Air Particulate Matter and Lichen**

Most air pollution studies employ sampling devices and other instruments, but since the 1970's, lichens have been increasingly used as bioindicators for air pollution and air particulate matter, with good results (*15,16*). Lichens are symbiotic organisms composed of fungi and photosynthetic algae. These organisms are quite suitable as biomonitors due to their ability to accumulate various trace metals, their wide geographical distribution, availability throughout the year, and having a morphology which shows no seasonal variations. Having no root system, lichens must obtain their mineral nutrients from airborne particulates and total deposition. Lichens accumulate particulate matter continuously from the atmosphere, and specific pollutant levels in a lichen sample represent an integrated exposure over time. Though lichens in some studies are being sampled to obtain general trace element maps of whole areas, a majority of studies on trace element accumulation in lichens have been done in areas with specific sources; *i.e.* power plants, smelters, or other industrial sources with specific emissions.

In the IAEA Co-ordinated Research Programme, "Applied Research on Air Pollution Using Nuclear-Related Analytical Techniques", the emphasis of the studies has been on harmonized, and therefore comparable, collection of air particulate matter using low-volume PM-10 stacked filter air samplers (*17*). Several of the participants have included the collection and analysis of these biomonitors as a supplementary part of their studies. Both passive (collection of organisms already in the study area) and active monitoring (using transplanted organisms) have been used to monitor air quality around known sources, to identify possible sources, and for general monitoring.

Transplantation and passive studies of the lichen, *Hypogymnia physodes*, have been carried out in areas of Slovenia at varying distances from the country's largest coal-fired power plant (*18*). The elements that were found to be elevated over that of a control area, U, Cd, Hg, Se, W, Sb, Mo, and Cr, are all considered to be characteristic of elements released during coal combustion. Quantitative relationships between the trace element concentrations in the lichen and those in air particulate matter and in deposition samples from the same sites are also being studied.

A systematic scheme for the collection of the lichen *Parmelia sulcata* in Portugal (*19,20*) has resulted in concentration maps for over 30 elements over the whole country. A study using transplanted lichens is also in progress, to study the trace element concentrations around several power plants. In addition to the lichen samples, complementary data is being obtained through air sampling and collection of total deposition samples.

Projects in additional participating institutes of the CRP are also currently studying or intending to include biomonitors (lichens or mosses) as part of their air pollution studies. These participants include those in Argentina, Bangladesh, Jamaica, Kenya and Thailand.

**Quality Assurance for Biomonitoring Programmes**

The IAEA is promoting the compatibility and comparability of the data produced in these studies, both through encouraging the establishment of quality assurance procedures at participating institutes, and through the development and characterization of reference materials.

A lichen reference material, IAEA-336, has been issued through a co-operation of the IAEA and the Instituto Tecnologico e Nuclear (ITN) in Sacavem, Portugal. Lichen collected from areas presumed to be of low contamination by trace metals was used in preparing the material, which was submitted to an international intercomparison exercise (*21*). Statistical evaluation of the data resulted in recommended values for 17 elements and information values for 15 elements. A second lichen material is planned that would have more elevated levels of several of the significant heavy metals, such as V, Cd, Pb, and Ni.

In order to validate measurements from studies of human hair analyses for mercury and methylmercury, two quality assurance materials with different levels of these analytes were developed. One is a human hair material at a natural low level, typical of a non-fish consuming population (IAEA-086), and the other is a material at an elevated level of methylmercury (IAEA-085), prepared by spiking half of the natural level material with MeHg. Production and preliminary characterization of the materials was accomplished through an international co-operation initiated by the IAEA, and has been described previously (*22*). An international intercomparison exercise has been completed, in which 65 participating institutes from 40 countries have contributed data on the amounts of Hg, MeHg, and other trace elements in the two materials.

The significance of studies using biomonitors in the programmes of the IAEA has been demonstrated, both through the results of the evaluation of methylmercury status in selected populations, and through the results of lichen analysis to evaluate air pollution. Because of the steps taken to ensure the reliability and comparability of the data, including development of suitable reference materials, the information from these programmes can be used as reference for related studies. Finally, the accomplishments of the programmes and activities described here emphasize that the importance of biomonitor samples in environmental studies will continue to increase in the future.

## Literature Cited

1.Weiss, B. In: Prenatal Exposure to Toxicants: Development Consequences  Editors, Needleman, H.L.; Bellinger, D.; Johns Hopkins Series in Environmental Toxicity, 1994,112-129.

2.Grandjean, P.; Weihe, P.; Nielsen, J.B. *Clin. Chem.* **1994**, *40*, 1395-1400.

3.Akagi, H.; Kinjo, Y; Branches, F.; Malm, O.; Harada, M.; Pfeiffer, W.C.; Kato, H. *Environ. Sci.* **1994** *3*, 25-32.

4.Srikumar, T.S.; Källgård, A.; Lindeberg, S.; Öckerman, P.A.; Åkesson, B. J. *Trace Elem. Electrolytes Dis.* **1994** *8*, 21-26.

5. Kjellström, T., Kennedy, P., Wallis, S., Mantell, C., Physical and mental development of children with prenatal exposure to mercury from fish. Stage 1. Preliminary tests at age 4. (Report 3080) Stockholm: National Swedish Environmental Board, 1986.

6. Kjellström, T., Kennedy, P., Wallis, S., Mantell, C., Physical and mental development of children with prenatal exposure to mercury from fish. Stage 2. (Report 3642) Stockholm: National Swedish Environmental Board, 1989.

7.Zelenko, V.; Kosta, L. *Talanta* **1973**, *20*, 115-123.

8.May, K.;Stoeppler, M.;Reisinger, M. *Fresenius Anal.Chem.* **1984**, *317*,248-251.

9.Liang, L.; Horvat, M.; Bloom, N.S., *Talanta* 1994, *41*,371-379.

10. Sherlock, J., Hislop, J., Newton, D., Topping G., Whittle, K., *Human Toxicol.* **1984**, *3*, 117-131.

11.International Programme on Chemical Safety. Methylmercury (Environmental Health Criteria 101). Geneva: World Health Organization 1990.

12.International Atomic Energy Agency, NAHRES-26, Report on the Third Research Co-ordination Meeting of the Co-ordinated Research Programme, "Assessment of Environmental Exposure to Mercury in Selected Human Populations as Studied by Nuclear and Other Techniques", Vienna, 1995, 87-100.

13.Vasconcellos, M.B.A.; Saiki, M.; Paletti, G.; Pinheiro, R.M.M.; Baruzzi, R.G., Spindel, R. *J. Radioanal. Nucl. Chem.* **1994**, *179*, 369-376.

14. Joint FAO/WHO Expert Committee on Food Additives, WHO Technical Report Series No. 776, Geneva: World Health Organization, 1989.

15. Olmez, I.; Cetin Guloval, M.; Gordon, G.E. *Atmos. Env.* **1985** *19*, 1663-1669.

16. Sloof, J.E., Ph.D. Thesis, Environmental Lichenology: Biomonitoring Trace-element Air Pollution, Delft Univeristy of Technology, 1993.

17. Zeisler, R., Haselberger, N., Makarewicz, M., Ogris, R., Parr, R.M., Stone, S.F., Valkovic, O., Vlakovic, V., Wehrstein, E., *J. Radioanal. Nucl. Chem.*, **1996**, in press.

18.Jeran, Z.; Jacimovic, J.; Smodis, B., In: Proc. 2nd Intl. Symp. and Exhibition of Environmental Contamination in Central and Eastern Europe, in press.

19.Freitas, M.C., Nobre, A.S. *J. Radioanal. Nucl. Chem.* in press.

20. Freitas, M.C., personal communication (Progress report), 1994.

21.Stone, S.F., Freitas, M.C.; Parr, R.M.; Zeisler, R. *Fresenius J. Anal. Chem.*, **1995**, *352*, 227-231.

22.Stone, S.F.;Backhaus, F.W.; Byrne, A.R.; Gangadharan, S.; Horvat, M.; Kratzer, K.; Parr, R.M.; Schladot, J.D.; Zeisler, R., *Fresenius J. Anal. Chem.* **1992**, *352*, 184-187.

# Chapter 5

# The Silver Content of the Ascidian *Pyura stolonifera* as an Indicator of Sewage Pollution of the Metropolitan Beaches of Sydney, Australia

T. M. Florence[1], J. L. Stauber[2], L. S. Dale[2], and L. Jones[1]

[1]Centre for Environmental & Health Science Pty. Limited, 112 Georges River Cres., Oyster Bay, New South Wales 2225, Australia
[2]Centre for Advanced Analytical Chemistry, CSIRO Division of Coal and Energy Technology, PMB 7, Bangor, New South Wales 2234, Australia

*Pyura stolonifera* is an ascidian which is common on the sandstone headlands of Sydney's metropolitan beaches. It strongly concentrates heavy metals from seawater, with an enrichment factor of $2 \times 10^5$ for silver. The concentration of silver is low (<1 ng/L) in coastal seawater but is much higher in Sydney's sewage outfall water (0.5-5 µg/L), which makes the concentration of silver in *Pyura* a useful indicator of the extent of sewage pollution of the beaches. Silver in *Pyura* was measured before and after the extension of Sydney's major sewage outfall pipes into deep offshore water. Samples of *Pyura* collected from pristine areas contained 0.05 µg/g silver wet weight, whereas samples of the ascidian taken near Sydney's beaches had 0.15-0.75 µg/g silver wet weight (contents of sack). Extension of the sewage outfalls did not significantly decrease the silver concentrations in *Pyura*, and it is suggested that this is because the ascidian is concentrating silver from suspended particles from contaminated nearshore sediments.

The ascidian *Pyura stolonifera* ("sea squirt", or the Aboriginal name "cunjevoi") populates the intertidal sandstone headlands around Sydney, Australia, in dense clusters. *Pyura* is found all along the New South Wales coast, and also on the western coasts of tropical South Africa and South America. Florence and Farrar (*1*) had found previously that *Pyura* is capable of concentrating several heavy metals from seawater. Preliminary studies (*2*) showed that, of 50 elements analysed in samples of Sydney's sewage outfall water, silver had the highest differential concentration between sewage water and seawater. Open ocean surface Pacific Ocean seawater contains 0.1 ng/L (*3*), while unpolluted coastal Pacific Ocean seawater has about 0.3 ng/L silver (*4,5,6*). By contrast, Sydney's sewage discharge water has 500-5,000 ng/L silver as a result of inputs from the film processing and electroplating industries. Samples of *Pyura stolonifera* collected from pristine areas along the New South Wales coast had silver concentrations of about 0.05 µg Ag/g wet weight (contents of sack), leading to an enrichment factor of $2 \times 10^5$.

Sydney has major ocean sewage outfalls in the metropolitan area, and the water, which contains both domestic sewage and controlled amounts of industrial waste, receives only primary treatment before discharge to the Pacific Ocean. Originally, the discharge pipes extended only a few hundred metres from the cliff face but, between October 1990 and August 1991, the three largest outfalls (Bondi, Malabar and North Head) were extended to 3 km offshore, so that the sewage is now discharged into deep water. Because *Pyura stolonifera* is so common on the sandstone headlands of Sydney's metropolitan beaches and because it concentrates silver so well, this animal should be a useful monitor of sewage impact on the beaches. Since sewage is the only significant source of anthropogenic silver in the seawater off Sydney, it was expected that analysis of *Pyura* samples before and after the installation of the deep ocean outfall pipes would reveal a substantial decrease in the silver content of *Pyura* after the outfall pipes had been extended.

**Sampling and Analysis**

The sampling points around Sydney for *Pyura stolonifera* and the sewage discharge points are shown in Figure 1, and some sewage discharge rates in Table I. Other sampling points were Mollymook Beach and Summercloud Bay (in Jervis Bay) on the New South Wales south coast, which were considered to be relatively unpolluted areas. Mollymook has a small (domestic only) sewage outlet near the beach. The sampling area was dominated by the effects of the East Australian Current, with currents running from north to south at about 10-20 cm/sec. Wind direction was predominantly from the south and south-east. A total of 4-8 animals were collected randomly over an area of approximately 50 $m^2$ from each locality and analysed individually.

The fleshy contents of the leathery sack or "tunic" (wet weight of contents was 5-35 g; mean, 14g) were removed with a stainless steel knife and homogenised in a blender with stainless steel blades. The average ratio of wet weight to freeze-dried weight was 6.5. About 1g wet wt. of homogenised flesh was weighed accurately into a polycarbonate vial and 10 mL conc. $HNO_3$ (Suprapur) was added. After standing overnight, the samples were digested in a microwave oven for 15 min on 20% power and for a further 15 min on 10% power. After cooling for 30 min, 1 mL of 30% $H_2O_2$ (Suprapur) was added and the samples again digested for 10 min on 10% power. After the addition of a 12 mg/L indium/gallium internal standard, the samples were diluted in Milli-Q water and then analysed by inductively-coupled plasma mass spectrometry (ICP-MS) using a Fisons Elemental PQ2 in peak jump mode for Ag-107 and In-115. Recovery of silver was checked using the IAEA MA-A-994 fish standard reference material. Standard additions of silver to the digested sample solutions gave recoveries of 92-105%. Copper, lead, cadmium, zinc and uranium concentrations in *Pyura* were determined on the same acid-digested samples using ICP-MS. The same analytical methods were used for all samplings.

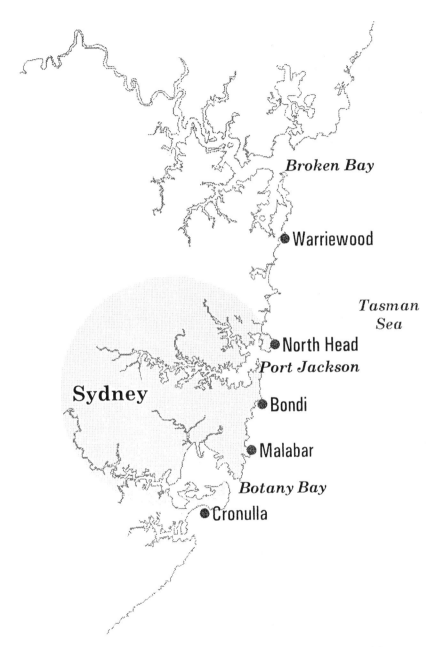

Figure 1. Sydney Environs and the Metropolitan Sewage Outfalls.

## Results and Discussion

Silver concentrations in samples of *Pyura stolonifera* collected before and after the installation of the deep ocean outfalls are shown in Table II. There was a large difference between the silver concentrations in *Pyura* samples collected near the major Sydney sewer outfalls and in Summercloud Bay, the location most remote from any settlement. However, there was no clear correlation between these silver concentrations and location of the Sydney sampling points, other than that, before the outfall extensions, *Pyura* from the major outfalls of North Head and Malabar had, with the exception of Palm Beach, the highest silver concentrations. The Palm Beach value for 1988 is anomalous, in that this location is well away from any outfall. However, the high silver values were maintained in samples collected during two separate sampling trips in 1988. Illegal dumping of trade wastes in this area during 1988 is a possible explanation. The expected low silver values were found in samples collected during 1992.

**Table I. Sydney Sewage Discharge Rates (1995)**

| Discharge location | Average discharge (ML/day) |
|---|---|
| North Head | 300 |
| Bondi | 141 |
| Malabar | 475 |
| Boat Harbour | 46 |
| Warriewood | 13 |

*Pyura stolonifera* has a lifetime of less than 2 years (mean, 9 months) (2) and, even 2-3 years after the 3-km extension of the North Head, Malabar and Bondi sewage outfalls (note that the smaller Warriewood and Boat Harbour were not extended), the silver concentration in *Pyura* sampled around Sydney beaches was still much higher than in *Pyura* from the pristine area of Summercloud Bay (Table II). This might be the result of a combination of factors, including the return to the beaches of some silver from the extended outfalls, contributions from the outfalls that were not extended, silver in stormwater drainage and, especially, silver mobilised from contaminated near-shore sediments. Higher concentrations of some other heavy metals such as copper and lead (but not cadmium and zinc) were also found in *Pyura* sampled near Sydney after the outfall extensions, compared to Summercloud Bay (Table III), and there was no significant decrease in the copper concentration of *Pyura* after the extensions were in place (Table IV).

There were significant differences in the silver concentrations of individual specimens of *Pyura* collected from a small area (~50 m$^2$) on a rock platform (Table V). These differences may be the result of differing ages of the animals, but there was no correlation between size of animal and silver concentration. However, crowding of the animals on the platform and competition for food could interfere with any size/age relationship (2).

Table II. Silver Concentrations in the Ascidian *Pyura stolonifera*. Samples
were analysed before and after the extension of the outfalls during October 1990
to August 1991. Silver concentrations correspond to the mean and standard
deviation of analyses on individual animals. Where no standard deviation is
shown, the result is for the analysis of combined individuals.

| Location and distance in km from nearest Sydney outfall | Date sampled | Silver, µg/g wet wt. | p* and no. (n) of samples |
|---|---|---|---|
| Palm Beach (4) | July 1988 | 0.76 ± 0.38 | 0.01 (n =6) |
| | November 1992 | 0.057 ± 0.012 | NS (n= 5) |
| North Head (0.2) | July 1988 | 0.51 ± 0.26 | 0.02 (n = 4) |
| | May 1993 | 0.42 ± 0.09 | 0.01 (n = 5) |
| Malabar (0.5) | May 1988 | 0.52 | n = 4 |
| | May 1993 | 0.26 ± 0.03 | 0.01 (n =5) |
| Oak Park (6) | May 1988 | 0.40 | n = 4 |
| | November 1992 | 0.21 ± 0.05 | 0.01 (n = 5) |
| Garie Beach (17) | July 1988 | 0.23 ± 0.08 | 0.02 (n = 4) |
| | May 1993 | 0.31 ± 0.06 | 0.01 (n =8) |
| Bondi Beach (0.4) | May 1993 | 0.25 ± 0.05 | 0.01 (n = 5) |
| Warriewood (0.5) | May 1993 | 0.24 ± 0.03 | 0.01 (n = 5) |
| Boat Harbour (0.2) | May 1988 | 0.20 ± 0.03 | 0.03 (n = 4) |
| | November 1992 | 0.31 ± 0.06 | 0.01 (n = 6) |
| Tamarama (1.2) | May 1988 | 0.15 | n = 4 |
| | May 1993 | 0.13 ± 0.06 | NS (n = 5) |
| Mollymook Beach (220) | November 1992 | 0.11 ± 0.07 | NS (n=8) |
| Summercloud Bay (180) | May 1988 | 0.025 | n = 4 |
| | November 1992 | 0.06 ± 0.02 | n = 5 |

* Significance of difference between the mean silver concentration of cunjevoi at a
site and the mean concentration of silver in cunjevoi sampled at Summercloud Bay
during November 1992.
NS= not significant (p = >0.05). Mann-Whitney U Test used for significance testing.

**Table III. Heavy Metals in the Ascidian** *Pyura stolonifera.* Samples were
collected between November 1992 and May 1993. Metal results correspond to
the mean and standard deviation of analyses on 5-8 individual animals. Where no
standard deviation is shown the result is for four individuals combined before
analysis. Results are in µg/g wet weight.

| Location | Copper | Lead | Cadmium | Zinc |
|---|---|---|---|---|
| North Head | 6.5 ± 1.7 | 0.32 | 0.03 ± 0.01 | 10 ± 3 |
| Malabar | 6.4 ± 1.6 | 2.3 ± 0.4 | 0.01 ± 0.00 | 11 ± 3 |
| Bondi | 6.6 ± 1.0 | 0.41 | 0.03 ± 0.01 | 9 ± 3 |
| Warriewood | 4.9 ± 0.6 | 0.18 | 0.01 ± 0.00 | 7 ± 1 |
| Boat Harbour | 13.1 ± 2.8 | 0.43 ± 0.12 | 0.02 ± 0.01 | 11 ± 1 |
| Mollymook | 4.0 ± 1.8 | 0.1 | 0.03 ± 0.01 | 11 ± 1 |
| Summercloud Bay | 1.6 ± 0.2 | 0.1 | 0.02 ± 0.01 | 9 ± 3 |

In an attempt to remove the differences in silver concentration between
individual *Pyura* specimens, the silver/uranium ratio was measured. Uranium is a
conservative, non-essential element in seawater, which contains 3 µg/L uranium (7),
compared with only 0.2 µg/L in Sydney sewage outfall water (Dale, unpublished
results). *Pyura stolonifera* concentrates uranium from seawater by a factor of about
ten, and it was hoped that the uranium concentration in *Pyura* would be an index of
the integrated volume of seawater pumped by the animal, so that the silver/uranium
ratio would then be independent of age or growth rate. However, this method
assumes similar accumulation and depuration rates for silver and uranium. Use of
silver/uranium ratio rather than silver concentration did not improve the uniformity
between different specimens of *Pyura* (Table V).

Silver in the mussels *Mytilus edulis* and *Mytilus californianus* and the clam
*Macoma balthica* have been used as indicators of anthropogenic pollution in San
Francisco Bay (*4,5,8*). Water in the south bay has silver concentrations as high as 27

**Table IV. Copper in** *Pyura Stolonifera* **Before and After the Outfall Extension.**
Mean copper concentration and standard deviation in µg/g wet weight (n = 5-8).
Where a standard deviation is not shown, the individual animals (n = 4) were
combined before analysis.

| Location | Before Extension | After Extension |
|---|---|---|
| Oak Park | 3.2 ± 0.3 | 7.1 ± 1.4 |
| Malabar | 7.2 ± 0.4 | 6.4 ± 1.6 |
| Warriewood | 5.2 | 4.9 ± .06 |
| Bondi | 7.0 | 6.6 ± 1.0 |
| Tamarama | 10 | 8.5 ± 1.3 |
| Palm Beach | 5 | 8.5 ± 1.2 |
| Summercloud Bay | 1.9 | 1.6 ± 0.2 |

Table V. Effect of Size of *Pyura stolonifera* on the Silver/Uranium Ratio. Results show the height and width (mm) of individual animals, and the silver and uranium concentrations in μg/g wet weight

| Location | Height | Mid-width | Silver | Uranium | Ag/U |
|---|---|---|---|---|---|
| North Head | 90 | 55 | 0.38 | 0.027 | 14 |
| 2 | 50 | 65 | 0.44 | 0.022 | 20 |
| 3 | 57 | 53 | 0.33 | 0.016 | 21 |
| 4 | 40 | 55 | 0.38 | 0.026 | 15 |
| 5 | 40 | 65 | 0.57 | 0.026 | 22 |
| Bondi | 50 | 40 | 0.22 | 0.029 | 8 |
| 2 | 60 | 35 | 0.24 | 0.020 | 12 |
| 3 | 65 | 35 | 0.19 | 0.032 | 6 |
| 4 | 60 | 35 | 0.26 | 0.023 | 11 |
| 5 | 40 | 50 | 0.33 | 0.037 | 9 |

ng/L, compared with 0.6 ng/L in the northern reach of the estuary, and 0.3 ng/L in pristine coastal seawater (5,8). Approximately 20 kg/day of silver is added to San Francisco Bay from industrial wastewater discharges, which have silver concentrations as high as 20,000 ng/L (5,8). Martin et al. (8) reported concentrations of silver in mussels from Royal Palms as high as 15 μg/g wet weight, compared with 0.02 μg/g wet weight silver in mussels from pristine coastal areas of California.

Smith and Flegal (5) found that most (~70%) of the silver in San Francisco Bay waters was in a non-filterable form. The most likely explanation for the insignificant change in the silver concentration in *Pyura* after the sewer outfall extension is that the animal is concentrating silver attached to particles of near-shore sediment contaminated with silver before the sewer extensions were made. This explanation is supported by the observation that after the extensions, copper and lead are still elevated in *Pyura* compared to background levels, whereas cadmium and zinc are not. Lead and copper adsorb much more strongly to sediment particles than do cadmium and zinc (9). *Pyura* appears to have a different mechanism for silver adsorption than other animals. Abbe and Sanders (10) found that the oyster *Crassostrea virginica* strongly concentrates dissolved, but not particulate, silver, and Connell et al. found a similar situation for grass shrimps (11).

**Conclusions**

1. *Pyura stolonifera* strongly concentrated silver from seawater, making it a useful biomonitor for this metal in aquatic systems.
2. Silver concentrations were elevated in *Pyura* collected from near Sydney's major sewage outfalls compared to *Pyura* collected from pristine areas.
3. Extension of Sydney's major sewage outfall pipes did not significantly decrease silver concentrations in *Pyura*, possibly because the ascidian concentrated silver from contaminated sediment particles.

**Literature Cited**

1.  Florence, T.M.; Farrar, Y.J. *J. Electroanal. Chem.* **1974**, *51*, 191-200.
2.  Jones, L. Thesis, University of Technology, Sydney, Australia, 1988.
3.  Martin, J.H.; Knauer, G.A.; Gordon, R.M. *Nature* **1983**, *305*, 306-9.
4.  Sanudo-Wilhelmy, S.A.; Flegal, A.R. *Environ. Sci. Technol.* **1992**, *26*, 2147-51.
5.  Smith, G.J.; Flegal, A.R. *Estuaries* **1993**, *16*, 547-58.
6.  Apte, S.C.; Batley, G.E.; Szymczak, R.; Rendell, P.S.; Lee, R.; Waite, T.D. *Aust. J. Mar. Fresh. Res.* (in press).
7.  Riley, J.P.;Skirrow, G. Chemical Oceanography, Vol 2, **1965**, Academic Press, N.Y., pp 429-432.
8.  Martin, M.; Stephenson, M.D.; Smith, D.R.; Gutierrez-Galindo, E.A.; Munoz, G.F. *Mar. Poll. Bull.* **1988**, *19*, 512-20.
9.  Florence, T.M. *Talanta* **1982**, *29*, 345-64.
10. Abbe, G.R.; Sanders, J.G. *Estuar. Coast. Shelf Sci.* **1990**, *31*, 113-23.
11. Connell, D.B.; Sanders, J.R.; Riedel, G.F.; Abbe, G.R. *Environ. Sci. Technol.* **1991**, *25*, 921-4.

# Chapter 6

# The Moose (*Alces alces* L.), A Fast and Sensitive Monitor of Environmental Changes

A. Frank[1] and V. Galgan[2]

[1]Centre for Metal Biology, P.O. Box 535, S–75121 Uppsala, Sweden
[2]National Veterinary Institute, Department of Chemistry,
P.O. Box 7073, S–75007 Uppsala, Sweden

Liver and kidney samples from 4,360 moose (*Alces alces* L.) were collected in the whole of Sweden in 1982 and analyzed for 14 essential and non-essential elements. The organ samples constitute a reference material, for use in future attempts to detect environmental changes with use of the moose as monitor. Changes in pH influence the mobility of metals. As a typical effect of acidification, decreasing pH in the soil results in increased uptake of cadmium, via plants, by the moose tissues. Accordingly, regional differences in the cadmium burden were observed within the reference material depending on the extent of acidification. By expressing the cadmium burden as the average renal cadmium uptake per year, different regions can be compared. In addition to comparing the cadmium burden between Swedish regions, this burden in Alaska and Canada was compared with Swedish levels by using the moose as monitor. With increasing pH, e.g. as a result of liming, levels of cadmium and other metals decrease, but molybdenum level increases, as found in the moose organs. A disturbed balance between copper and molybdenum can cause severe copper deficiency in the animals, leading to death, as reported from Sweden in the late 1980s. An environmental change within a period of only 5 years can be recognized using the the moose as monitor.

Monitors in the environmental context have been defined as organisms in which changes in known characteristics can be measured in order to assess the extent of environmental contamination, so that conclusions regarding the health implication for other species of the environment as a whole can be drawn (*1*). Monitors can give information about the environmental concentration both of essential and non-essential metals, in that they are able to demonstrate deficiency and toxicity, and over a period of time can also show changes in the concentrations of the metals in the environment. Environmental contamination may include contamination by metals, the origin of which may be natural and/or anthropogenic. The metals may affect life or disturb life processes on different trophic levels, which may influence the choice of suitable monitors (2). Too low or too high a concentration of essential elements, and too high a concentration of toxic elements in the bedrock or soil can cause a state of deficiency or toxicity and thus become a direct risk for human and animal health (3, 4, 5).

Atmospheric deposition of oxides of sulfur and nitrogen (acid rain) may influence the weathering of bedrock and soils. A decreasing pH causes mobilization of metals in the upper soil layer, leading to greater availability of metals, via plants, to grazing animals. Elements which are easily mobilized are Ca, Mg, K, Na, Al, Cd, Mn, Ni and Zn, and to a lesser extent Hg, Pb and Cu. When the buffering capacity of the soil is insufficient, acid rain results in leaching and eluting of metals essential for plants and animals, thereby causing deficiencies of these metals in plants and, via plants, in herbivorous animals.

*Certain essential trace elements such as Se and Mo become less soluble in an acidic environment, which means that their availability for plants will decrease. In the case of an increasing pH, the chemical behavior of these elements will be the reverse.* A change in uptake by herbivorous animals via plants may result in a change in metal concentrations in the animal tissues as well as a change in the relationships between the concentrations of different elements in the tissues, with severe consequences for the health status of grazing animals. Selenium uptake by animals via plants is influenced by different parameters (6) such as the geochemical background, aerial deposition of selenium from natural and anthropogenic sources, the buffer capacity of the soil, and the chemical form of the selenium. In the western parts of Sweden, the atmospheric transport of organic selenium compounds from the marine environment compensates to some degree for the loss of selenium bioavailability by acidification (7).

In a search for monitors of cadmium among the species of the wild Swedish fauna during 1973 - 1976, the moose (*Alces alces* L.) was found to be useful for this purpose (8, 9, 10) according to the above definitions (1). The moose, a large wild ruminant, is found in most parts of Sweden. It is a relatively stationary animal and its range of seasonal migration seldom exceeds 50 - 60 km. Its age is easy to determine. Tissues for investigation can be obtained through cooperation with hunters' organizations during regular hunting seasons in the autumn. Moreover, the moose acts as an integrating "sample collector" of essential and non-essential elements, browsing a great variety of plants. The moose lives in the northern hemisphere - Scandinavia, Russia, Poland, Canada, Alaska, and some northern parts of the contiguous United States.

From 4,360 moose, samples of liver and kidneys, and also of the left mandible for age determination, were collected in the whole of Sweden during the hunting season in 1982. All 24 counties but one participated in the sample collection. Tissue samples were pretreated by automatic wet digestion (11, 12) and submitted to simultaneous multielement analysis with respect to 13 metals using a direct current plasma-atomic emission spectrometer (DCP-AES) (13). The metals assayed were Al, Ca, Cd, Co, Cr, Cu, Fe, Mg, Mn, Ni, Pb, V and Zn. Molybdenum was determined by

ICP-AES. Selenium was measured by flow-injection hydride generation atomic absorption spectrometry (FI-HG-AAS) (*14*).

The assays yielded *metal concentrations in the liver and kidney samples, constituting a reference material for the year 1982* (*7, 9, 15*).

In the present study **regional** differences were found, reflecting differences in the *availability* of metals, via plants, to the moose. The concentrations of essential and non-essential elements in the liver and kidneys provided information about normal variations in the different regions, which are important from toxicological and nutritional points of view. Comparisons of values obtained in future investigations with reference values from the same geographic region from a previous year will demonstrate changes in the environment.

Among different parameters influencing the mobility of metals and their uptake by animals via plants, the **pH of the soil** appears to be the most important. This seems to be more significant than, for example, the decreasing rate of atmospheric deposition of heavy metals. Data of soil pH in Sweden are compiled in maps (16). At a low pH of the soil, e.g. 3.8-4.2, most of the metals, including cadmium, become mobilized, but molybdenum becomes bound to the soil. *Increasing pH, e.g. when liming, has the opposite effect, molybdenum now being mobilized - as mentioned above.* An increase in pH from 5.0 to 5.5 results in an approximately 45 percent increase in Mo uptake by grasses and clovers (*17*). The chemistry of Se in the soil is complex (*6*). Selenium in soils is present in the form of different molecular species and states of oxidation. The water soluble selenate is available for plants, but the transformation of the different compounds into selenate by a chain of chemical reactions is a slow pocess and is influenced by the redox potential of the soil. In a reductive, acid environment the bioavailability of Se slowly decreases, and similarly, when the pH increases the transformation of Se compounds to selenate takes time and becomes delayed (6).

## A Map of the Cadmium Burden of Sweden, Corresponding to a Map of Acidification

At the time of sample collection, the main interest was focused on the cadmium burden in the different geographic regions of Sweden. The critical organ for cadmium is the kidney, which is also the storage organ, although this meta accumulates to a lesser extent in the liver as well. The accumulation of cadmium is age-dependent. To some extent, the rate of accumulation depends on the total amount of cadmium present in the environment. However, influencing factors include the geochemical background, the atmospheric transport and deposition of particulate matter as well as of oxides of sulfur and nitrogen (acid rain), and anthropogenic activities, all of which seem to contribute to the flux of cadmium in the biosphere.

A number of statistical models were tested to express the cadmium burden of a population of animals. Earlier investigations have demonstrated very high cadmium burdens in the south and south-west of the country (*9, 15*). The cadmium burdens of the moose and other animals are presented in different ways in the literature. Comparisons between different regions are difficult. In the present study, therefore, to be able to compare the cadmium burdens of different regions, a new concept was developed. A statistical model allowing such comparisons has been found most useful

(*18*) and may be described briefly as follows: A quotient is formed by dividing the cadmium concentration of the organ in question with the estimated age class of the animal in the region investigated. The median quotient is calculated for every age class and is weighted for age distribution and the number of individuals in percent in each age class. The resulting constant is characteristic for the region in question and represents the **average renal or hepatic cadmium uptake per year** (mg Cd/kg tissue wet weight). The burden is expressed by calculating the **yearly average cadmium uptake by the moose kidney**. Using this model, the cadmium burdens in the investigated regions were calculated. A shaded map expressing these burdens in 25 regions of 23 counties of Sweden in 1982 is shown in Figure 1. (No moose on the island of Gotland (I).)

The counties represented by letters in Figure 1 are as follows: AB, Stockholm; C, Uppsala; D, Södermanland; E, Östergötland; F, Jönköping; G, Kronoberg; H, Kalmar; I, Gotland; K, Blekinge; L, Kristianstad; M, Malmö; N, Halland; O, Göteborg & Bohus; P, Älvsborg; R, Skaraborg; S, Värmland; T, Örebro; U, Västmanland; W, Kopparberg; X, Gävleborg; Y, Västernorrland; Z, Jämtland; AC, Västerbotten; BD, Norrbotten.

When this cadmium map as shown in Figure 1 was compared with 1) a map of atmospheric deposition of cadmium, measured by moss analysis (*19*), 2) a bedrock geological map, 3) a map of acidification (*20*), and 4) a biogeochemical map (*21*), the best correspondence was found with the latter two maps. Biogeochemical maps are based on metal determinations in biological material consisting of roots of aquatic plants from stream banks and aquatic mosses.

Although atmospheric deposition of particulate matter contributes to the cadmium burden, the main effect in the southern and south-western parts of Sweden is attributable to metal mobilization by acid rain. Thus, the cadmium burden is highest in regions closest to the pollution-emitting countries west and south of Scandinavia, and declines in the eastern and northern parts of Sweden as illustrated in Figure 1. The relationship between counties can be as high as 3:1 (*9, 15*).

The advantage of the method described is that it permits **comparison of the cadmium burden between geographical regions on a standardized basis both nationally and internationally.** By using the model of average renal cadmium uptake per year, a rough comparison can be made with the cadmium burdens in Alaska 1986 and in Canada in 1985. Moose in one region of Alaska had a 30% lower (Yakutat) and in another region a 30% higher (Galena) cadmium burden than the lowest and highest values in Sweden, respectively (Frank, A.; Zarnke, R.L., unpublished data). The moose in Huntsville, Ontario, Canada, had in 1985 a cadmium burden three times higher than the highest Swedish values ( Frank, A.; Addison, E.M., unpublished data).

As mentioned above, the mobility of cadmium and its uptake in the moose via plants is pH dependent. In the present study a correlation was found between the degree of acidification and the cadmium burden of the moose in individual regions. Thus, *the moose is a good monitor* in reflecting the relationship between acidification and cadmium burden. In a similar manner it may be possible in the future to reveal environmental changes as a function of acidification (*9, 15*).

## Environmental Changes Demonstrated by Monitoring after Only 4 - 5 Years

During the 1980s a previously unrecognized disease of the moose, with an unknown etiology, was reported in one of the most acidified south-west parts of Sweden, in the county of Älvsborg, "P" in the map shown in Figure 1. Tissue metal concentrations were again measured in 1988, 1992 and 1994, in the same manner as before, and were related to the metal concentrations of the reference material of 1982.

The results showed changes in the metal concentrations in the investigated tissues (liver and kidneys) of the moose during the period 1982 - 1994, as illustrated for Mo, Cu and Cr in Figure 2.

Decreased hepatic copper concentrations, relative to the reference values, were noted in 1988 (reduction by 30%), in 1992 (50%) (*22*), and in 1994 (50%). Chromium declined continuously up to 1994, to the level of the detection limit. Decreases in the concentrations of other essential and non-essential metals were noted in the liver. The concentrations of Fe, Zn, Pb, and Cd were also decreased. In the kidneys, reductions in Ca, Mg, and Mn indicated severe metabolic disturbances (*22*). The cadmium concentration in the kidneys decreased by about 30% between 1982 and 1988, which is the opposite of what could be expected during acidification.

Another unexpected finding in this strongly acidified environment was an increase in molybdenum in the liver by 20 - 40% during the study period, as shown in Figure 2. Molybdenum is present in the environment in anionic form, in contrast to the other metals, which are found in the cationic form. When the pH in the soil decreases, molybdenum becomes less available to plants and its uptake diminishes. As mentioned above, at pH 5 a pH increase of 0.5 causes an increase in molybdenum uptake by plants of about 45 percent by grass and clover (*17*). Our findings of a decreasing level of copper and an increasing level of molybdenum in the liver, evidently indicate an increased pH in the environment of the moose.

Wetlands, lakes, fields, pastures, and to a lesser extent forest areas of this acidified region have been *limed* since the beginning of the 1980s. The liming activities were intensified during the second half of the 1980s, concomitantly with the occurrence of the moose disease. In addition, as a result of a changing flora due to decades of acidification, the moose seems to have altered its browsing habits. The animals now also frequently browse on cultivated fields of pasture (clover and grass), oats, and rape, all of which fields are heavily limed by the farmers. An *increased pH* in the environment of the moose is suggested as a plausible explanation for the increased hepatic molybdenum concentrations and for the concomitant decrease in, for example, cadmium.

Increasing molybdenum relative to the copper content in the feed reportedly causes copper deficiency in ruminants (17). Cattle, especially, have shown a high susceptibility to copper deficiency (*17, 23*). The clinical signs in the cattle are closely similar to those observed in the moose disease (*24, 25, 26, 27*). Thus, **secondary copper deficiency, chromium deficiency and a disturbance of trace element metabolism** have been suggested as the cause of the disease (*22*).

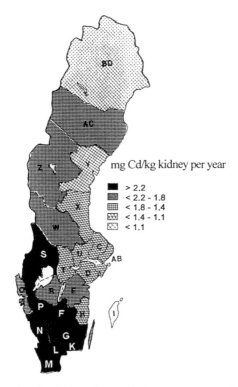

Figure 1. Cadmium burden in Sweden, with the moose as monitor. Average renal cadmium uptake per year (mg/kg w.w.). The highest cadmium burden was found in the southern and western parts of the country. The darker the shading, the higher cadmium burden. For explanation of letters, see text.

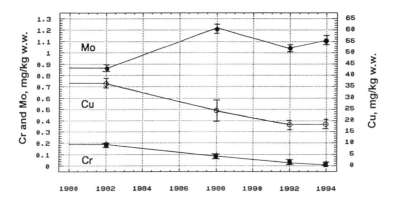

Figure 2. Molybdenum, copper and chromium in moose liver, yearlings. Median concentrations $(\tilde{x} \pm SD)$ are expressed on a wet weight basis. Southern district of the county of Älvsborg "P", 1982-1994.

## Conclusion

The regional differences in the cadmium burden, as evident from the differences in the cadmium uptake in the moose, are highly related to the pH-dependent mobility/bioavailability of the metal. In addition, after a period of only 4 - 5 years, in 1988, a short time in the history of nature, significant changes, such as a 30% reduction in the copper and cadmium contents of the liver and kidneys, respectively, were displayed by the moose. Moreover, after ten years, in 1992, the copper content was *reduced* by 50 %, cadmium by 35 %, chromium by about 80 %, and lead by 60 - 70 %, but that of molybdenum *increased* by 20 - 40 %. Changes in the tissue concentrations of molybdenum, copper and chromium clearly reflect alterations in environmental conditions.

It is evident from the above two examples that *the moose is a useful, fast and sensitive monitor of environmental changes.*

## Acknowledgments

The authors whish to thank the Swedish Environmental Protection Agency, Solna, the Swedish Hunters' Association, Spånga and "Stiftelsen Rädda Älgstammen", Borås, Sweden for supporting these studies financially.

## Literature Cited

1. O´Brien, D.J.; Kaneene, J.B.; Poppenga, R.H. *Environ. Health Perspect.* **1993**, *99*, 351-68.
2. Wren, C.D. *Review of occurrence and toxicity of metals in wild animals*; Ref. no. KN107-2-4609. Canadian Wildlife Service, 100 Gamelin Bvd. Hull, Quebec, 1983; p 158.
3. Låg, J. In: *Geomedicine*; Låg, J., Ed.; CRC Press: Boca Raton, Ann Arbor, London, 1990; pp 1-24.
4. Låg, J. In: *Human and animal health in relation to circulation processes of selenium and cadmium*; Låg, J., Ed.; The Norwegian Academy of Science and Letters, 1991; pp 9-21.
5. Thornton, I. In: *Cadmium in the environment*; Mislin, H.; Ravera, O. Eds.; Birkhäuser Verlag, Basel, 1986; p 8.
6. Kabata-Pendias, A.; Pendias, H. *Trace elements in soil and plants*; 2nd ed., CRC Press: Boca Raton, Ann Arbor, London, 1992; pp 217-225.
7. Galgan, V.; Frank, A. *Sci. Total Environ.* **1995**, *172*, 37-45.
8. Frank, A. *Sci. Total Environ.* **1986**, *57*, 57-65.
9. Frank, A.; Petersson, L.R. *Fresenius' Z. Anal. Chem.* **1984**, *317*, 652-53.
10. Frank, A.; Petersson, L.R.; Mörner, T. *Svensk Veterinärtidn.***1981**, *33*, 151-56.
11. Frank, A. *Fresenius' Z. Anal. Chem.* **1976**, *279*, 101-2.
12. Frank, A. In: *Trace element analytical chemistry in medicine and biology;* Brätter, P.; Schramel, P., Eds.; Proceedings of the Fifth International Workshop, Walter de Gruyter, Berlin, New York, 1988, Vol. 5; pp 78-83.
13. Frank, A.; Petersson, L.R. *Spectrochim Acta* **1983**, *38 B*, 207-20.

14. Galgan, V.; Frank, A. In: *Trace element analytical chemistry in medicine and biology*; Brätter, P.; Schramel, P., Eds.; Proceedings of the 5th International Workshop, Walter de Gruyter, Berlin, New York, 1988; Vol. 5; pp 84-89.

15. Frank, A.; Petersson, L.R.; Mörner, T. In: *Das freilebende Tier als Indikator für den Funktionszustand der Umwelt*; Proceedings. Symposium. nov. 1984; Institut für Wildtierkunde, Savoyenstrasse 1, A-1160, Wien, Austria.; pp 83-94.

16. *The Swedish National Atlas*; Environment; Bernes, C.; Grundsten, C., Eds.; Bokförlaget Bra Böcker: Höganäs, Sweden, 1991; p 42.

17. Mills, C.F.; Davies, G.K. In: *Trace elements in Human and Animal Nutrition*; Mertz, W., Ed.; Academic Press, Inc.: Orlando San Diego, New York, Austin, London, Montreal, Sydney, Tokyo, Toronto, 1987, Vol.1; pp 429-57.

18. Eriksson, O.; Frank, A.; Nordkvist, M.; Petersson, L.R. *Rangifer* 1990, Special Issue No. 3. pp 315-331.

19. Rühling, Å.; Skärby, L. *National survey of regional heavy metal concentrations in moss*; PM. 1191; Swedish Environmental Protection Agency, Solna, Sweden, 1979.

20. *Monitor 1981*; Swedish Environmental Protection Agency, Solna, Sweden, 1981; p 131.

21. Selinus, O.; Frank, A.; Galgan, V. In: *Environmental Geochemistry and Health in Developing Countries*; Appleton, D.; Fuge, R.; McCall, J. Eds.; Special Publication No. 113. Geological Society of London; Chapman and Hall, 1996; pp 81-89.

22. Frank, A.; Galgan, V.; Petersson, L.R. *Ambio*, 1994, *23*, 315 - 17.

23. Blood, D.C.; Radostits, O.M. *Veterinary Medicine;* 7th ed. Baillière Tindall, London, Philadelphia, Sidney, Tokyo, Toronto, 1989; pp 1160-73.

24. Feinstein, R.; Rehbinder, C.; Rivera, E.; Nikkilä, T.; Stéen, M. *Acta Vet. Scand.* 1987, *28*, 197-200.

25. Rehbinder, C.; Gimeno, E.; Belák, S.; Stéen, M.; Rivera, E.; Nikkilä, T. A. *Vet. Rec.* 1991, *129*, 552-54.

26. Stéen, M.; Diaz, R.; Faber, W.E. *Rangifer* 1993, *13*, 149-156.

27. Stéen, M.; Frank, A.; Bergsten, M.; Rehbinder, C. *Svensk Veterinärtidn.* 1989, *41*, 73-77.

# Chapter 7

# Use of Manganese Concentration in Bivalves as an Indicator of Water Pollution in Japanese Brackish Lakes

Yoshiko Yano

Tokyo National College of Technology, 1220–2 Kunugidacho
Hachioji, Tokyo (193), Japan

A bivalve, *Corbicula japonica* or *Yamatoshijimi*, lives in brackish water and its habitats are widespread in Japan. This paper concerns Mn which is accumulated in the soft tissue of *Corbicula japonica*. Recently very high amount of Mn was found in *Corbicula japonica* living in Lake Shinjiko. A lake, located in a densely populated district, often suffers from deficiency of dissolved oxygen because of eutrophication. Under such circumstances Mn(II) can be liberated into water from sediment and it can be taken by aquatic life. Lake Shinjiko is surrounded by a big city, and it has an eutrophication problem. The maximum level of Mn ever observed was more than 2000 $\mu$g/g of specimen from Lake Shinjiko on dry weight basis. Typical amounts in Lake Hachirogata and Lake Jusanko were 140 and 80 $\mu$g/g, respectively. Population density decreases around the three lakes in the above order. By contrast, *Corbicula japonica* at the mouth of the Tonegawa showed only 10-20 $\mu$g/g in spite of its location in densely populated district. The flow may help in enriching the river water with dissolved oxygen. Manganese in the tissue showed a distinct increase during 15 months, 9 to 130 $\mu$g/g, by transfer from the Tonegawa to Lake Shinjiko. It was shown that the concentration of Mn in *Corbicula japonica* and other related species could be an indicator of oxygen deficiency at the very bottom of hydrosphere.

Trace element content in the body of living creatures is largely affected by the surroundings where they live, and they are accumulated on the physiological demands of them. This is why biological systems are useful for monitoring various environmental conditions. However, relative abundance ratio of trace elements in organisms may not necessarily be same as those in the surroundings. There have been several observations (*1-5*) that some types of organisms seem to accumulate a particular element through bioconcentration.

Several years ago during a study of the contents of minerals as nutrients in foods, the author noticed that the concentration of Mn in the bivalve, called *Shijimi* in Japan, varied largely from sample to sample which was purchased at a store. The finding seemed important for nutrition because it had been proposed to add Mn (6) into food component table in Japan as an item. Also the finding led the author to speculate that Mn concentration in the clam might reflect environmental conditions of the place of its origin. The author hypothesized that the concentration of Mn could reflect conditions of brackish hydrosphere since *Shijimi* clam is often a representative creature in brackish water. The quality of the brackish lake waters, in contrast with lakes in the mountains, would indicate more exactly the effect of humans on the environment.

## Materials and Methods

**Sample Collection and Treatment**. *Corbicula japonica*, or *Yamatoshijimi* in Japanese, is 10 - 30mm in shell length and 2 - 10 g in whole body weight including shells. Among several related species, it was used in the present study. It lives and breeds in brackish hydrosphere. Places from where these bivalves were taken to be analyzed are shown in Figure 1 ; (a) Lake Shinjiko, Shimane Prefecture, (b) Lake Hachirogata, Akita Prefecture, (c) Lake Jusanko, Aomori Prefecture, and (d) the mouth of the Tonegawa, on a boundary between Ibaraki and Chiba Prefectures. Usually 2 kg of the clams was collected at once. Living clams collected were kept in fresh water for one day and at least 300 g of them was randomly taken and then frozen.

About 10 kg of living clams of *Corbicula japonica* were collected at the mouth of the Tonegawa near the end of April 1994. Most of the clams were immediately transferred to Lake Shinjiko. Then they were divided into four portions of about 2 kg each and each portion was put into a cage of 70W x 50D x 40H in cm. The four cages with the clams inside were sunk and half-buried into the bottom ground where native *Corbicula japonica* of Lake Shinjiko lives. Five months later the first cage was pulled up and the clams were re-taken as a sample. Native clams of Lake Shinjiko were also collected to be compared with the transferred clams on the same date at the same sampling point. Further four and ten months later the sampling was done in the same way.

Also, *Corbicula japonica* from Lake Shinjiko were kept in artificial brackish water without any feed in an aquarium with aeration. About 300 g of each of the clams was taken from the aquarium once a week and frozen.

The frozen clams were thawed and shelled. The soft tissue part was homogenized with a blender and freeze-dried to obtain powdered samples. Contamination from stainless steel cutter of the blender was negligible. Also, the varying operation period of the blender had no effect on the analytical results.

The powdered sample prepared was further homogenized by an agate ball mill and kept at 85 °C until a constant weight was attained, and then about 0.3 g of the powder was weighed. The weighed sample was decomposed by overnight digestion with 3 mL of nitric acid followed by heating with additional 1 mL of nitric acid and fuming treatment with 1 mL each of hydrofluoric and perchloric acid at a temperature of 190 °C.

The residue was diluted to 10 mL with pure water and 0.5 mL of concentrated nitric acid.

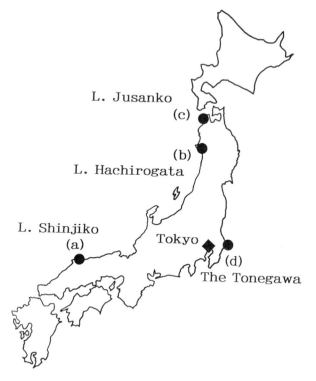

L. Jusanko
(c)

(b)
L. Hachirogata

L. Shinjiko
(a)

Tokyo

(d)
The Tonegawa

**Figure 1. Sampling Locations for *Corbicula japonica*.**

The acids used for decomposition were reagents specified for the trace analysis of heavy metals. Ultra-pure water produced by Yamato WQ500 was used throughout all experimental procedures.

Two types of laboratory wares were utilized for the decomposition procedure mentioned above. One was a PFA (perfluoroalcoxy fluororesins) test tube (18 i.d., 21 o.d. and 130 mm h.). An aluminium block heater was used to heat the test tube. The other was a pressurized PFA container in a PTFE outer vessel with a stainless steel jacket. No difference was observed in analytical results between the two, though the latter container had been expected more effective to complete decomposition because of the pressure caused by heating.

**Analytical Methods.** Inductively coupled plasma atomic emission spectrometry (ICP-AES) was used for determining the amount of trace elements Cu, Fe, Mn, and Zn in the soft tissue of *Corbicula japonica*. A Jarrell-Ash Model 96-975 ICP-AES instrument was operated under the usual condition. In order to confirm the validity of the analytical method including pretreatment and ashing procedures, the biological reference material, mussel tissue powder (7) prepared by National Institute for Environmental Studies (NIES), Japan Environment Agency, was analyzed.

When Mn was the only element to be analyzed, flame atomic absorption spectro-metry (FAAS) was applied. A Seiko SAS 7500 instrument was used with an attached micro-volume PTFE cell (8) for supplying sample solution into the acetylene-air flame. Interferences from co-existing salts were examined in advance by comparison between the method of standard additions and conventional calibration. The linear calibration curve method was found to be satisfactory since the difference between them did not exceed 5%.

## Results and Discussions

Usually 3 to 5 portions of each homogenized sample powder were taken and analyzed repeatedly. A mean value, $\bar{x}$, with standard deviation (SD) and a median, $\tilde{x}$, with quartile deviation (QD) were obtained. All data of metal concentrations in the present paper are expressed by $\mu g$ per g of soft tissue on dry weight basis, which is abbreviated as "$\mu g/g$ tissue, d.w.". Statistical significances between metal concentrations of two or more samples were calculated mainly according to the Kruskal Wallis test (9). The Wilcoxon two-sample test (9) was also used when a small size of one statistical sample, i.e. $n \leq 5$, was compared with a similar size of the other.

Analytical values of trace elements concentrations found in the biological reference material agreed with certified ones as shown in Table I. The results indicated that our analytical method was appropriate.

**Table I. Concentrations of trace elements in a biological reference material, NIES No.6 (mussel)**

All values are given in $\mu g/g$ tissue, d.w.

|  |  | Cu | Fe | Mn | Zn |
|---|---|---|---|---|---|
| Found (n=7) | $\bar{x} \pm$ SD | $5.3 \pm 0.4$ | $153 \pm 17$ | $13.8 \pm 2.2$ | $112 \pm 16$ |
|  | Range | $4.9 - 6.1$ | $120 - 170$ | $10.1 - 17.3$ | $86 - 135$ |
| Certified |  | $4.9 \pm 0.3$ | $158 \pm 8$ | $16.3 \pm 1.2$ | $106 \pm 6$ |

**Concentration of Trace Elements in *Corbicula japonica* from Different Habitats.** Table II shows the analytical results of Cu, Fe, Mn, and Zn in soft tissue. The results of the statistical calculations were also listed in the Table. Manganese content differed largely from habitat to habitat. The maximum concentration of Mn ever observed was 2800 $\mu g/g$ in a powdered sample from Lake Shinjiko. Slight statistical dependence on habitats was found for Zn concentrations. Both the Cu and Fe concentrations showed no dependence on habitats.

Figure 2 shows the impact of population densities around the four habitats of *Corbicula japonica* on the Mn content of the clams. The denser the human population near a lake habitat, the higher the risk of pollution. When the pollution causes deficiency of dissolved oxygen, Mn(II) ion can be liberated into water (10) from

**Table II    Metal Concentrations in the Soft Tissue of *Corbicula japonica***

All values of mean ($\bar{x}$), median ($\tilde{x}$), and range are given in μg/g, d.w.

| | Habitats | n | $\bar{x}\pm$SD | $\tilde{x}\pm$QD | Range | Significances | |
|---|---|---|---|---|---|---|---|
| Cu | The Tonegawa | 19 | 30.7±6.6 | 30.2±53[a] | 20.2–40.3 | n.s. | ab ac ad bc bd cd |
| | L. Jusanko | 3 | 34.3±11.3 | 38.3±5.3[b] | 21.6–42.9 | | |
| | L. Hachirogata | 5 | 30.4±6.6 | 29.2±4.8[c] | 22.7–38.7 | | |
| | L. Shinjiko | 13 | 37.3±19.1 | 33.9±4.0[d] | 21.2–99.3 | | |
| Fe | The Tonegawa | 19 | 526±308 | 431±169[a] | 185–1342 | n.s. | ab ac ad bc bd cd |
| | L. Jusanko | 3 | 852±310 | 1011±401[b] | 495–1050 | | |
| | L. Hachirogata | 5 | 652±387 | 398±338[c] | 321–1080 | | |
| | L. Shinjiko | 13 | 621±234 | 576±117[d] | 299–1210 | | |
| Mn | The Tonegawa | 22 | 19.1±9.4 | 15.9±5.3[a] | 7.5–49.4 | *** | ac ad |
| | L. Jusanko | 3 | 74.5±27.2 | 79.8±13[b] | 45.1–98.7 | ** | ab bd cd |
| | L. Hachirogata | 5 | 137±69 | 145±43[c] | 54.0–227 | n.s. | bc |
| | L. Shinjiko | 18 | 814±697 | 534±353[d] | 133–2820 | | |
| Zn | The Tonegawa | 19 | 147±21 | 144±11[a] | 108–190 | * | ad |
| | L. Jusanko | 3 | 167±9 | 163±4[b] | 161–178 | n.s. | ab ac bc bd cd |
| | L. Hachirogata | 5 | 178±35 | 177±21[c] | 130–220 | | |
| | L. Shinjiko | 13 | 173±33 | 184±25[d] | 117–224 | | |

n : number of samples.

Significances :  n.s. : not significant,  * : $P < 0.05$,  ** : $P < 0.01$,  *** : $P < 0.001$ .

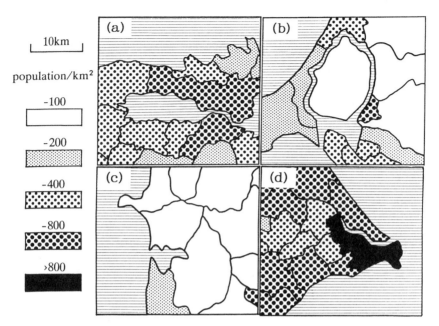

**Figure 2. Population Densities around the Four Habitats of**
*Corbicula japonica.* (a) L. Shinjiko, (b) L. Hachirogata, (c) L. Jusanko,
(d) The Tonegawa.

sediment and it can be easily taken by aquatic life directly and/or through a food chain
in the hydrosphere.    In fact, Lake Shinjiko and Lake Hachirogata are both listed as
eutrophic and Lake Jusanko as mesotrophic (*11*). At the mouth of the Tonegawa, the
river flow and tide facilitate the mixing of oxygen into the water.

In conclusion, it may be possible to use the concentration of Mn in *Corbicula
japonica* as an indicator of oxygen deficiency at the very bottom of the hydrosphere.

**Seasonal Effect on Mn Content in *Corbicula japonica*.**    In order to study the
seasonal effect on Mn content, *Corbicula japonica* was collected monthly at the mouth
of the Tonegawa from March 1990 to March 1991 and analyzed.    The results in Figure
3 show that there was no significant change in the concentration of any of the elements
investigated with change of the season.    Thus it is confirmed that the Mn concentration
in *Corbicula japonica* is not affected by physiological or any other seasonal factors.

**Effect of transfer of *Corbicula japonica* from the Tonegawa to Lake
Shinjiko.**    The sample of clams, which was collected at the mouth of the Tonegawa
and transferred to Lake Shinjiko, had originally very low Mn concentration, 8.8 $\mu$g/g
tissue, d.w.    The Mn level gradually increased as time passed and a value of 128 $\mu$g/g
was observed 488 days later.    During the course of the increase, as shown in Figure
4, Mn concentration decreased for once from 150 to 60 $\mu$g/g.    It may be that the Mn
once taken in by *Corbicula japonica* could be partly discharged with a change of

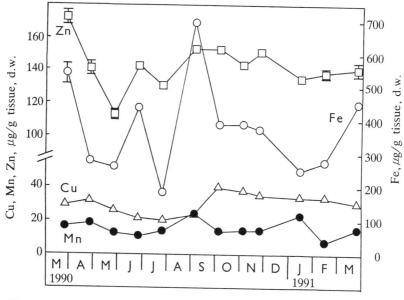

**Figure 3. Monthly Change in Concentration of Cu, Fe, Mn, and Zn in *Corbicula japonica* at the Mouth of the Tonegawa.**

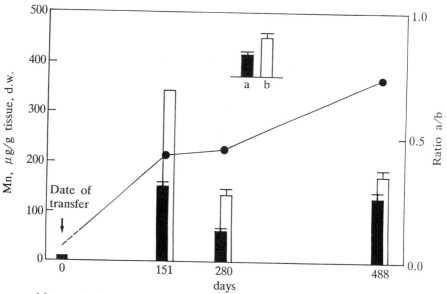

a: Mn, $\mu$g/g, tissue, d.w. of transferred sample from the Tonegawa + SD.
b: Mn, $\mu$g/g, tissue, d.w. of native sample at L.Shinjiko + SD.

**Figure 4. Effect of Transfer from the Tonegawa to Lake Shinjiko on Concentration of Mn in Soft Tissue Part of *Corbicula japonica*.**

Table III   Change in concentration of Cu, Fe, Mn, and Zn in
            *Corbicula japonica* with keeping time in an aquarium.
            All values are given in $\mu$ g/g tissue, d.w.   Number of samples
            was 3 for each mean and median value.   Statistical significances
            for metals between different durations are reported.

| | duration, weeks | $\bar{x} \pm SD$ | $\tilde{x}$ | Range | Significances |
|---|---|---|---|---|---|
| Cu | 0 | 30.1±1.5 | 30.7[a] | 28.4−31.1 | |
| | 1 | 29.4±0.4 | 29.5[b] | 28.9−29.7 | ab  n.s. |
| | 3 | 30.5±0.6 | 30.7[c] | 29.8−30.9 | bc  n.s. |
| | 4 | 29.2±1.1 | 28.6[d] | 28.5−30.6 | cd  n.s. |
| Fe | 0 | 695±36 | 702[a] | 656−727 | |
| | 1 | 384±11 | 381[b] | 375−395 | ab  * |
| | 3 | 333± 8 | 334[c] | 325−341 | bc  * |
| | 4 | 314± 3 | 315[d] | 313−318 | cd  * |
| Mn | 0 | 903±47 | 912[a] | 853−945 | |
| | 1 | 881±50 | 867[b] | 839−937 | ab  n.s. |
| | 3 | 788±10 | 783[c] | 782−799 | bc  * |
| | 4 | 670± 4 | 670[d] | 669−676 | cd  * |
| Zn | 0 | 188±3 | 190[a] | 185−190 | |
| | 1 | 199±2 | 199[b] | 197−201 | ab  * |
| | 3 | 293±9 | 292[c] | 285−303 | bc  * |
| | 4 | 278±2 | 277[d] | 275−280 | cd  n.s. |

n.s.: not significant,   * : P< 0.05 .

environmental condition of the lake.   A similar decrease in Mn concentration was observed in the native clams of Lake Shinjiko at the same sampling point where the clams transferred from the Tonegawa had lived.   It was also noticed that Mn level in the native Shinjiko clams was lower over the period of investigation of effect of the transfer.   The median value and the range of the Mn concentration in the period were 274 and 133-373 $\mu$g/g tissue, d.w., and those observed in 2.5 years prior to the period were 903 and 221-2820 $\mu$g/g tissue, d.w.   The lower level may be due to an improvement of the eutrophic condition of the lake.

Comparison of Mn concentrations was made between the transferred Tonegawa clams and the native Shinjiko clams on the same date.   The results, given in Figure 4 as ratios of Mn concentration in the transferred Tonegawa clams to that of the native Shinjiko ones, evidently increased reaching a maximum of 0.8.   The change might indicate the adaptation process of the transferred *Corbicula japonica* to the environmental condition of the new habitat, Lake Shinjiko.

**Discharge of Mn from *Corbicula japonica*.**   Table III indicates that Mn was slowly discharged from soft tissue when *Corbicula japonica* was kept in a synthetic brackish water aquarium.   Though it was not clear whether the aeration affected the dissolution of Mn(II) in the water, it seemed that discharge of Mn in the soft tissue of *Corbicula japonica* took some time.   Therefore if there was any eutrophication problem in a lake, the Mn concentration would be an indicator of low dissolved oxygen.

**Distribution of Metals in *Corbicula japonica*.**   Distributions of Cu, Fe and Mn in the soft tissue of *Corbicula japonica* from Lake Shinjiko were determined by the technique of two dimensional X-ray fluorescence analysis (*12*) using Mo K$\alpha$ line. Sample preparation for the analysis was as follows; raw soft tissue of several clams was pressed to 1.5 mm in thickness between two glass plates and then it was freeze-dried.   The clams used were picked up from the lot of clams which gave an averaged Mn concentration of 1300  $\mu$ g/g tissue, d.w.   One of results is shown in Figure 5 as chemical images of Mn and Fe.   Note that Mn was highly concentrated in a very small area, possibly the kidney of marine bivalves (*2,4,13*).   Distribution of Fe is similar to that of Mn whereas that of Cu is independent of both Mn and Fe.   Copper was found to be distributed over the whole soft tissue part.

In order to find out elements accompanying Mn other than Fe,   scanning electron microscopy with energy dispersive X-ray spectroscopy (SEM-EDX) was applied. From the pressed and dried specimen of *Corbicula japonica*, small area where Mn was highly concentrated, was cut off, and was ground and dispersed into the water.   Brown colored heavy granules were precipitated and were isolated by washing off the floating matter away.   Results of SEM-EDX suggested that the granule consisted of several kinds of chemical compounds.   It was interesting that Mn was always accompanied by Fe, Ca and high concentration of P.   No other element seemed to coexist as a major component with Mn.   A typical example of an X-ray spectrum containing Mn is shown in Figure 6.   It strongly suggested that Mn, together with Fe, was embedded into calcium phosphate.   Such embedding of Mn onto phosphate had been already reported in the marine bivalve, *Cyclosunetta menstrualis* (*2,4,13*).   There were some other spectra showing Ca and S to suggest calcium sulfate, or showing only Ca as a major

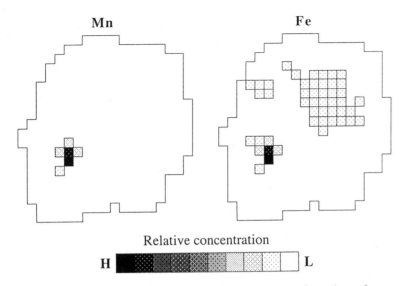

Figure 5. Two Dimensional X-ray Fluorescence Imaging of *Corbicula japonica* from L. Shinjiko Showing Distribution of Mn and Fe. Each square spot corresponds 1mm × 1mm step.

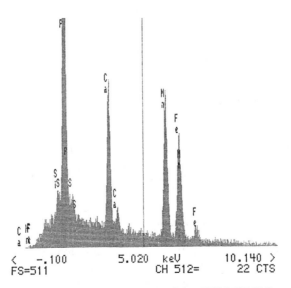

Figure 6. X-ray Spectrum Obtained by SEM-EDX for a Fine Granule from *Corbicula japonica*.

component to suggest calcium carbonate, or showing Si and Al to suggest some aluminosilicate, etc.

## Conclusion

The concentration of Mn in *Corbicula japonica* reflected environmental conditions of brackish lake in Japan. In a eutrophic lake, Mn was accumulated in soft tissue of clams probably as phosphate. It was shown that Mn content in *Corbicula japonica* and in its related species can be a useful indicator of low dissolved oxygen, since their habitats are widespread not only in the lakes and rivers in Japan, but also in China, Korea and Taiwan.

## Acknowledgments

The author expresses her thanks to Mr. Z. Saruta for providing *Corbicula japonica* from various places; to Mr. T. Hara of Shinjiko Fishermen's Association for assistance in taking care of *Corbicula japonica* transferred from the Tonegawa; Ms. Y. Takeuchi for the ICP-AES measurements; and Dr. I. Nakai and Ms. C. Numako for carrying out the two dimensional X-ray and SEM-EDX measurements. The bulk of the analytical data presented here could not have been obtained without help of Ms. H. Oohashi, Ms. A. Kougo, Ms. M. Kataniwa, Ms. S. Hata, Ms. Y. Oda, Ms. T. Oomori, Ms. A. Sakai, Mr. T. Kawashima, Ms. T. Ogasawara, and Ms. M. Ikema. The author is also grateful to Mr. Y. Nakamura, an official of Choshi City, Dr. H. Hashitani, former professor of The University of Shimane, and Dr. I. Sakamoto of Shimane Medical University, for their helpful discussions.

## Literature Cited

1. Henze, M. *Hoppe-Seyler's Z. Physiol. Chem.* **1911**, *72*, 494-501.
2. Ishii, T.; Otake, T.; Hara, M.; Ishikawa, M.; Koyanagi, T. *Bull. Jpn. Soc. Sci. Fish.* **1986**, *52*, 147-154.
3. Ishii, T.; Nakai, I.; Numako, C; Okoshi, K; Otake, T. *Naturwis.* **1993**, *80*, 268-270.
4. Numako, C; Nakai, I.; Okoshi, K; Ishii, T. *Jpn. J. Appl. Phys.* **1993**, *32*, Suppl. 32-2, 556-558.
5. Nakai, I.; Numako, C; Okoshi, K; Ishii, T. *Jpn. J. Appl. Phys.* **1993**, *32*, Suppl. 32-2, 559-560.
6. *Standard Table of Food Composition in Japan -Minerals-* ; Resources Council, Science and Technology Agency, Japan, Ed.; The Printing Bureau of the Ministry of Finance : Tokyo, 1991; 76.
7. Okamoto, K. *Fresenius' Z. Anal. Chem.* **1988**, *37*, 524-527.
8. Uchida, T.; Isoyama, H.; Tatematsu, M.; Kojima, I.; Goto, K.; Matsubara, M. *Bunseki Kagaku* **1988**, *37*, T232-T237 (Jpn).
9. Yanagawa, T. *Nonparametric Hou (Method)*; Baifukan : Tokyo, 1982; 41-42, 118-127 (Jpn).
10. Davison, W.; Woof, C. *Water Res.* **1984**, *18*, 727-734.

11. *Chronological Scientific Tables* ; National Astronomical Observatory, Japan, Ed. ; Maruzen: Tokyo, 1995; 672-673 (Jpn).
12. Nakai, I.; Mochizuki, A.; Iida, A; Taguchi, I.; Yamasaki, K. *Bull. Natl. Museum Jpn. History* **1992,** *38,* 145-166 (Jpn).
13. Numako, C; Nakai, I.; Okoshi, K; Ishii, T. *Bull. Inst. Oceanograph., Monaco* **1994,** *special 14,* 105-113.

# Chapter 8

# Mutagenesis and Acute Toxicity Studies on Saliva-Leached Components of Chewing Tobacco and Simulated Urine Using Bioluminescent Bacteria

**Shane S. Que Hee**

Department of Environmental Health Sciences and Center for Occupational and Environmental Health, School of Public Health, University of California—Los Angeles, 10833 Le Conte Avenue, Los Angeles, CA 90095–1772

The task of minimizing expensive chemical analyses is usually attempted through bioassay techniques or tracers to signal the presence of compounds of interest before complete chemical analysis. The *Salmonella typhimurium* reverse mutation assay has been the chief bioassay used. However only point mutations are determined, and metals and organics genotoxic through other mechanisms are not detected. The use of bioluminescent bacteria in acute (*Photobacterium phosphoreum*) and genotoxic (*Vibrio fischeri* dark mutant) assays has been investigated for compounds in chewing tobacco that are available to simulated saliva and whose metabolites may be present in urine (investigated in simulated urines). Gas chromatography/mass spectrometry revealed that nicotine was the major compound leached into simulated saliva. Nicotine was not genotoxic in the reverse mutation test using a dark mutant, unlike its metabolite, cotinine. Recommendations on the conditions that need to be controlled are provided for these tests.

Chemical analyses of samples containing unknown compounds is expensive. Often toxicity information is the desired endpoint although monitoring for regulated pollutants is often assumed to suffice. There are millions of nonregulated pollutants whose toxicity may be potent, and, if not so, may be present at high enough concentrations to pose threats to biota. This is so for many wastewaters, landfill leachates, and polluted environments. Biological media are even more complex mixtures of endogenous compounds, xenobiotics, and their metabolites and adducts.

Tracers have been used to show that pollutant discharges contribute to environmental quality. Tracers also inform on the degree of contribution of a specific point source to a sample. However, no information is provided on toxicity. One approach is to define any medium having the tracer to be toxic no matter the dilution as the U.S.

EPA does for a diluted listed hazardous waste. For animal bioassays, radiolabels have enabled investigation of toxicokinetics and toxicodynamics of xenobiotics without identifying the labeled species, thus avoiding tedious chemical analyses where endogenous chemicals may interfere. Attempts to identify an unknown compound lead to having to cope with the background of endogenous chemicals.

The best approach is to determine toxicity directly. There are many kinds of toxicities to humans: acute (effects for up to one week after exposure), subacute (effects for up to a month or so), subchronic (effects beyond one month to a year), and chronic (effects beyond a year). In addition ecotoxicological effects to biota have different time frames and susceptibilities. In many cases non-human biota are more at risk to pollution than are humans. The most logical approach is to use chemical tests and microbial assays before plant, invertebrate, and vertebrate bioassays. The Toxicity Characteristic Leaching Procedure (TCLP) measures human bioaccessability of specific organics and inorganics of a solid waste that may potentially pollute ground and surface waters used for human drinking water. Bioassays like the *Salmonella typhimurium* reverse mutation assay (the "Ames Test"} are the most familiar of the microbial assays. The Ames test detects organics that induce point mutations but not metal or organic genotoxins that operate by other mechanisms($1$). A battery screening approach is ultimately necessary if human health effects are to be predicted. Such an approach is also essential for ecotoxicological screening.

There is still need for a general genotoxicity test that not only detects point mutations, but also such DNA effects as large and small deletions, inhibition of synthesis, crosslinking, and intercalation. The reverse mutation assay of a dark mutant of the bioluminescent bacterium *Vibrio fischeri* to regain the bioluminescent state detects all these mechanisms sensitively ($2$). The same instrumentation is used to evaluate acute and chronic toxicity using bioluminescent *Photobacterium phosphoreum* ($3,4$). The effective concentration at which the luminescence is 50% that of the control ($EC_{50}$) is the diagnostic parameter for the acute test. The 5 to 30 min $EC_{50}$ for the acute test (the Microtox test) is correlated to Draize Test results for eye irritation, the acute $LD_{50}$ for *Salmo gairdneri, Daphnia, Spirillum*, and the 14-day $LD_{50}$ for guppies (*Poecilia reticulata*) ($4$). The chronic test 24-h Lowest Observed Effective Concentration (LOEC) is correlated to 7-day *Ceriodaphnia dubia* data ($3$).

The aims of the present paper are to review the author's research with bioluminescent bacteria in assessing the acute toxicity of simulated biological media like urine and saliva in 5.0-mL volumes using *Photobacterium phosphoreum* (*4-8*) and genotoxicity with a dark mutant of *Vibrio fischeri* in O.200 mL volumes (*2*). The measurements were performed on the Microtox Model 500 Analyser (Microbics Corp., Carlsbad, CA) at 15°C for the acute test, and at 27°C for the genotoxicity test.

**Acute Toxicity Experiments**

**Test Media.** The acute test dilution medium was 2.0% saline. Extensive testing (4) showed that the optimum bioluminescence at 5, 15 and 25 min occurred over the sodium chloride range of 1.9-3.3% (ionic strengths of 0.32-0.58 M) with $EC_{50}$ values of 1.3% and 5.0%. Thus sample sizes in toxicity testing should be kept small relative to growth medium. A saline content of 2% discourages growth of many bacterial contaminants, and allows nonsterile conditions to be used during toxicity testing. All light was quenched at pH values below 3.20, with the $EC_{50}$ being pH 5.34. The pH optimum was between 6.5-7.5. These results showed that pH control was essential and that acid media were toxic. To demonstrate whether media truly contained toxic compounds instead of being toxic through pH or osmotic effects, the pH was kept constant in the 6.5-7.5 range and the saline concentration between 2.0-3.0% for environmental and biological media acute toxicity evaluations.

The biological test media examined for acute toxicity were four standard freeze-dried urines and one simulated saliva. Two urines were from the National Institute for Standards and Technology (Gaithersburg, MD) SRM 2670 [normal (I) and spiked (II) concentrations of toxic metals]; two others were from Instrumentation Laboratory (Orangeburg, NY) with Level 1 being unspiked (I) and Level II containing patented amounts of organics and metals. Each liter of simulated saliva contained: 555 mg sodium as sodium chloride; 500 mg potassium as potassium chloride; 100 mg calcium as calcium chloride; 150 mg phosphorus as sodium dihydrogen phosphate; 25 mg magnesium as magnesium chloride; 2700 mg of mucin Type III; 88 mg urea; 200 mg D-glucose; 100 units of amylase; 700 units of lysozyme; and 4 units of phosphatase. The medium was adjusted to pH 7.0 with hydrochloric acid and sodium hydroxide. The additional substrates tested were nicotine (*4*), cotinine (*4*), straight chain aldehydes and carboxylic acids up to $C_{14}$ (*5*), glyoxal and glyoxylic acid, pyruvaldehyde and pyruvic acid, and various ketones, especially methyl ketones that were straight chain up to $C_{14}$ (*8*).

**Aldehydes, Ketones, and Carboxylic Acids.** Quantities in the ng/mL-mg/L range of aldehydes, ketones, and carboxylic acids have been detected as ozonolysis bypro-ducts of drinking water (*9*). Since the luciferin responsible for light emission in *Photobacterium phosphoreum* is *n*-tetradecanal, competitive inhibition of luciferase by aldehydes and carboxylic acids was expected. To make the insoluble compounds bioavailable, aldehydes beyond $C_3$, carboxylic acids beyond $C_7$, and ketones beyond $C_9$ were solubilized in 0.45% methanolic growth medium.

Table I gives the 25 min $EC_{50}$ values for each class of compound (*5*). From $C_1$ to $C_7$, the toxicity for compounds of the same number of carbon atoms decreased in the order carboxylic acid, aldehyde, and ketone. Similarly beyond $C_7$, the order was aldehyde, ketone and then carboxylic acid. The toxicities for formic, acetic, and pyruvic acids were about the same. However formaldehyde was far more toxic than acetaldehyde or butyraldehyde. In contrast acetone was far less toxic than methyl ethyl ketone. Pyruvaldehyde was far more toxic than glyoxal but not as toxic as formaldehyde. Both the $C_{14}$ acid and aldehyde caused stimulation of bioluminescence proving that both can act as luciferins. A cocktail of aldehydes and carboxylic acids at their highest concentrations found in drinking water disinfected with ozone (*9*) was not toxic, indicating that the Microtox test was not sufficiently sensitive. The data show that methyl ketones also inhibit luciferase competitively just as aldehydes and carboxylic acids do. The carbonyl group in the latter is not situated too differently from that in carboxylic acids. Another indication of the importance of the position of the carbonyl group is evidenced from the fact that 3-pentanone is about three times less toxic than 2-pentanone.

Table I. $EC_{50}$ (25-Min) Values (in mM) for Some Aldehydes (ALD),
Ketones (KET), and Carboxylic Acids (CAR) of Different Chain
Lengths ($C_n$) in the Acute Toxicity Test (5,8)

| $C_n$ | ALD | CAR | $KET^a$ |
|---|---|---|---|
| 1 | 0.227 (0.014) | 0.172 (0.0076) | Not applicable |
| 2 | 6.88 (0.51) | 0.160 (0.0096) | Not applicable |
| 3 | Not done | Not done | 243 (19) |
| 4 | 1.374 (0.011) | 0.196 (0.0070) | 51.9 (2.7) |
| 5 | Not done | Not done | 5.30 (0.36) |
| 7 | 0.101 (0.0083) | 0.133 (0.0028) | 0.360 (0.047) |
| 8 | Not done | Not done | 0.0781 (0.011) |
| 10 | 0.0186 (0.0042) | 0.0523 (0.0052) | 0.0275 (0.0026) |
| 11 | 0.0297 (0.0037) | Not done | Not done |
| 12 | Not done | 0.0163 (0.0011) | Not done |
| 13 | 0.00486 (0.00051) | 0.0216 (0.0062) | 0.0070 (0.0014) |
| 14 | Stimulation | Stimulation | Not done |
| 2-Di | 7.40 (0.44) | 0.127 (0.0029) | Not applicable |
| 3-Pyr | 0.540 (0.017) | 0.151 (0.0039) | Not applicable |

[a] Methyl ketones with the carbonyl group at the 2-position.
ALD beyond $C_3$, CAR beyond $C_7$, and KET beyond $C_9$ are dissolved in 0.45%
methanolic saline growth medium.
Di, two carbonyls or carboxylic acid groups.
Pyr, pyruvaldehyde or pyruvic acid.
The values in parentheses are standard deviations of the triplicate measurements.

**Urine Experiments.** The two SRM urines on reconstitution had pH values of 4.5-
4.8, and were toxic because of the acidity. These urines were henceforth adjusted to
pH 6.5. This caused a twenty-fold toxicity decrease. However the supposedly more
toxic Level II SRM was less toxic (25-min $EC_{50}$ of 11.6 +/-0.94 g/L) than the
unspiked urine (25-min $EC_{50}$ of 9.44 +/-0.52 g/L), correcting for creatinine.
Inductively coupled plasma-atomic emission spectroscopy ICP-AES (10)
showed that zinc was present in higher concentrations in the unspiked SRM (1.316
mg/L) than in the spiked (0.494 mg/L) sample, offsetting the higher concentrations of
copper, lead and selenium in the spiked SRM. The zinc is a contaminant from the
vial stopper. Copper (365 ng/mL) accounted for the toxicity of the spiked urine.
The other two reference urines had pH values of 5.8-6.5 and therefore did not
require pH adjustment. Upon correction for creatinine, the unspiked urine had a 25
min-$EC_{50}$ value of 10.2+/-0.98 g/L whereas the spiked urine had a value of 2.57+/-
0.16 g/L, or four times the toxicity. ICP-AES confirmed that heavy metals were
generally higher in the spiked urine than in the unspiked, especially cadmium (30
ng/mL), copper (421 ng/mL), lead (447 ng/mL), selenium (824 ng/mL), and zinc (2.7
mg/L). Ionic strengths of all urines fell within the optimum range so that any effects
could not be caused by this parameter.

Nicotine (25-min $EC_{50}$ of 7.39+/-1.46 x $10^{-4}$ M) was more acutely toxic than cotinine (25-min $EC_{50}$ of 11.8+/-1.35 x $10^{-4}$ M) in 2% saline solution. When nicotine and cotinine were spiked into the urines to produce concentrations found in the urines of tobacco users, the urine matrix was the major toxic contribution rather than the added compound so that toxicities did not differ from those of unspiked urines. Thus unless urines are dilute enough to allow the toxicity of nicotine and cotinine to be manifested, the acute test is more sensitive to the urine matrix than to the nicotine or cotinine. Thus it is unlikely that the acute test will be a sensitive nonspecific screening test for nicotine and cotinine in urine.

**Saliva Testing of Chewing Tobacco Leachates.** Smokeless tobacco is the chief tobacco-based alternative to smoking. Different 3 oz brands of ground chewing tobacco of weight 2 g (Redman long cut; Beechnut wintergreen; and Levi Garett) were extracted with artificial saliva at 37°C for 1 h in a shaking water bath to simulate mastication. After centrifugation and filtration through 5.0-, 2.0-, and 0.45 -µm filters, the pH was adjusted to 7.0 (NaOH), and the toxicity evaluated correcting for color (7). The 25-min $EC_{50}$ values were in mg/mL: Redman, 5.30+/-0.54; Beechnut, 6.46+/-0.91; and Levi Garett, 4.80+/- 0.52, based on dissolved solid dry weights.

An average of 37+/-2% of the original mass was extracted. An EPA type extraction scheme firstly at pH 7.0 (hexane; 3 extractions), then at pH 2.0 (diethyl ether; 3 extractions), and then at pH 10.0 (hexane; 3 extractions) was then performed. The major toxic compounds were extracted at pH 7.0 (4-5 fold increase in $EC_{50}$). Gas chromatography/mass spectroscopy (GC/MS) of the pH 7.0 hexane extract revealed that the major component was nicotine, then dihydroactiniolide, 1H-indole-3-acetonitrile, and an unidentified trace peak. A quantitative nicotine extraction study showed it was >90% extracted with 5 extractions at pH 10.0, but at pH 7.0 even 10 extractions were only 60% efficient in spite of an asymptote for cumulative mass extracted. The quantitative extraction of nicotine at pH 10.0 removed the major part of the toxicity. ICP-AES analysis revealed that the residual toxicity in the aqueous layer was probably caused by iron (2.77 mg/L), the 15-min $EC_{50}$ of ferric chloride being 0.8 mg/L. Thus the analysis of nicotine in saliva is feasible since the saliva medium is not toxic and the effect of nicotine dominates unlike for urine.

A further study investigated whether carbonyl compounds contributed to the residual toxicity after hexane extraction at pH 10.0. This was accomplished by reaction with O-(2,3,4,5,6)-pentafluorophenylhydroxylamine hydrochloride at pH 2.0 before hexane extraction of oximes (11). Carbonyl compounds did not cause the residual toxicity nor significant toxicity in the original leachates (6).

**Genotoxicity of Nicotine and Cotinine**

The genotoxicity experiments were carried out over 40 hours in a very complex medium (2) also containing 0.47% dimethyl sulfoxide with and without rat S9 fraction (2). Positive controls were: 2-aminoanthracene (2AA) as a S9-point mutagen; N-methyl-N'-nitro-N-nitrosoguanidine (MNNG) as a direct acting point mutagen; and phenol as a direct acting DNA intercalator which also served as the positive control for the acute test. Nicotine and cotinine and their mixtures, with nicotine added to the positive controls were evaluated. Twofold dilutions of analytes were examined over the nicotine concentration range 0.07-1.6 mg/mL observed for the saliva of tobacco chewers. The same concentration range for cotinine was examined also. Genotoxicity is signalled by the reappearance of bioluminescence from a dark mutant of *Vibrio fischeri* Strain NRRL-B-11177 at 27°C after incubation for a specified time and at a given concentration.

Nicotine was not genotoxic in the absence or presence of S9. However cotinine was a direct acting genotoxin at 1.24 mg/mL at 13-15 h incubation, and at 2.50 mg/mL between 18-30 h. Addition of S9 caused genotoxicity only at 1.25 mg/mL after 9 h. The effects of nicotine on positive controls were complex: potentiation occurred with 2AA with and without S9, phenol without S9, and cotinine with and without S9 by shifting the incubation times and concentrations for bioluminescence; and nicotine antagonized MNNG. Nicotine acts as a potentiator and antagoniser.

Since nicotine (half-time, 2-3 h) is quickly metabolized to cotinine and other compounds, cotinine adverse genotoxicity may be of great importance relative to genotoxic tobacco human chronic effects. Fortunately exposure to tobacco is usually expressed through cotinine marker in biological fluids. The correlations with cotinine however assume that it is less biologically active than nicotine. At least relative to genotoxicity this is not so, and lower birth weights are correlated with increasing cotinine in body fluids (2). The effect may be direct. The role of the tobacco-specific nitrosamines (TSNA), the major known genotoxins in smokeless tobacco, is unknown with regard to their interactions with nicotine and cotinine.

## Conclusions

The acute toxicity bioluminescence (Microtox) test cannot be used to screen nicotine and cotinine in urine, but can be used for saliva. Growth media containing environmental and biological samples must be adjusted to pH 6.5-7.5 for toxicity testing. Cotinine is a direct acting genotoxin in the *Vibrio fischeri* dark mutant (Mutatox) test but nicotine is not active with or without S9. Nicotine has a complex interactive effect on potent genotoxins.

## Literature Cited

1. Gee, P.; Maron, D.M.; Ames, B.N. *Proc. Natl. Acad. Sci. U.S.A.* **1994**, *91*, 11606-11610.
2. Yim, S.H.; Que Hee, S.S. *Mutat. Research* **1996**, *336*, 275-283.
3. Mazidji, C.N.; Koopman, B.; Bitton, G.; Voiland, G.; Logue, C. *Tox. Assessment* **1990**, *5*, 265-277.
4. Chou, C.C.; Que Hee, S.S. *J. Biolumin. Chemilumin.* **1993**, *8*, 39-48.
5. Chou, C.C.; Que Hee, S.S. *Ecotoxicol. Environ. Safety* **1992**, *23*, 355-363.
6. Chou, C.C.; Que Hee, S.S. *J. Agric. Food Chem.* **1994**, *42*, 2225-2230.
7. Chou, C.C.; Que Hee, S.S. *Environ. Toxicol. Chem.* **1994**, *13*, 1177-1186.
8. Chen, H.F.; Que Hee, S.S. *Ecotoxicol. Environ. Safety* **1995**, *30*, 120-123.
9. Yamada, H.; Somiya,I. *Ozone Sci. Eng.* **1989**, *11*, 125-141.
10. Que Hee, S.S.; Boyle, J.R. *Anal. Chem.* **1988**, *60*, 1033-1042.
11. Cancilla, D.A.; Chou, C.C.; Barthel, R.; Que Hee, S.S. *J. Assoc. Offic. Anal. Chem. Internat.* **1992**, *75*, 842-854.

# Chapter 9

# Distribution of Polychlorinated Biphenyl Congeners in Bear Lake Sediment

Min Qi, S. Bierenga, and J. Carson

Chemistry Department and Water Resources Institute,
Grand Valley State University, Allendale, MI 49401

Contamination of the Bear Lake sediment with polychlorinated biphenyls (PCBs) and the distribution of PCB congeners as a function of core depths were determined using a freeze corer sampling technique. The maximum level of PCBs was found in the 15 - 25 cm core section from the surface, then decreased with the depth of core section to the bottom. Congeners #105 and 118 showed a considerable dechlorination pattern compared with the Aroclor mixture which was initially introduced to the environments. However, congeners #101, 138, and 180 exhibited very small or almost no alterations in relative amount. The PCB concentrations were very homogeneous in three studied sites along Bear Lake, which suggests that the contamination of Bear Lake with PCBs is probably due to the atmospheric deposition. This study would provide very useful information for envorinmental and biomonitoring activities in west Michigan.

Polychlorinated biphenyls (PCBs) are compounds that have a highly lipophilic nature and are ubiquitous, persistent environmental contaminants. As a consequence, PCBs bioaccumulate in fatty tissue and induce a variety of biological and adverse health effects that can be congener specific and differ between species. The non-ortho and mono-ortho substituted co-planar PCBs have been shown to exhibit "dioxin-like" effects; they bind to the aryl hydrocarbon receptor, and are potent inducers of aryl hydrocarbon hydroxylase (1, 2). However, the neurotoxic effects of PCBs seem to be associated with some degree of ortho-chlorine substitution (3), therefore the toxicity of each congener must be determined. The reproductive failure of fish populations and the high prevalence of skin and liver tumors that have been reported in the Great Lakes region are very likely related to the contamination of PCBs, DDE, and other polynuclear aromatic hydrocarbons (PAHs) (4).

Muskegon Lake is a 4,150 acre lake and discharges to Lake Michigan through a channel. A major tributary of Muskegon Lake is Bear Lake, with an average annual discharge of 1.0 $m^3$/s. The historical contamination of Bear Lake with PCBs from industrial discharges has not been reported; however, fish taken recently from Bear Lake have shown elevated levels of PCBs and mercury, the study was conducted by Michigan Department of Natural Resources (MDNR) (5). The source for fish contamination could be related to water pollution, atmospheric deposition, contaminated food consumption, and pollutants released from the sediment. To assess whether the pollution is due to the local historical accumulation or global cycling, it will be necessary to determine the distribution of PCB congeners in the lake sediment. Sediment samples from Bear Lake were collected in 1988 and 1990 for the determination of PCBs. However, the analytical results were inconclusive because of interference from high levels of sulfur and an insufficient method detection limit (5).

The objective of this study was to use the freeze corer sampling technique (6, 7, 8) for collecting lake sediment samples, to measure the concentrations of PCBs and DDE within specified depth intervals of bottom sediments in Bear Lake, and to establish a model of pollutant distribution. The study will be beneficial to management decision makers in development, remediation, and long term exposure risk assessment activities in the West Michigan ecosystem.

## Sample Collection and Analytical Method

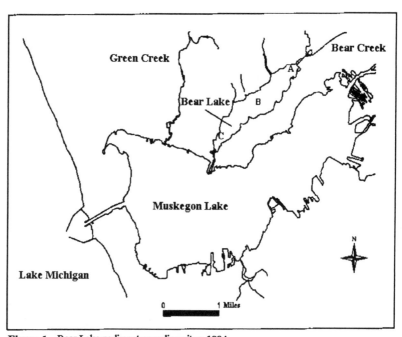

**Figure 1.** Bear Lake sediment sampling sites, 1994

**Sample Collection**: Sites A, B and C as shown in Figure 1 were selected for sample collection and analysis in this study. Site A was near the Bear Creek inlet, site B was near the middle of Bear Lake, and site C was near the outlet draining into Muskegon Lake where contamination has been reported previously (5). Two cores were collected from site A and C in November of 1994, one core was collected from site B in June of 1995. The freeze core technique (8) was used in sample collection. The corer is a heavy brass tube, 129 cm in length and 7.6 cm in diameter. The tube was first filled with dry ice and alcohol to keep the temperature of the device below - 50 °C. The device was then dropped into the lake quickly to penetrate sediments. Water and sediment were frozen on the outside of the tube *in situ*. After about 10 minutes, the device was taken out from the lake and the frozen sediment cores were thawed off from the device by adding hot water to the inside of the tube. The cores were segmented every 10 cm down to 95 cm and kept at - 20 °C until the time of analysis.

**Total Volatile Solids (TVS) Determination:** Total volatile solids were determined using the method described in "Standard Methods for the Examination of Water and Wastewater" (9). About 5 g of wet sediment samples were weighed out and dried in the oven at 105 °C for 12 hours. After the percentage of water in the sediment samples was determined, the dried samples were transferred into the muffle furnace and heated at 550 °C for 1 hour. The TVS in the samples was then calculated based on the weight loss after heating (Table 1).

### Table 1.  Total Volatile Solids in Bear Lake Sediments

| Site | Depth (cm) | % Water | % TVS |
|---|---|---|---|
| A | 0-10 | 95.8 | 29 |
| | 10-20 | 90.5 | 32.1 |
| | 20-30 | 91.5 | 34.8 |
| | 30-40 | 87.5 | 34.3 |
| | 40-55 | 92.3 | 37.2 |
| | Mean ± Std.dev | 91.5 ± 3.0 | 33.5 ± 3.1 |
| B | 0-10 | 93.2 | 39.9 |
| | 10-20 | 92.1 | 39.8 |
| | 20-30 | 92.1 | 40.7 |
| | 30-40 | 91.5 | 40.7 |
| | 40-50 | 90.6 | 41.9 |
| | 50-60 | 90.2 | 44.3 |
| | 60-70 | 89.9 | 44.2 |
| | Mean ± Std.dev | 91.4 ± 1.2 | 41.6 ± 1.9 |
| C | 0-15 | 94.6 | 33.5 |
| | 15-25 | 92.2 | 32.2 |
| | 25-35 | 91.7 | 33.6 |
| | 35-45 | 90 | 34.7 |
| | 45-55 | 90 | ND |
| | 55-65 | 90.5 | ND |
| | 67-75 | 89.8 | ND |
| | Mean ± Std.dev | 91.2 ± 1.7 | 33.5 ± 1.0 |

Note: ND = Sample Lost.

**Solvent-Solvent Extraction**: All of the solvents used were pesticide grade and purchased from E.M. Science (Gibbstown, New Jersey). PCB standards were purchased from Accu Standards Inc. (New Haven, CT). About 100 g of wet sediment were weighed out and left in the hood overnight before extraction. Chlorinated biphenyl (CB) congeners # 46 and 142, which are not found in the environment, were chosen as surrogate standards and they were added to all of the samples before any analytical procedures. An appropriate amount of anhydrous sodium sulfate (10 - 60 mesh, E.M. Science), which was heated at 600 °C for 3 hours before application, was added to the samples. Sediments were extracted using a

Braun-Sonicu homogenizer with 50 mL of acetone and 100 mL of hexane. The extraction was repeated with additional hexane. Copper turnings (E.M. Science) were used to remove the sulfur. The combined extracts were concentrated with a slow stream of $N_2$ and were cleaned using a 2% deactivated Florisil column (E.M. Science). Finally the hexane solution was exchanged for iso-octane.

**Method Detection Limits** (MDL): MDL, the constituent concentration that produces a signal with 99% probability that it is different from the blank, was determined following the EPA procedure (10). An uncontaminated soil sample collected from Allendale, Michigan, was used as the blank and spiked with 12 CB congeners. The spike concentration was 0.17 ng/g for congeners 6, 44 and 52; and 0.075 ng/g for congeners 101, 138, 153, 180, 205, and DDE. Seven spiked sediment samples were extracted and method detection limits were determined. A comprehensive quality assurance and quality control program included a reagent blank, sediment blank, spiked matrix sample and a duplicate sample was incorporated with every 10 samples. All data reported in this paper are based on dry sediment weights and have not been corrected for recoveries with the surrogate standards. Total PCB levels were calculated as the sum of individual congener concentrations.

**Table 2. Method Detection Limit for Selected CB Congeners**

| Congener | Spiked Amount ng/g | Mean (n=7) ng/g | Stds | MDL pg/g |
|---|---|---|---|---|
| 6 | 0.17 | 0.175 | 0.037 | 120 |
| 46 | 0.17 | 0.101 | 0.018 | 56 |
| 52 | 0.17 | 0.153 | 0.014 | 44 |
| 44 | 0.17 | 0.139 | 0.013 | 41 |
| 101 | 0.083 | 0.079 | 0.0097 | 30 |
| DDE | 0.083 | 0.052 | 0.0038 | 12 |
| 77 | 0.083 | 0.055 | 0.033 | 100 |
| 142 | 0.083 | 0.046 | 0.0032 | 10 |
| 153 | 0.083 | 0.061 | 0.0032 | 10 |
| 138 | 0.083 | 0.058 | 0.0043 | 14 |
| 180 | 0.083 | 0.069 | 0.0032 | 10 |
| Mirex | 0.083 | 0.068 | 0.32 | 100 |
| 169 | 0.083 | 0.035 | 0.016 | 52 |
| 205 | 0.083 | 0.063 | 0.0027 | 8.5 |

**GC-EC and GC-MS Analysis:** A Perkin-Elmer Auto-System GC with $Ni^{63}$ electron capture detector and RTX-5 capillary column (30 m x 0.25 mm x 0.25 μm, Restek, Bellefonte, PA) were used for PCB analysis. The column temperature was programmed as follows: 80 °C for 2 minutes, 10 °C/min to 160 °C, 1.5 °C/min to 190 °C, 2 °C/min to 256 °C, and the column was held at that temperature for 6 minutes. Helium and Nitrogen were applied as the carrier gas and the makeup gas, respectively. The injector temperature was 260 °C and the detector temperature was 330 °C. A GC-MS (Perkin-Elmer Q-mass 910) with the same type of the capillary column and temperature program as in GC analysis was used for the chemical confirmation. CB congeners #30 and 204 were used as internal standards, #30 for the congeners with retention times smaller than that of congener 101, and #204 for all other congeners including 101. They were added to each of the samples before the GC analysis. Our preliminary studies indicated 43 individual CB congeners present in Bear Lake sediment; therefore, they were used in our calibration standards. Structural

**Table 3. Structure and Retention Time of 44 Congeners Analyzed in the Sediment Samples**

| IUPAC# | Structure | Retention Time (min.) |
|---|---|---|
| 6 | 2,3' | 16.21 |
| 8 | 2,4' | 16.47 |
| 30* | 2,4,6 | 17.96 |
| 18 | 2,2',5 | 18.96 |
| 28 | 2,4,4' | 22.22 |
| 46** | 2,2',3,6' | 24.50 |
| 52 | 2,2',5,5' | 24.66 |
| 44 | 2,2',3,5' | 26.36 |
| 64 | 2,3,4',6 | 27.71 |
| 94 | 2,2',3,5,6' | 29.69 |
| 95+66 | 2,2',3,5',6+2,3',4,4' | 30.55 |
| 101 | 2,2',4,4',6 | 32.59 |
| 99 | 2,2',4,4',5 | 33.22 |
| 97 | 2,2',3',4,5 | 34.71 |
| 87 | 2,2',3,4,5' | 35.18 |
| *p,p'*-DDE | | 35.97 |
| 77+110 | 3,3',4,4'+2,3,3',4',6 | 36.16 |
| 149 | 2,2',3,4',5',6 | 38.38 |
| 118 | 2,3',4,4',5 | 38.56 |
| 142** | 2,2',3,4,5,6 | 39.72 |
| 153 | 2,2',4,4',5,5' | 40.54 |
| 105 | 2,3,3',4,4' | 40.94 |
| 141+179 | 2,2',3,4,5,5',+2,2',3,3',5,6,6' | 41.85 |
| 138 | 2,2',3,4,4',5' | 43.02 |
| 129 | 2,2',3,3',4,5 | 43.90 |
| 126 | 3,3',4,4',5 | 44.00 |
| 187 | 2,2',3,4',5,5',6 | 44.78 |
| 183 | 2,2',3,4,4',5',6 | 45.24 |
| 189+128 | 2,3,3',4,4',5,5'+2,2',3,3',4,4' | 45.61 |
| 174 | 2,2',3,3',4,5,6' | 46.86 |
| 204* | 2,2',3,4,4',5,6,6' | 48.38 |
| 180 | 2,2',3,4,4',5,5' | 49.27 |
| 169 | 3,3',4,4',5,5' | 51.32 |
| 170 | 2,2',3,3',4,4',5 | 52.00 |
| 201 | 2,2',3,3',4,5',6,6' | 52.82 |
| 195 | 2,2',3,3',4,4',5,6 | 55.85 |
| 194 | 2,2',3,3',4,4',5,5' | 57.69 |
| 205 | 2,3,3',4,4',5,5',6 | 58.13 |
| 206 | 2,2',3,3',4,4',5,5',6 | 60.90 |
| 209 | 2,2',3,3',4,4',5,5',6,6' | 63.54 |

* internal standard    ** surrogate standard

information of these congeners and their retention time on the RTX-5 column can be found in Table 3, the toxicity information of these congeners can be obtained elsewhere (11). The relative response factors to be used for determining the concentration of each of the congeners in the samples were calculated by using the above internal standards. The congeners were identified using GC-MS and by comparing their retention times with Mullin's data (12).

## Results and Discussion

Sediments obtained from Bear Lake were analyzed for the percentage of water and the total amount of volatile solids, which was proportional to the total amount of organic carbon. As shown in Table 1, the percentage of water in sediment sections ranged from 89 in the bottom (85-95 cm down), to 95 at the surface. The Bear Lake sediments were composed primarily of sand (more than 70%), and there was no significant variation in the TVS observed in the samples. TVS is largely associated with the fine-grained portion of sediments such as silt and clay (13), and was negatively correlated with the sand content of the sediment. Due to the sandy nature of Bear Lake sediments, variations in PCB concentrations were not related to the TVS as shown in Table 4. This pattern might be indicative of the influence of PCB atmospheric deposition in Bear lake.

The recovery for spiked congeners, as shown in Table 2, ranged from 60 to 104%, and the MDL ranged from 120 pg/g for congener 6 and 8.5 pg/g for congener 205. PCB concentrations in the cores generally increased with depth up to the 10 to 20 cm section, and then decreased in the bottom core sections. The results of congener and DDE concentrations are summarized in Table 4. The highest total PCB (67 ng/g) and DDE (23 ng/g) concentrations were observed at site C with the depth 15 - 25 cm. At 50 - 60 cm from the surface, total PCB concentration decreases to 4.55 ng/g. A similar depth profile was also found at sites A and B. Based on an average sedimentation rate of 0.02 g/cm$^2$/yr in several Lake Michigan basins (14), the 20 cm sediment strata from Bear Lake, amounted to about 50 years of deposition. The decrease in PCB concentration from top to bottom was quite different from the profiles reported for the Hudson River in New York State (15), and for the Newark Bay Estuary in New Jersey (16), where the highest PCB concentration was determined in the bottom core sections, but was similar to that observed for the Sheboygan River sediment in Wisconsin (17).

The concentrations of PCB congeners among the three sites studied were only two to three times different; therefore, the contamination in Bear Lake with the PCBs was very likely due to atmospheric deposition rather than point source accumulation. Atmospheric PCB concentrations in Michigan monitored during 1990 to 1991 (18) were found to range from 603.3 pg/m$^3$ to 92.9 pg/m$^3$. Hermanson *et al* (14) investigated atmospheric and sedimentary PCB fluxes in Lake Michigan and concluded that about 55 - 100 % of PCBs accumulated in the sediment were from atmospheric inputs to the lake.

A typical chromatogram of sediment extract is shown in Figure 2. The numbers associated with the peaks are the IUPAC numbers of the congeners. This chromatogram was compared with Aroclors 1016, 1221, 1242, 1254, and 1260. The best perceivable match could be found only with the Aroclor mixture of 1254 and 1260. The distribution of homolog groups with different core sections showed that the penta group was dominant at the surface and near the surface sections, and then dropped as shown in Figure 3. All groups except tri- and octa-, displayed a maximum at the 15 - 25 cm section.

Figure 4 shows a comparison of the congener concentration ratio over total PCBs with those existing in the original mixture of Aroclors 1016, 1221, 1254 and 1260, also the changes in the concentration ratio with the core depth. The Aroclors mixtures were chosen because they covered the majority of the congeners detected in the sediments (15). Notable changes in the relative distributions of CB congeners, possibly due to reductive dechlorination, were found in the corer samples (Figure 4). Figure 4 clearly shows

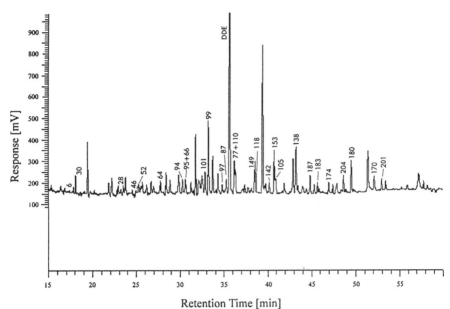

Figure 2. GC Chromatogram of Sediment from Site C at 15-25 cm Depth

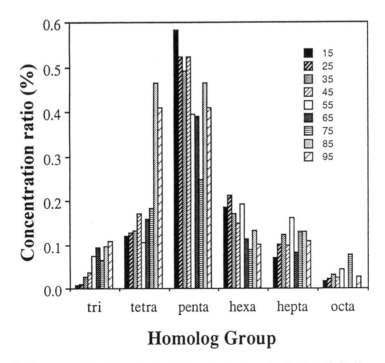

**Figure 3. Concentration Ratio of PCB Homolog Group in the Site C Sediment with Different Depth (cm)**

**Table 4. Summary of CB congeners concentration detected in Bear Lake sediments**

Congeners (ng/g)

| Site | Depth (cm) | 8+5 | 18 | 28 | 52 | 44 | 64 | 94+74 | 66+95 |
|---|---|---|---|---|---|---|---|---|---|
| A | 0-10 | ND | ND | 0.04 | 0.14 | 0.16 | 0.05 | ND | 0.20 |
| A | 10-20 | ND | ND | 0.26 | 0.50 | 1.11 | 0.39 | 1.95 | 0.22 |
| A | 20-30 | ND | 0.38 | 0.37 | 0.67 | 1.19 | 0.59 | 1.72 | 1.37 |
| A | 30-40 | ND | 0.25 | 0.24 | 0.96 | 1.05 | 0.53 | 1.91 | 1.42 |
| A | 40-50 | ND | ND | 0.26 | 0.55 | 0.93 | 0.37 | 1.03 | 0.92 |
| B | 0-10 | ND | 0.26 | 0.35 | 1.48 | 0.95 | 0.74 | 0.02 | 0.98 |
| B | 10-20 | 0.25 | 0.51 | 0.33 | 1.22 | 0.75 | 0.66 | 0.12 | 0.82 |
| B | 20-30 | ND | 0.25 | 0.25 | 0.87 | 0.48 | 0.46 | 0.09 | 0.54 |
| B | 30-40 | ND | 0.39 | 0.31 | 1.25 | 0.50 | 0.57 | ND | 1.29 |
| B | 30-40 | ND | 0.41 | 0.28 | 0.98 | 0.41 | 0.50 | 0.10 | 1.12 |
| B | 40-50 | ND | 0.26 | 0.22 | 0.92 | 0.35 | 0.39 | ND | 0.83 |
| B | 50-60 | 0.14 | 0.34 | 0.10 | 0.58 | 0.41 | 0.18 | ND | 0.28 |
| B | 50-60 | 0.17 | 0.37 | 0.14 | 0.59 | 0.56 | 0.27 | ND | 0.47 |
| B | 60-70 | 0.25 | ND | 0.13 | 0.91 | 0.63 | ND | 0.4 | 0.18 |
| B | 70-75 | 0.35 | ND | 0.15 | 0.75 | 0.55 | 0.55 | 0.16 | 0.46 |
| C | 0-15 | ND | ND | 0.36 | 2.02 | 3.21 | 1.14 | 3.58 | 2.66 |
| C | 15-25 | ND | ND | 0.61 | 3.07 | 3.53 | 1.66 | 0.34 | 3.19 |
| C | 25-35 | 0.31 | 0.40 | 0.29 | 1.53 | 1.36 | 0.63 | ND | 1.70 |
| C | 35-45 | ND | 0.51 | 0.35 | 1.98 | 1.49 | 0.63 | 1.84 | 1.65 |
| C | 45-55 | 0.34 | 0.42 | 0.25 | 0.24 | 0.59 | 0.13 | 0.65 | 0.41 |
| C | 55-65 | 0.53 | 0.15 | 0.26 | 0.12 | 0.4 | 0.16 | 0.61 | 0.23 |
| C | 65-75 | 0.96 | 0.22 | 0.27 | 0.11 | 0.81 | 0.46 | 0.73 | 0.30 |
| C | 75-85 | ND | 0.12 | 0.22 | 0.18 | ND | 0.07 | 0.26 | 0.19 |
| C | 85-95 | ND | 0.32 | 0.36 | 0.43 | 0.33 | 0.09 | ND | 0.15 |

Congeners (ng/g)

| Site | Depth (cm) | 101 | 99 | 97 | 87 | 77+110 | 149 | 118 | 153+132 |
|---|---|---|---|---|---|---|---|---|---|
| A | 0-10 | 0.21 | 0.29 | 0.15 | 0.11 | ND | 0.04 | 0.15 | 0.16 |
| A | 10-20 | 0.95 | 2.98 | 0.31 | 0.56 | 4.2 | 0.79 | 1.28 | 1.27 |
| A | 20-30 | 0.90 | 1.85 | 0.30 | 0.49 | 3.57 | 0.64 | 1.08 | 1.04 |
| A | 30-40 | 1.01 | 0.48 | 0.31 | 0.57 | 3.99 | 0.74 | ND | 0.96 |
| A | 40-50 | 0.95 | 1.26 | 0.31 | 0.47 | 3.23 | 0.56 | ND | 0.81 |
| B | 0-10 | 2.76 | 6.34 | 0.84 | 1.13 | 2.03 | 0.45 | 1.56 | 1.53 |
| B | 10-20 | 1.91 | 4.16 | 1.03 | 0.90 | 1.61 | 0.45 | 1.48 | 1.28 |
| B | 20-30 | 1.36 | 1.29 | 0.53 | 0.57 | ND | 0.25 | 0.79 | 0.76 |
| B | 30-40 | 1.57 | 1.34 | 0.80 | 0.66 | 0.21 | 0.26 | 0.84 | 0.83 |
| B | 30-40 | 1.71 | 1.53 | 0.74 | 0.66 | 0.15 | 0.29 | 0.64 | 0.88 |
| B | 40-50 | 1.41 | 0.25 | 0.79 | 0.54 | 1.26 | 0.23 | 0.55 | 0.82 |
| B | 50-60 | 0.43 | ND | 0.47 | 0.16 | 0.58 | ND | 0.18 | 0.48 |
| B | 50-60 | 0.50 | 0.39 | 0.46 | 0.2 | ND | 0.08 | 0.20 | 0.25 |
| B | 60-70 | 0.18 | ND | 0.58 | 0.11 | 0.25 | ND | 0.04 | 0.26 |
| B | 70-75 | 0.30 | ND | 0.23 | 0.15 | 0.42 | ND | ND | 0.51 |
| C | 0-15 | 3.38 | 6.84 | 0.90 | 1.37 | 8.15 | 1.63 | 2.49 | 3.12 |
| C | 15-25 | 3.16 | 9.66 | 0.89 | 1.36 | 10.5 | 2.08 | 3.16 | 4.39 |
| C | 25-35 | 1.55 | 2.69 | 0.62 | 0.99 | 4.94 | 1.08 | ND | 1.69 |
| C | 35-45 | 1.35 | 0.47 | 0.41 | 0.68 | 4.48 | 0.88 | 1.01 | 1.25 |
| C | 45-55 | 0.83 | ND | 0.43 | 0.48 | 0.47 | 0.56 | ND | 0.30 |
| C | 55-65 | 0.22 | 0.11 | 0.13 | 0.13 | 0.11 | 0.15 | 0.12 | 0.11 |
| C | 65-75 | 0.26 | 0.18 | 0.15 | 0.09 | 0.16 | 0.26 | ND | 0.15 |
| C | 75-85 | 0.40 | 0.34 | ND | 0.19 | 0.29 | 0.10 | ND | 0.21 |
| C | 85-95 | 0.25 | 0.50 | 0.37 | 0.31 | 0.39 | 0.12 | 0.58 | 0.29 |

**Table 4.** *Continued*

**Congeners (ng/g)**

| Site | Depth (cm) | 105 | 141+179 | 138 | 187 | 183 | 180 | 170+190 |
|---|---|---|---|---|---|---|---|---|
| A | 0-10 | 0.03 | ND | 0.19 | 0.02 | 0.02 | 0.04 | 0.03 |
| A | 10-20 | 0.63 | 0.35 | 1.61 | 0.48 | 0.25 | 0.74 | 0.43 |
| A | 20-30 | 0.48 | 0.50 | 1.27 | 0.38 | 0.14 | 0.63 | 0.35 |
| A | 30-40 | 0.54 | 0.33 | 1.25 | 0.37 | ND | 0.58 | 0.29 |
| A | 40-50 | 0.40 | 0.11 | 0.96 | 0.28 | ND | 0.54 | 0.21 |
| B | 0-10 | ND | ND | 2.23 | 0.62 | 0.22 | 0.91 | 0.64 |
| B | 10-20 | ND | ND | 2.15 | 0.36 | 0.23 | 0.73 | 0.52 |
| B | 20-30 | ND | 0.06 | 1.34 | 0.29 | 0.16 | 0.44 | 0.31 |
| B | 30-40 | 0.07 | 0.07 | 1.41 | 0.24 | 0.20 | 0.43 | 0.31 |
| B | 30-40 | ND | 0.08 | 1.54 | 0.22 | 0.27 | 0.38 | 0.36 |
| B | 40-50 | 0.07 | ND | 1.34 | 0.32 | 0.17 | 0.44 | 0.31 |
| B | 50-60 | 0.08 | ND | 0.46 | 0.07 | 0.10 | 0.13 | 0.07 |
| B | 50-60 | 0.12 | ND | 0.45 | 0.11 | 0.08 | 0.12 | 0.08 |
| B | 60-70 | 0.25 | 0.06 | 0.24 | ND | 0.13 | ND | ND |
| B | 70-75 | 0.44 | ND | 0.31 | 0.08 | ND | ND | ND |
| C | 0-15 | 1.20 | 0.36 | 3.89 | 0.84 | 0.39 | 1.84 | 0.79 |
| C | 15-25 | 1.52 | 1.70 | 5.36 | 1.23 | 0.51 | 2.73 | 1.48 |
| C | 25-35 | 0.64 | 0.42 | 1.82 | 0.50 | 0.21 | 1.0 | 0.62 |
| C | 35-45 | 0.66 | 0.34 | 1.43 | 0.36 | 0.17 | 0.06 | 0.35 |
| C | 45-55 | 0.09 | 0.25 | 0.88 | 0.27 | 0.1 | 0.33 | 0.28 |
| C | 55-65 | ND | ND | 0.21 | 0.15 | ND | 0.1 | ND |
| C | 65-75 | ND | ND | 0.27 | ND | 0.21 | 0.22 | 0.54 |
| C | 75-85 | ND | ND | 0.18 | 0.15 | 0.13 | 0.08 | ND |
| C | 85-95 | ND | ND | 0.22 | 0.20 | 0.17 | 0.12 | 0.17 |

**Congeners (ng/g)**

| Site | Depth (cm) | 201 | 194 | 205 | 206 | 209 | Total | DDE |
|---|---|---|---|---|---|---|---|---|
| A | 0-10 | ND | ND | ND | ND | ND | 2.03 | 0.85 |
| A | 10-20 | 0.22 | 0.17 | ND | 0.16 | 0.46 | 22.3 | 8.09 |
| A | 20-30 | ND | ND | ND | 0.14 | ND | 20.1 | 9.66 |
| A | 30-40 | ND | ND | ND | ND | 0.32 | 18.1 | 8.79 |
| A | 40-50 | 0.11 | ND | ND | 0.13 | 0.43 | 14.8 | 6.63 |
| B | 0-10 | 0.28 | 0.32 | ND | 0.15 | 1.05 | 27.8 | 6.42 |
| B | 10-20 | 0.19 | 0.16 | ND | 0.10 | 0.27 | 22.2 | 1.93 |
| B | 20-30 | 0.13 | 0.10 | ND | 0.88 | 0.17 | 12.4 | 1.33 |
| B | 30-40 | 0.14 | ND | ND | 0.1 | 0.12 | 13.9 | 1.68 |
| B | 30-40 | 0.15 | 0.10 | ND | 0.12 | 0.21 | 13.8 | 0.85 |
| B | 40-50 | 0.13 | 0.12 | ND | 0.08 | 0.15 | 12.0 | 1.78 |
| B | 50-60 | 0.05 | ND | ND | ND | 0.08 | 5.37 | 2.01 |
| B | 50-60 | ND | 0.06 | ND | ND | 0.08 | 5.75 | 2.13 |
| B | 60-70 | ND | ND | ND | 0.08 | ND | 4.68 | 0.42 |
| B | 70-75 | ND | ND | ND | 0.07 | ND | 5.48 | 0.35 |
| C | 0-15 | 0.44 | 0.34 | 0.14 | ND | ND | 51.1 | 23.6 |
| C | 15-25 | 0.67 | 0.50 | 0.23 | 0.49 | ND | 64.0 | 21.4 |
| C | 25-35 | 0.35 | 0.23 | 0.12 | 0.19 | 0.99 | 26.9 | 5.28 |
| C | 35-45 | 0.28 | 0.15 | ND | 0.17 | 0.53 | 24.2 | 4.47 |
| C | 45-55 | 0.18 | 0.11 | 0.11 | ND | 0.70 | 9.40 | 3.15 |
| C | 55-65 | ND | ND | ND | ND | 0.29 | 4.29 | 1.43 |
| C | 65-75 | 0.36 | ND | 0.23 | 0.11 | ND | 7.05 | 1.5 |
| C | 75-85 | ND | ND | ND | ND | ND | 3.11 | 1.35 |
| C | 85-95 | ND | ND | 0.17 | 0.07 | ND | 5.61 | 1.49 |

(ND: Not-Detectable)

that a) relative concentrations of congeners were changing compared to aroclor mixture; b) the changing rate of individual congeners was extraordinarily different; and c) these changes were the results of environmental dechlorination. The change in concentration ratios of CB congeners # 28, 52 and 44, which have less chlorine substitutions, with the corer depth compared to the quantity in the Aroclor mixture was insignificant, and such a phenomenon could not be explained by common physico-chemical dechlorination models. Because these congeners are more soluble and volatile than most of the other congeners, their concentration would decrease more quickly because of evaporation or decomposition. Analogous amounts of those congeners in different core sections can only be interpreted by the "stepwise" dechlorination pattern for highly chlorinated CB congeners, with each change in step occuring when the concentration of substrate declined to appropriate levels (15, 19, 20). Similar observations were also reported by David et al (17), Quensen et al (21) and Brown (22).

Figure 4 shows the considerable decrease in relative abundance of mono-ortho substituted co-planar CB congeners #105 and 118. When core sections reached 25 cm deep, concentrations of congener 118 dropped close to method detection limit. The steric structure of these co-planar PCBs, and the toxic effects evoked by them are similar to that of TCDD (23). Even though their concentration in the sediments was much lower than other ortho-substituted CB congeners, their affinities for the aryl hydrocarbon receptor are two or three magnitude higher (11), and thus, they may be of greater concern to the environment. For CB congeners #149 and 153, only surface core sections showed a

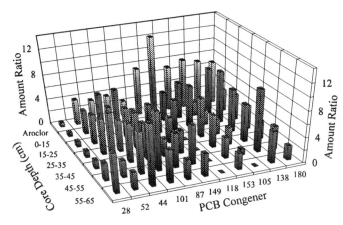

**Figure 4.** PCB Amount Ratios in Aroclor Mixture (1016, 1221, 1254 and 1260) and Sediment Cores by Depth. (The bars with solid black caps are the relative amounts of congeners in Aroclor mixture.)

large decrease in relative abundance. On the other hand, CB congeners # 101, 138 and 180 seemed to have potential resistance to the dechlorination process and their relative concentrations remained almost the same as in the Aroclor mixture. Although the sediments were from different sites, Figure 5 shows similar concentration ratio changes for the congeners discussed in Figure 4; thereby signifying that the congener distributions for the surface sediments throughout the lake were relatively uniform.

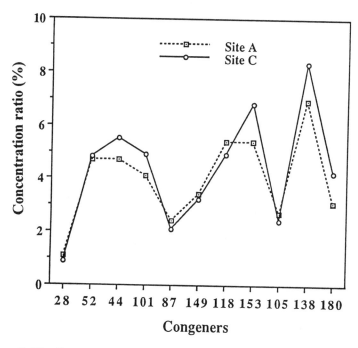

**Figure 5.** The Congener Distribution for Site A and Site C with the Core Depth at 10 cm.

Preferential dechlorination of meta- and para- substituted congeners (#105 and 118) observed in the present study was most likely indicative of microbial processes (22). Different bacteria are probably responsible for the dechlorination. This hypothesis is also supported by other studies (24, 25, 26).

## Conclusion

Our survey of total PCBs and PCB congeners in 3 sediment cores from Bear Lake showed that the vertical distribution of PCBs were highly heterogeneous in all the sites studied. The PCB contamination reached a peak at the 15 -25 cm strata, and declined down to the bottom. The variation of surface concentrations of PCBs in the three sites were very insignificant, and we could conclude that PCB contamination in Bear Lake was due to atmospheric deposition.

## Acknowledgments

The authors wish to thank J. Cooper and B. Miller for the sample collection, R. Denning for the figures presented in the manuscript, R. Rediski and E. Baum for valuable discussion, and R. Ward, Director of Water Resources Institute, for the support. The work was supported by The Grand Rapids Foundation and the Science and Math Division at Grand Valley State University.

## References

*1.* Safe, S. Toxicology 1990, 21(1): 15-88.
2. Patterson, D. G. Environmental Health Perspectives Supplements 1994. 102 Suppl I, 195-204.
*3.* Seegal, R.F., Bush, B. and Shain, W. Toxicology and Applied Pharmacology 1990, 106, 136-144.

4. NCASI. LTI, Limno-Tech, Inc 1993.
5. MDNR: Michigan Department of Natural Resources. 1994 Update.
6. Patts, G.E., Thomas, M.C., Brittan, K and Atkin, B. The Science of the Total Environment 1989, 84, 259-272.
7. Rymer, l. and Neale,J. Australian J. of Ecology 1981, 6, 123-126.
*8.* Salisbury, J. *Publication # MR-93-5*, Water Resources Institute at Grand Valley State University. 1993.
9. Greenberg, A.E. Standard Methods for the Examination of Water and Waste Water 18th Edition. 1992, 2-57
10. FRNARA. *Published by the Office of the Federal Register National Archives and Records Administration.* 1993, 569.
11. Kafafi, S.A.; H.Y. Afeefy; A.H. Ali; H.K. Siad; I.S. Adb-Elazem; A.G. Kafafi. Carcinogenesis 1993, 14, No. 10, 2063-2071.
12. Mullin, M.D. and Cynthia, M.P. Environmental Science & Technology 1984, 18, 468-476.
13. Latimer, J.S. *"PhD Thesis, University of Rhode Island, Kingston, RI.* 1989.
14. Hermanson, Mark H.; Erik R. Christensen; Dale J. Buser; Li-Mei Chen. Journal of Great Lakes Resources 1991, 17(1) 94-108.
*15.* Bush, B., L.A. Shane; M. Mahlen. Chemosphere 1987. 16, No. 4, 733-744.
16. Iannuzzi, T.J.; S.L. Huntly; N.L. Bonnevie; B.L. Finley; R.J. Wenning. Archives Environmental Contamination & Toxicology. 1995, 28, 108-117.
*17.* David, M., L.M. Brondyk; William Sonzogni. Journal of Great Lakes Research 1994. 20(3), 510-522.
18. Monosmith, C.L. International Association for Great Lakes Research 1995, 69.
19. Lake, J.; R.J. Preull; F.A. Osterman. Organic Substances and Sdiments in Water 1991, 3, 173-197.
20. Sokol, R. C.M. Bethoney; G.Y. Rhee. Chemosphere 1994, 45, 1735-1743.
21. Quensen, J.F.; S.A. Boyd; J.M. Tiedje. Science 1988, 242, 752-754.
22. Brown, J.F. Jr.; D.L. Bedard; M.J. Brenan; J.C. Carnahan; R.J. Science 1987, 236, 709-712.
*23.* Dewailly, E. Environmental Health Perspectives Supplements 1994, 102, Suppl I 205-209.
24. Brown, J.F.Jr. Ninth Progress Report, GE Co. 1990,87-90.
25. Rhee, G.Y.; B. Bush; C.M. Bethoney; A. DeNucci; H-M Oh; R.C. Sokol. Environmental Toxicology and Chemistry 1993b. 12 1033-1039.
26. Rhee, G.Y. Environmental Toxicology and Chemistry 1993a, 12, 1025-1032.

EXPOSURE ASSESSMENT

Chapter 10

# Challenges to Health from the Major Environmental Chemical Contaminants in the Saint Lawrence River

S. Raman[1] and K. S. Subramanian[2,3]

[1]Department of Epidemiology and Community Medicine, University of Ottawa School of Medicine, 451 Smyth Road, Ottawa, Ontario K1H 8M5, Canada
[2]Environmental Health Directorate, Postal Locator 0800B3, Health Canada, Tunney's Pasture, Ottawa, Ontario K1A 0L2, Canada

The International Joint Commission (IJC) has identified polychlorinatedbiphenyls (PCBs) and mercury as the major chemical contaminants in the St. Lawrence River system. Since 1979, PCBs have been repeatedly identified in sediment, forage, sport fish, water fowl, snapping turtles and native mussels at various locations in the St. Lawrence River from Cornwall, Ontario to Valleyfield, Quebec. PCB concentrations in the river have rarely been below 1 ng/L, the Ontario Water Quality Objective. Since 1988, mercury levels in the Cornwall-Massena area of the St. Lawrence River have been found to be below the analytical detection limit of 10 ng/L. However, there are no systematic population monitoring of the health effects of the contaminants in the St. Lawrence river. One way of assessing the impact is through the National Vital Statistics on mortality and hospital morbidity. This paper discusses the limitations of such conventional epidemiologic studies in assessing the impact of environmental chemical contaminants on population health.

"Environmental degradation inevitably causes a degradation of human life, even when it does not interfere with fundamental biological activities" (1). This simple yet poignant statement implies that mankind is dependent on clean water, fresh air and unpolluted land for continued sustenance. Nevertheless, human beings have been degrading the environment at an alarming rate within the past 50 years, thereby causing considerable damage to human and ecosystem health. A prime example of this is the St. Lawrence River. This is a major river system that constitutes the outflow of the largest reservoir of fresh water on earth. It serves

[3]Corresponding author

as a source for a number of vital services, including the provision of drinking water, transportation, sport and commercial fisheries, tourism and a host of industrial activities. Historically, the River has served as a rich habitat for a variety of wild life, and a means of livelihood for people living on its shores. However, the river has reached a state of both chronic and acute ill-health owing to nearly two centuries of human habitation and commercial exploitation relative to its pristine paleolimnological picture *(2)*. In recent decades, the River has exhibited increasing symptoms of ecological pathology. The high levels of organic and inorganic pollutants in the water sediment and biota have wrought drastic and generally undesirable changes to the lives of many people whose traditional livelihood depended on the river. This has resulted in legitimate concerns about risks to human health.

This chapter will focus on epidemiologic aspects of the challenges - especially in predicting the relationship between exposure to a given contaminant (or contaminants) and the consequent development of an adverse effect - to health from some of the major environmental contaminants in the St. Lawrence River. The investigation is part of our mandate to study the gaps in knowledge in relation to the ecological impact of the River on the health of the Cornwall and Akwasasne communities *(3)*. The final report of this 3-year (1993-1996) multidisciplinary study will identify the course of action required for the ecologic rehabilitation of the degraded River system with a focus on the sustainable re-development of the Cornwall region.

## History of Chemical Contaminants in the St. Lawrence

In epidemiologic parlance the *chemicals* represent the *agent*, exposure to some of which have adverse health consequence. In 1981 the IJC compiled an inventory of over 1000 chemical substances; 360 of which were applicable to the Great Lakes system and about 100 of these compounds have potentially adverse health effects in humans *(4, 5)*. Environment Canada and the Quebec Ministry of Environment have conducted surveys of over 100 sampling stations in the St. Lawrence River and collected information on 45 variables which have been entered into a national database for the 1978-1988 decade *(6)*. In a 1979-81 sediment survey conducted by Environment Canada, only PCBs and mercury were found at elevated concentrations: PCBs ranging from 1.0 to 1900 ng/g with a median of 170 ng/g and mercury levels ranging from 0.5 to 1472 ng/g with a median of 133.5 ng/g; the Ontario guidelines for PCBs and mercury are 50 ng/g and 300 ng/g, respectively, in dredged sediments *(7)*.

## Human Health Effects of Some Major Contaminants

There is general consensus that the pollution of the Great Lakes and the St. Lawrence River creates health risks for the people living in the region and also for the fish and wildlife that are dependent on these waterways *(8)*. Cancer, infertility, birth defects, immune system impairment, thyroid abnormalities and impaired learning ability in children are some of the possible major health effects causing the most concern *(8)*. Although there is a decline in pollutant levels from their peak values recorded in the 1970s, the humans and animals living in the

Great Lakes ecosystem still carry a "body burden" of the persistent bioaccumulative toxic chemicals such as PCBs and methyl mercury.

Several recent studies have examined the impact of a variety of contaminants present in the Great Lakes Basin ecosystem on the extent and nature of risk to human health (9). Although these studies show that the Great Lakes environment contains many harmful chemicals which have been detected in human serum, fat and breast milk, PCBs and mercury have been identified as the major contaminants in the St. Lawrence ecosystem. Acute exposure to specific PCB congeners and related compounds are known to cause adverse reproductive effects, carcinogenicity and immuno and hepato toxicity (10). Exposure to methyl mercury leads to neurotoxicity, impairment of vision and in some cases fatality (11). Developmental deficits in infants have been associated with methyl mercury levels as low as 30 $\mu$g/g in hair and 100 $\mu$g/L in blood (12). The IJC (13) has concluded that "persistent toxic substances in the Great Lakes Basin ecosystem pose serious health threats to living organisms including humans".

**Chemical Contaminants as Agent**

Since the chemical contaminants provoke the health outcome they are equivalent to the *agent* in the context of the communicable diseases. The policy strategy for an effective containment of the health problem will be through the control or elimination of the *agent*. We shall consider step by step whether we have relevant and adequate information on the *agent:*

(i) Do we have a complete temporal and spatial information on the contaminants? The answer is a qualified "yes" for some of the contaminants at least for the time being. Several millions of dollars have been spent in the past both in U.S. and Canada to characterize the spatial and temporal trends of important chemicals in the Great Lakes and St. Lawrence River system (14). But the process should be on-going. Due to recurring budget cuts the activities are being curtailed, and future implications are not known.

(ii) Do we know the health effects from the different contaminants? The answer is "yes" for some of the well known chemicals wherein human exposure levels and health information are available from occupational and toxicity studies. These results are in most cases acute toxicity studies and the long-term chronic effects are unknown in a number of cases due to lack of cohort follow up. Also new chemicals continue to be introduced into the system and we do not have an adequate monitoring database to alert us to the potential health hazards. Indeed, there is no systematic monitoring of exposed cohorts in Canada to the best of our knowledge.

(iii) Do we know the dose-response effect? There is a significant information gap in this area. The dose levels are established based on animal experiments and the extrapolation to human effects is fraught with considerable uncertainty. Also, information on chronic low dose exposure effects is highly limited because accurate measurement of subtle effects requires large numbers of animals over long periods of time, and this is often not possible. Again, test animals may be different in size, life span, and sensitivity to contaminants (*8*). Ingenious mathematical models have been proposed but epidemiologic studies are generally non-existent. The basic problem is that an animal which shows  adverse health effects from exposure to pollution does not usually react the same way that humans do. For example, animals process or metabolize contaminants differently, and they are exposed to contaminants differently from humans. Furthermore, laboratory studies with animals usually involve exposure to just one chemical at a time, and as a result interactions among contaminants might be overlooked (*8*).

(iv)  Do we know what is a safe dose? The designation of a given dose as safe depends on the probability of adverse effect at that exposure level. The probability is pegged usually at 1 in a million or 1 in 100,000. Hence contrary to public understanding a safe dose is not a risk free dose. There are also limitations on the  detection capabilities of analytical instrumentation. Inability to detect doses below the instrumental detection limits does not imply safety. At present the safe level of human exposure to many of the persistent contaminants including PCBs and methyl mercury are not known (*15, 16*). Also, many gaps remain in our understanding of the ways by which these pollutants cause the adverse health effects.

(v)  Are we measuring the right type of chemical contaminants? In many cases, the answer is no.  The health effects are highly dependent on the particular chemical species and its concentration. Chemical speciation is very important but is not fully understood. Extensive and intensive studies are needed in this area. Some of the University of Ottawa teams are investigating the chemical speciation of the major contaminants present in the St. Lawrence ecosystem.

(vi) Do we know the interaction effects among different chemicals?  There is a paucity of information in this area. Exposure data are invariably based on exposure of adult males to a single pollutant. How can we extrapolate these studies to the vulnerable segments of the population such as women (especially pregnant women), children (especially the developing fetus), the elderly and other special groups who may be exposed to a mixture of environmental pollutants at different levels, in different ways, for different periods of time ?" (*8*). For example, lead and mercury, both of which are neurotoxic, can produce cumulative effects which may be  unacceptable, though both may be present at safe levels as individual contaminants. More basic research is needed in  this area.

(vii) Do we know the body burden of the contaminants?  The answer is no for a number of chemicals. Even baseline values are not known for a number of chemical contaminants. In the context of increasing pollution loads from exotic chemicals it is essential to establish baseline levels for all of them to set national and international standards.

(viii) Are we measuring the contaminants at the right locations? We do not know whether the sampling stations truly represent the distribution levels of the contaminants for the purpose of estimating human exposure. That would depend on the flow characteristics of the river and the location of the sources of pollutants. The hydrogeology team of the ecoresearch project has produced sophisticated computer software to investigate models of pollution distribution and dispersion which could form the basis for the selection of optimal sampling stations (*17*).

(ix) Are there gaps in the knowledge with respect to our understanding of the health effects of contaminants? The answer is definitely yes. For example the hormone disrupter effect of PCBs was not anticipated but was discovered serendipitously (*18*). It has been claimed that there is a causal link with breast and testicular cancer. Much of this new knowledge is arising from current developments in molecular epidemiology and computer modelling.

**Environmental Factors**

Let us now focus our attention on problems relating to the *environment* which is the *medium* through which the *agent* is transmitted to the *host*.

(i) Do we know the mode of human exposure to the major contaminants found in the River?
**A. Drinking Water.** The river water serves as a source of drinking water to the communities residing along its route. Drinking water is one of the means of exposure to the contaminants. In general, the contaminant levels are so low in water that the hazard of toxic accumulation is also considered to be low. Yet the big question remains: what is a safe lifetime exposure level, which will not cause birth defects, cancer or chronic illness? For example, the drinking water contribution to the total daily intake of lead increases from 10% to 53% for a two-year-old Canadian child weighing 13.6 Kg as the lead level in water increases from 5 to 50 $\mu$g/L. The corresponding values for adults weighing 70 Kg are 12 and 57% (*19*). Under these conditions, drinking water becomes a major pathway for lead intake by humans. One may not be safer by switching to other sources of drinking water. Bottled water is not free from harmful chemicals either (*20*). Another source of exposure is the indirect use of water secondary to the preparation of other foods. We do not have data on the secondary use of water as part of food consumption. It is tacitly assumed that such use of water is just as safe as the use of water for drinking. To the best of our knowledge, no survey of drinking water consumption pattern for the Cornwall area seems to have been conducted, let alone its use in beverage preparation.

**B. Fish Consumption.** The important source of exposure to contaminants of the population arises from the consumption of fish found in the River. While the contaminant levels in the water may be safe, the levels in fish may be a hundred-fold due to bioaccumulation constituting a major hazard to humans. The Ontario Guide to Eating Sport Fish lists the Ontario waterbodies alphabetically specifying how much of fish caught in a given waterbody can be safely consumed according to type and length of fish. However there is no consistency in the recommendations between Ontario and New York State sport fish eating guidelines (8). Epidemiologic surveys have shown that nearly 50% of the population are not aware of the fish eating guidelines (21). Even sports fishermen more often disregard the guidelines (22). It is part of the risk-taking behaviour of the population. We need more data on the pattern of fish consumption in Cornwall to assess the hazard to the general population. The health science team (the present authors) of the ecoresearch group is collecting information on this aspect which will be made available as part of the final report.

(ii) Do we know the pattern of exposure in the human population?   No. Occupational exposure is monitored by law. Community exposure patterns can be assessed only through community surveys. The Canada Health Survey, and the recent Ontario Health Survey do not have the statistical power to assess exposure at local levels. Human exposure depends on a variety of characteristics: age, education, socioeconomic status, recreational activity, lifestyle factors etc. There are a number of variables to be investigated to disentangle the role of a particular exposure variable, and this can be done only through a carefully planned epidemiological study.

(iii) Are there susceptible groups who may be sensitive to low exposure levels? A quick answer is more studies are needed. Traditionally children, elderly and pregnant women are known to be highly susceptible groups. Sport fish eating population is definitely a high risk group. The case of the native population is worth investigation. They are genetically a distinct group and should be monitored separately though their numbers are too small to derive statistically significant conclusions. PCB levels in breast milk among natives have been falling since they adhere strictly to the fish eating guidelines. However the shift from traditional native fish diet to other foods may have consequences that should also be considered which leads to the interesting paradox: " What constitutes a Cause?".

**Host Factors**

Do we know what are the health effects from the contaminants?   In a number of cases we do. But the list is growing with the accumulation of scientific knowledge. There are both short-term effects and long-term effects. The short term effects can be assessed from quick cross sectional surveys. The long-term effects are more difficult to assess due to the lag time. It is also necessary to adjust for multiple risk factors and the confounding effect of variables. The indicators most commonly used to monitor population health are sickness absenteeism, spontaneous abortions, congenital anomalies, infant mortality and still births, infertility, sex ratio of new born, toxicological studies of chord blood/placenta, contaminants in breast milk, height, weight and nutritional status or growth and development studies, learning disabilities, standardized neurological testing of preschool children, cancer and neurological disorders. The national disease notification system is insensitive to detecting problems due to contaminants (except for major outbreaks). The public health system has not been primed to detecting a number of the listed problems. The family medical practitioner has neither the training nor the time to do the detective work.

### Epidemiologic Strategies

Can traditional epidemiologic studies help? There are a number of study design options depending on the magnitude of the effects to be measured, the nature of the health outcome and the availability of resources (23). There are four common approaches: (i) ecologic; (ii) cohort: (iii) case control; and (iv) nested case control. An **ecologic** study, also called an aggregate study, is one in which the unit of analysis is a group, most often defined geographically. It can often be carried out using national or provincial vital statistics system. In the case of chemical contaminants, it lacks statistical power since the group differences are usually small. Further it does not identify the problem cases; nor will it be possible to adjust for other exposure factors. **Cohort** studies are population follow-up studies and are excellent for identifying the risk to the exposed group. Also they provide a direct estimate of the risk to the exposed individual after adjusting for relevant co-factors. The major disadvantage is the long duration needed to observe effects as well the cost of monitoring the groups for a long time. **Case control studies** provide a good compromise to cohort studies. They can occasionally be expensive, but provide quick results for implementing policy actions. Case control studies require good participation from the selected public to avoid bias. Though they do not provide direct estimate of risk, it is possible to estimate the relative risk of the exposed population with respect to the unexposed population. Such estimates may be adequate for the evaluation of the population risk. **Nested case control studies** combine the advantage of both cohort and case control approaches. Since a direct estimate of risk is possible, we avoid the problem of bias that can invalidate the conclusions in case control studies.

Finally, the high degree of logistical complexity required for identifying the adverse health effects arising from contaminants found in the river, calls for the use of a variety of scientific talents in the field of analytic chemistry, biology, biochemistry, ecology, epidemiology and medicine. Such a well-knit multidisciplinary team is essential in order to arrive at any meaningful conclusions on the relationship between the chemical contaminants in the St. Lawrence River and the health profiles of the communities residing along its shores.

## Literature Cited

(1)   Dubois, R. *The Mirage of Health;* Harper: New York, NY, 1973.
(2)   Reavie, E. D.; Smol, J. P. *Abstract.* International Conference on the St.Lawrence Ecosystem, Cornwall, ON, Canada, 1995, p 16.
(3)   Crabbe, P. *Ecosystem Recovery on the St.Lawrence: A Proposal Submitted to the Eco-Research Grants Program.* National Research Council of Canada, Ottawa, ON, Canada, 1993.
(4)   International Joint Commission. *Status Report on Organic and Heavy Metal Contaminants in Lake Erie, Michigan, Huron and Superior Basins.* IJC, Windsor, ON, Canada, 1978.
(5)   International Joint Commission. *Committee on the Assessment of Human Health Effects of Great Lakes Water Quality: Annual Report.* IJC, Windsor, ON, Canada, 1981.
(6)   NAQUADAT. *Quality of Water for Direct Human Consumption: St.Lawrence Update.* Environment Canada, Ottawa, ON, Canada, 1992.
(7)   Remedial Action Plan. *Choices for Cleanup: Deciding the Future of a Great River.* St.Lawrence RAP Team and Public Advisory Committee, 1994.
(8)   Boyer, B. *No Place to Hide? Great Lakes Pollution and Your Health;* The Baldy Center for Law and Social Policy, State University of New York at Buffalo, Buffalo, NY, 1991.
(9)   Flint, R. W.; Vena, J. *Human Health Risks from Chemical Exposure: The Great Lakes Ecosystem;* Lewis Publishers: Michigan, IL, 1991.
(10)  Agency for Toxic Substances and Disease Registry. *Toxicological Profile for Polychlorinated Biphenyls. Report TP-92/16;* U.S. Department of Health and Human Services, ASTDR, Atlanta, GA, 1993.
(11)  Suzuki, T.; Imura, N.; Clarkson, T. W. *Advances in Mercury Toxicology;* Plenum Press: New York, NY, 1991, pp 1-32.
(12)  Choi, B. H. *Prog. Neurobiol.* **1989,** *32,* 447-470.
(13)  International Joint Commission. *Committee on the Assessment of Human Health Effects of Great Lakes Water Quality: Annual Report.* IJC, Windsor, ON, Canada, 1990.
(14)  *Toxic Chemicals in the Great Lakes and Associate Effects;* Environment Canada: Ottawa, ON, Canada, 1991; Vol. 1.
(15)  Goyer, R. A. *Environ. Health Persp.* **1993,** *100,* 177-187.
(16)  Subramanian, K. S. In *Quantitative Trace Analysis of Biological Materials;* McKenzie, H. A.; Smythe, L. E., Eds; Elsevier: Amsterdam, 1988, pp 589-603.
(17)  Morin, J.; Boudreau, P.; Leclerc, M. *Abstract.* International Conference on the St.Lawrence Ecosystem, Cornwall, ON, Canada, 1995, p 13.
(18)  *Hormone Disruptors in the Environment;* Conference on Human and Wildlife Health Issues, Nazareth College of Rochester, New York, NY, March 1995.

(19)    Subramanian, K. S.; Sastri, V. S.; Elboujdaini, M. E.; Connor, J.; Davey, A. B. C. *Water Res.* **1995**, *29*, 1827-1836.

(20)    Engel, J. *The Complete Canadian Health Guide;* University of Toronto Press: 1993, pp 346-350.

(21)    Jordan-Simpson, D.; Sherman, G.; Walsh, P.; Grant, D.; Hills, B.; Feeley, M.; Gilman, A.; Kearney, J. *Ontario Cohort Study;* Science Workshop Document, Health Canada: Ottawa, ON, Canada, 1993.

(22)    Kearney, J.; Cole, D. C.; Haines, D. S. *Great Lakes Anglers Pilot Exposure Assessment Study;* Science Workshop Document, Health Canada: Ottawa, ON, Canada, 1993.

(23)    Kleinbaum, D. G.; Kupper, L. L.; Morgenstern, H. *Epidemiologic Research: Principles and Quantitative Methods;* Lifetime Learning Publications: Belmont, CA, 1982.

Chapter 11

# Body Burden Concentrations in Humans in Response to Low Environmental Exposure to Trace Elements

Jane L. Valentine

School of Public Health, University of California—Los Angeles,
10833 Le Conte Avenue, Los Angeles, CA 90095-1772

Few studies have been published relating low trace element environmental exposures and resulting body burdens. This paper presents an evaluation of concentrations of arsenic and selenium in human urine when low concentrations, less than 30 μg/L, in the drinking water were consumed. Five communities for arsenic exposures and three communities for selenium exposures were selected. A good dose/response curve for selenium was obtained but the correlation coefficient, $r = 0.9508$, $n = 3$, was found not to be statistically significant ($p > 0.10$). A similar dose/response curve was obtained for arsenic and found to be statistically significant, $r = 0.9898$, $n = 5$, $p < 0.002$.

Human body burden has been shown to reflect trace element intakes in the cases of both deficient and adequate dietary exposure to the trace element examined. An example of such a relationship can be made with the element selenium. Oster and Prellwitz (*1*) have shown that dietary selenium intake correlated well with selenium excreted in urine. This was determined using urine and dietary Se intake estimates from studies conducted by various researchers in China, New Zealand, Italy, France, West Germany, Japan, and Canada. Dietary intakes ranged from 11 μg/day to 750 μg/day with a seemingly good correlation as well at the low intake range of 11 μg/day to 50 μg/day. The lowest intake level reported by Oster and Prellwitz was taken from the study of Yang *et al.* (*2*). At this intake level, blood, hair, and urine concentrations of 2.1 μg/100 mL, 0.074 μg/g, and 7 μg/L were reported respectively.
 The study of Oster and Prellwitz (*1*) was followed by numerous reports of a linear correlation between dietary selenium intake and levels of selenium in whole blood and its components and in urine (*3-8*). In contrast to the above studies, Thimaya and Ganapathy (*9*) found that hair selenium levels and dietary intake were not correlated in their study of 115 subjects. Selenium intakes ranged from 25 to 204 μg/day with a mean of 87±3 μg/day. However the extreme low value of 25 μg/day was reported for only one individual. Hair selenium levels in this study averaged 0.64±0.02 μg/g.
 Similar evaluations for environmental exposure/body burden relationships have been conducted for the assessment of biological monitoring utility and for studies of potential toxicity. Creason *et al.* (*10*) evaluated the relationship of trace elements in hair and exposure to dustfall, house dust, and soil. They found Ni, Cr, Sn, Ba, Hg, Pb,

and V environmental exposure gradients in dustfall and house dust to be significantly associated with the same elements in scalp hair. Selenium showed no gradient in the dust samples and arsenic was not evaluated. Valentine *et al.* (*11,12*) studied the dose/response relationship for two trace elements, arsenic and selenium, when exposure was via tap water. Levels of selenium in water ranged from 26 µg/L to 1800 µg/L for family units on individual wells (*11*). Arsenic levels in water reported as mean arsenic concentrations in water ranged from < 6.0 µg/L to 393 µg/L for community groups (*12*). Concentrations of arsenic and selenium in hair and urine of populations showed good correlations with drinking water concentrations. The exposures in these studies were mostly at very high levels, greater than 100 µg/L.

Few studies have been conducted to investigate the effects of low environmental exposures on body burden. Biological monitoring at low levels is needed to reflect accurately trace element exposures at normal, presumably non-life-threatening concentrations.

Andrews *et al.* (*13*) studied the relationship between low exposures to selenium through drinking water and body burden. They found no correlation between levels of selenium in tap water and erythrocyte and plasma selenium levels in any of three counties in Georgia studied either individually or when data from the three counties were combined. Selenium levels in 20 tap water samples were reported to be less than 0.01 µg/L. No values for selenium in blood were reported other than the statement that concentrations were < 15 µg Se/dL in all subjects. A review by Valentine (*14*) concerning health assessments for U.S. and Canadian populations evaluated 11 published reports on participants residing in communities from the states of California, Alaska, Nevada, and Utah, and from Canada (*12,15-19*). A dose/response relationship for reported values of arsenic in tap water *vs.* urinary arsenic was observed. Although this dose/response relationship included high exposure to arsenic, concentrations of arsenic in tap water at the lower range were less than 8 µg/L. Similar dose/response relationships were reported for tap water arsenic *vs.* arsenic concentrations in hair (*14*). In that study no attempt was made to evaluate the relationships to body burden for the case of low exposure only. No other studies of the relationship between arsenic exposure and body burden for low arsenic exposures from drinking water have been published.

In this report an evaluation of human body burden is presented, as indicated by urine concentrations of arsenic and selenium when very low concentrations in drinking water were consumed. This evaluation takes the form of a review of major studies conducted in the United States to obtain values of low environmental exposures to the two elements where body burden data are also given for the exposed individuals. In addition, a reassessment of our previously collected data for individuals is provided.

## Methods

**Population Data Sets**  A data set on individuals exposed to various elements in water has been maintained by Valentine (*11,12,19,20*). This data set consists of concentrations of a particular element in drinking water and levels of the respective element in hair, blood, and urine. Such data were gathered from volunteer participants from various communities experiencing low or high exposures to arsenic or selenium in their drinking water. Previous publication of this data set has been done (*11,12,19,20*). From these previous publications information on length of residence, age, gender, etc., can be obtained. Each participant in these studies was required to have consumed from a single water supply for at least one year.

For the purposes of this paper selected data sets were compiled from selenium and arsenic exposures and their body burden estimates grouped to determine whether low exposure situations can be related to body burden estimates. Urine values were employed due to the good dose-response relationship when drinking water is consumed (*11,12*). Robberecht and Deelstra (*21*) have stated that urine would serve as a good indicator since the kidneys play an important role in homeostasis and since urinary excretion is the most important route of elimination. Biological monitoring using urine is advantageous due to its ease of collection, non-invasive nature, and volume of sample obtained.

Three community groups of participants representing Casper (Wyoming), Grants (New Mexico), and Sun Valley (Nevada), were selected for the study of selenium. The five community groups of participants with exposure and body burden data for arsenic were Casper(Wyoming), Fairfax and Edison (California), and Sun Valley (Nevada). Concentration values of selenium and arsenic in urine were selected from individuals whose drinking water exposures were less than 50 μg/L. The current EPA maximum concentration limit (MCL) is 50 μg/L for both elements.

**Laboratory Methodology**  For our data set the laboratory methodology of hydride generation employing sodium borohydride ($NaBH_4$) as the reductant was used. The reductant was added to acid digests to generate either arsine ($AsH_3$) or hydrogen selenide ($SeH_2$) gases. Details of the methodology have been published (*11,12,22*). The hydrides were decomposed and arsenic and selenium were determined using an atomic absorption spectrophotometer (Perkin-Elmer Model 603). A variety of other methods were used in other studies whose results are referred to here.

**Statistical Analysis Techniques**  Statistical evaluations of the data involved determination of linear correlation coefficients using the mean values of participant exposures *vs.* participant group body burdens to assess a dose/response trend for the elements selenium or arsenic. In addition data from the literature were gleaned to apply to this dose/response evaluation in so-called "normal — low exposure" situations.

## Results

**Selenium body burden *vs.* low water selenium**  The mean values of selenium in urine as reported in the literature for supposedly non-exposure, non-disease-state populations range from 37 to 76 μg/L (*21,24*). Robberecht and Deelstra (*21*) have stated that normally urine selenium concentrations in non-exposed circumstances are below 30 μg/L. In our surveys, selenium levels in urine averaged less than 49 μg/L for persons in a high selenium area (soil and ground water) but consuming low-selenium water (*11*) and less than 33 μg/L for participants residing in low selenium areas (*20*), as shown in Table I.

Hadjimarkos and Bonhorst (*24*) noted that the mean selenium levels determined in urine for their study, 76 μg/L, were indicative of selenium intake. However, selenium levels in drinking water were low in their study area (*25*). Longnecker *et al.* (*6*) determined urine values of 2.14 μmol/d (168.97 μg/day) for persons residing in high selenium areas of the U.S. Water concentrations in their study were not reported. Values as low as 7 μg/L selenium in urine have been reported for populations from selenium-deficient areas in China (*2*). Good reviews of the renal excretion and medical implications of selenium have been presented by Oster and Prellwitz (*1*) and by Robberecht and Deelstra (*21*).

The data set we have maintained on selenium was also evaluated to ascertain the relationship of urinary excretion of selenium to its drinking water intake (Figure 1). The relationship between exposure levels and body burden was found to be statistically insignificant ($r = 0.95$, $p > 0.10$, n=3) when mean values for three community groups of participants' water selenium were related to their group mean values for urinary selenium excretion. The correlation coefficient was derived including participants from Grants (New Mexico) whose wells were not exposed. However, some wells of other residents of Grants were exposed to high selenium (*11*) and their data are not used. Grants participants included in this study had levels of selenium in their water supplies of less than 30 μg/L. Another comparison was made using a "dose" value. This was computed from each participant's reported water consumption and their individual concentration of selenium in drinking water. Dose values of 1 µg/day, 4.8 µg/day, and 66 µg/day were used. This correlation to body burden also was not statistically significant.

In evaluating the relationship of selenium concentrations in water for very low intakes (< 50 µg/L), no correlation could be determined when literature values of others (Table II) and mine (Table I) for low exposures were employed. A correlation coefficient, r = 0.08 for n=7, which was not statistically significant at the p > 0.10 level was obtained.

The results seem to show a weak dose/response relationship at low exposures to selenium in drinking water. However, the relationship is not statistically significant possibly due to the small sample size used in our data set. Also the earlier reports on selenium in water and body burden may have employed analytical methods that were not sensitive enough to produce a definitive conclusion. It would appear that no statistically significant relationship to selenium exposure via drinking water and body burden can be ascertained with existing data when levels of exposure are less than 50 µg/L.

The relationship of selenium levels in drinking water to concentrations in urine for all exposures evaluated in the author's laboratory was examined. The data set is given in Table I for low and high exposures. A correlation coefficient of 0.99 for n = 5, p < 0.002, was obtained. This correlation coefficient excludes the data of Grants (New Mexico) as the mean values for water and urine are not meaningful due to the extreme range of exposures in that community.

**Arsenic body burden *vs*. low water arsenic**  The water and urinary arsenic values for each community group are given in Table III. The mean arsenic concentrations in drinking water for each community ranged from 0.55 to 15.5 µg/L. The highest value in the water samples was less than half that of the current EPA standard.

Mean urinary arsenic concentrations ranged from 3.4 to 18.7 µg/day. The values were somewhat lower than normal limits reported by Wagner and Weswig (26). Their study gave values ranging from 24 µg/day to 201 µg/day for normal urinary arsenic excretion when 24-hour samples were obtained. The urinary arsenic value for Edison as surveyed in 1986–87 was somewhat higher than that of the Fairfax community (1986–87) in comparison to their water arsenic exposures. This fact cannot be explained. However, the Edison community used low-arsenic water for a year and a half prior to our survey in 1986–87. In the years preceeding, arsenic in their water supplies were 393 µg/L in 1979 with greatly elevated urine arsenic concentrations (12). It is possible that a prior arsenic exposure produced the comparatively higher urinary arsenic excretion for Edison when compared to Fairfax.

The plot of mean urinary arsenic against mean water arsenic at very low concentrations by community groups is shown in Figure 2. An increase in urinary excretion with increase in arsenic concentration in water was observed. The correlation coefficient for this comparison was r=0.99 at p < 0.002 and was statistically significant.

A comparison of low exposure body burden estimates with those of high has been reported (14) and repeated here in Tables IV and V. Correlation coefficients ≥ 0.87 were obtained and found to be statistically significant (p < 0.001).

**Discussion**

Concentrations of arsenic and selenium in drinking waters of the United States are generally very low, less than 50 µg/L. A recent study by Engel and Smith (28) documented the mean arsenic concentrations for the period 1968–1984 in 30 U.S. counties. The concentrations ranged from 5.4 µg/L to 91.5 µg/L. Only one county

Table I  Urine *vs.* tapwater mean selenium concentrations for 6 community samples (Valentine *et al.*)

| Tapwater Se (μg/L) | Urine Se (μg/L) | Community | Reference |
|---|---|---|---|
| Low exposures: | | | |
| trace * | 24.3 | Sun Valley | Valentine *et al.* (*20*) |
| 2 | 33.5 | Casper | Valentine *et al.* (*20*) |
| 28 | 49.1 | Grants, non-exposed | this study |
| High exposures: | | | |
| 90 | 615.3 | Grants** | this study |
| 194 | 100.3 | Jade Hills | Valentine *et al.* (*20*) |
| 494 | 299.6 | Red Butte | Valentine *et al.* (*20*) |

* Approximated at 0.1 μg/L for purposes of linear regression analysis.
** Includes values for participants whose drinking water concentrations were greater than 30 μg/L.

Table II  Urine *vs.* tapwater mean selenium concentrations for 4 community samples (other researchers)

| Tapwater Se (μg/L) | Urine Se (μg/L) | Community | Reference |
|---|---|---|---|
| trace * | 37 | Klamath | Hadjimarkos (*25*) |
| 1 | 74 | Jackson | Hadjimarkos (*25*) |
| 2 | 76 | Josephine | Hadjimarkos (*25*) |
| < 17 | 56** | rural Colorado | Tsongas and Ferguson (*23*) |

* Approximated at 0.1 μg/L for purposes of linear regression analysis.
** Overall mean value computed from x=71, n=22 for males and x=44, n=27 for females.

Table III  Urine *vs.* tapwater mean arsenic concentrations for 5 community samples for low exposures

| Tapwater As (μg/L) | Urine As (μg/day) | Community | Reference |
|---|---|---|---|
| 0.6 | 3.4 | Casper, Wyoming | Valentine *et al.* (*18*) |
| 2.3 | 6.8 | Sun Valley, Nevada | Valentine *et al.* (*18*) |
| < 6 | 10.9 | Fairfax, California | Valentine *et al.* (*12*) |
| 8 | 12.5 | Edison, California* | Valentine *et al.* (*19*) |
| 15.5 | 18.7 | Fairfax, California* | Valentine *et al.* (*19*) |

*Participants in these groups were surveyed in 1986-1987 (*19*). At this time the Edison participants had received low arsenic water for 1-1/2 years. Fairfax participants were also surveyed earlier for the 1979 report (*12*).

**Table IV**   Hair *vs.* tapwater mean arsenic concentrations for 14 community samples

| Tapwater As (μg/L) | Hair As (μg/g) | Community | Reference |
|---|---|---|---|
| 2 | < 0.10 | Sun Valley, Nevada | Valentine *et al.* (*18*) |
| < 6 | 0.15 | Fairfax, California | Valentine *et al.* (*12*) |
| 8 | 0.06 | Edison, California | Valentine *et al.* (*19*) |
| 15 | 0.2 | Fairfax, California | Valentine *et al.* (*19*) |
| 80 | 0.51 | Ester Dome, Alaska | Harrington *et al.* (*15*) |
| 100 | 0.48 | Virginia Foothills, Nevada | Valentine *et al.* (*12*) |
| 100 | 0.57 | Fallon, Nevada | Valentine *et al.* (*12*) |
| 100 | 0.6 | Nova Scotia, Canada | Hindmarsh *et al.* (*16*) |
| 120 | 0.5 | Hidden Valley, Nevada | Valentine *et al.* (*12*) |
| 180 | 1.21 | Hinkley, Utah | Southwick *et al.* (*17*) |
| 270 | 1.09 | Deseret, Utah | Southwick *et al.* (*17*) |
| 390 | 1.16 | Edison, California | Valentine *et al.* (*12*) |
| 390 | 3 | Nova Scotia, Canada | Hindmarsh *et al.* (*16*) |
| 400 | 3.29 | Ester Dome, Alaska | Harrington *et al.* (*15*) |

SOURCE: Adapted from ref. 14.

**Table V**   Urine *vs.* tapwater mean arsenic concentrations for 11 community samples

| Tapwater As (μg/L) | Urine As (μg/L) | Community | Reference |
|---|---|---|---|
| < 6 | 8.4 | Fairfax, California | Valentine *et al.* (*27*) |
| 8 | 9.6 | Edison, California | Valentine *et al.* (*19*) |
| 16 | 18.7 | Fairfax, California | Valentine *et al.* (*19*) |
| 100 | 45 | Ester Dome, Alaska | Harrington *et al.* (*15*) |
| 100 | 48.5 | Virginia Foothills, Nevada | Valentine *et al.* (*27*) |
| 100 | 54.2 | Hidden Valley, Nevada | Valentine *et al.* (*27*) |
| 100 | 61.5 | Fallon, Nevada | Valentine *et al.* (*27*) |
| 180 | 175 | Hinkley, Utah | Southwick *et al.* (*17*) |
| 270 | 211 | Deseret, Utah | Southwick *et al.* (*17*) |
| 390 | 176 | Edison, California | Valentine *et al.* (*26*) |
| 400 | 170 | Ester Dome, Alaska | Harrington *et al.* (*15*) |

SOURCE: Adapted from ref. 14.

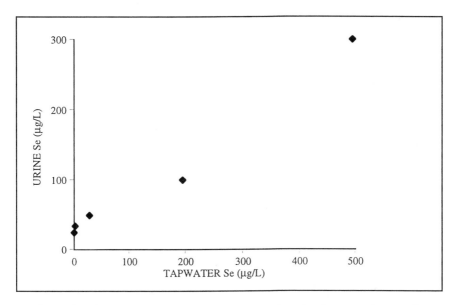

**Figure 1** Urine *vs.* tapwater selenium concentrations for 5 community samples with low and high exposure.

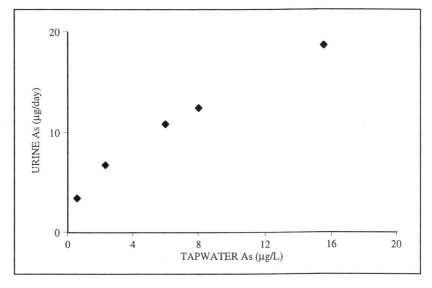

**Figure 2** Urine *vs.* tapwater mean arsenic concentrations for 5 community samples with low exposure.

showed mean arsenic concentrations for the 16 year period that were above 50 µg/L. Borum and Abernathy (29) estimated that 9% of the U.S. population are exposed to arsenic in drinking water at concentrations greater than 10 µg/L. Isolated incidents occur where arsenic concentrations of 0.50 ng/L to as much as 2.5 mg/L have existed in small public systems and individual wells (14). However, very low arsenic concentrations and very low selenium concentrations represent the norm in the USA. Selenium levels, less than 10 µg/L, are usually reported for most community water supplies (Valentine et al., unpublished data).

Previous studies by Valentine et al. (11,12,14,20) have shown that selenium and arsenic levels in urine samples increase with increasing concentrations in drinking water in cases of high exposures. Urine concentrations were found to be significantly correlated with blood and hair values for both elements (11,12).

The suitability of biological monitoring at low levels has not been previously investigated. It was commented upon by Andrews et al. (13) for selenium levels in drinking waters in the state of Georgia, United States.

From our investigation it appears that very low levels of 1 µg/L to 17 µg/L selenium in drinking water can show a dose/response relationship to body burden. However, the range of selenium concentrations in our present study was relatively narrow. In addition only three distinct data points, community means, were available. The lack of a correlation between body burden and water exposure for literature data of all researchers may be due to differences in the analytical methods employed.

Several studies have shown a relationship of exposure to body burden but only for cases of high environmental exposures (11,20,23). In some cases a lack of a dose/response relationship for drinking water at nanogram concentrations (13) may be related to dietary intake. At such levels in drinking water, dietary intake may overshadow any intake attributed to the environment. Andrews et al. (13) state that since the drinking water in their study contributed very little to the total daily intake of selenium, it was not surprising to find the absence of significant correlations between selenium levels in blood and selenium levels in drinking water. Robberecht and Deelstra (21) corroborated this viewpoint in their review of selenium in human urine by commenting on the marked but variable response in urine to dietary intake of the element. They cited the work of Clementi et al. (30) and Griffiths (31) where no correlation between diet intake at very low levels and urinary excretion could be ascertained. They argued that since intake levels were so low the urine values represented body stores and blood levels rather than intake. Levander et al. (32) on the other hand found good correlations between selenium in urine and Se intake at very low levels, 33–36 µg/day through the diet, where water intake was negligible. Similarly, Robberecht et al. (33) have stated that in most studies published on the daily intake of selenium, the role of drinking water's contribution is neglected. According to them drinking water would be responsible for a significant fraction of the total intake in some countries.

It has been stated that the mean dietary intake for U.S. adults amounts to approximately 108 µg/day (34). This level is quite high in comparison to human exposure from environmental sources, most notably water, where in many instances at most trace to 10 µg/L are experienced which would provide trace to 20 µg/day intakes from a water source if two liters of water are consumed daily.

Deficient Se intakes from diet have been recorded at <20 µg/day (4,35). In such cases high water intakes would be considerable under low dietary exposure conditions and could provide twice the intake value of food. Robberecht et al. (33) estimated that in Belgium and England drinking water would account for at most 4.8% and 6% respectively in an individual's total intake. Using our dose values drinking water would contribute about 1% to 61% of the total intake of selenium. Further studies on this issue are needed.

Arsenic levels in drinking water and dose response show much more promise during low exposures to the element, as in general the levels of arsenic in food are lower. Borum and Abernathy (29) gave a range of 10 μg/day to 14 μg/day intake of inorganic arsenic from food. They further stated that total arsenic food intake (inorganic plus organic) could amount to 21.5, 27.6, and 52.6 μg/day for the average infant (6 months old), toddler (2 years old), and adult respectively. We have determined that water could contribute as much as 30 μg/day with a two liter intake, representing 57% of the adult total food intake.

We have determined that arsenic body burden responds very well to very low arsenic concentrations, <50 μg/L. Also at these levels when evaluated along with high exposure levels — those above 100 μg/L — responsive evaluation is also obtained. These comparisons seem to hold for studies performed in the United States and when values of other countries are considered. Quarterman (36) stated that the diagnosis of low-level chronic metal toxicity in individuals has proved extremely difficult and must be the aim of a great deal of research in the future. In addition there is a need for biological monitoring at low levels where further research is needed.

## Literature Cited

1. Oster, O., and Prellwitz, W. *Biological Trace Element Research* **1990**, *24*:119–146.
2. Yang, G., Wang, S., Zhou, R., and Sun, S. *Amer. J. Clin. Nutr.* **1983**, *37*:872–881.
3. Schrauzer, G.N., and White, D.A. *Bioinorg. Chem.* **1978**, 8:303-318.
4. Stewart, R.D.H., Griffiths, N.M., Thomson, C.D., and Robinson, M.F. *Br. J. Nutr.* **1978**, *40*:45–94.
5. Palmer, I.S., Olson, O.E., Ketterling, L.M., and Shank, C.E. *J. Amer. Diet. Assoc.* **1983**, *82*:511–515.
6. Longnecker, M.P., Taylor, P.R., Levander, O.A., Howe, M., Veillon, C., McAdam, P., Patterson, K.Y., Holden, J.M., Stampfer, M.J., Morris, J.S., and Willet, W.C. *Amer. J. Clin. Nutr.* **1991**, *53*:1288–1294.
7. Gebre-Medhin, M., Gustavson, K.-H., Gamstorp, I., and Plantin, L.-O. *Acta Paediatr. Scand.* **1985**, *74*:886–890.
8. Westermarck, T. *Acta pharmacol. et toxicol.* **1977**, *41*:121–128.
9. Thimaya, S., and Ganapathy, S.N. *The Science of the Total Environment* **1982**, *24*:41–49.
10. Creason, J.P., Hinners, T.A., Bumgarner, J.E., and Pinkerton, C. *Clin. Chem.* **1975**, *21*:603–612.
11. Valentine, J.L., Kang, H.K., and Spivey, G.H. *Environ. Res.* **1978**, *17*:347–355.
12. Valentine, J.L., Kang, H.K., and Spivey, G. *Environ. Res.* **1979**, *20*:24–32.
13. Andrews, J.W., Hames, C.G., Metts, J.C., Waters, L., Davis, J.M., and Carpenter, R. *J. Environ. Pathol. and Toxicol.* **1980**, *4*:313–318.
14. Valentine, J.L. In: W.R. Chappell, C.O. Abernathy, and C.R. Cothern (eds.) *Arsenic Exposure and Health.* Science and Technology Letters, Northwood, 1994.
15. Harrington, J.M., Middaugh, J.P., Morse, D.L., and Housworth, J. *Amer. J. Epidemiol.* **1978**, *108*:377–385.
16. Hindmarsh, J.T., McLetchie, O.R., Heffernan, L.P.M., Hayne, O.A., Ellenberger, H.A., McCurdy, R.F., and Thiebaux, H.G. *J. Anal. Toxicol.* **1977**, *1*:270–276.
17. Southwick, J.W., Western, A.E., Beck, M.M., Whitley, T., Issacs, R., Petajan, J., and Hansen, C.D. In: William H. Lederer and Robert J. Fensterheim (eds.), *Arsenic: Industial, Biomedical, Environmental Perspectives.* Van Nostrand Rheinhold Co., New York, 1983.

18. Valentine, J.L., Reisbord, L.S., Kang, H.K., and Schluchter, M.D.    In: C.F. Mills, J. Bremner, and J.K. Chesters (eds.), *Trace Elements in Man and Animals – TEMA 5*. Commonwealth Agricultural Bureaux, Aberdeen, 1985.

19. Valentine, J.L., Bennett, R.G., Borok, M.E., and Faraji, B.    In: B. Momcilovic (ed.) *Trace Elements in Man and Animals – TEMA 7*. IMI Institute for Medical Research and Occupational Health, University of Zagreb, Zagreb, Yugoslavia, 1991.

20. Valentine, J.L., Reisbord, L.S., Kang, H.K., and Schluchter, M.D.    In: G.F. Combs, Jr., J.F. Spallholz, O.A. Levander, and J.E. Oldfield (eds.). *Selenium in Biology and Medicine*. Van Nostrand Rheinhold Co., New York, 1987.

21. Robberecht, H.J., and Deelstra, H.A.    *Clinica Chimica Acta* **1984**, *136*:107–120.

22. Kang, H.K., and Valentine, J.L.    *Anal. Chem.* **1977**, *49*:1829–1831.

23. Tsongas, T.A., and Ferguson, S.W.    In: M. Kirchgessner (ed.) *Trace Element Metabolism in Man and Animals – 3. Proceedings of the 3rd International Symposium*. Weihenstephan, 1978.

24. Hadjimarkos, D.M., and Bonhorst, C.W.    *J. Pediat.* **1958**, *52*:274–278.

25. Hadjimarkos, D.M.    *Arch. Environ. Health* **1965**, *10*:893–899.

26. Wagner, S.L., and Weswig, P.    *Arch. Environ. Health* **1974**, *28*:77–79.

27. Valentine, J.L., Campion, D.S., Schluchter, M.D., and Massey, F.J.    In: J.McC. Howell, J.M. Hawthorne, and C.L. White (eds.) *Trace Element Metabolism in Man and Animals*. Australian Academy of Science, Canberra, 1981.

28. Engel, R.R., and Smith, A.H.    *Arch. Environ. Health* **1994**, *49*:418–427.

29. Borum, D.R., and Abernathy, C.O.    In: W.R. Chappell, C.O. Abernathy, and C.R. Cothern (eds.) *Arsenic Exposure and Health*. Science and Technology Letter, Northwood, 1994.

30. Clementi, G.F., Rossi, L.C., Santaroni, G.P.    *J. Radioanal. Chem.* **1977**, *37*:549–558.

31. Griffiths, N.M.    *Proc. Univ. Otago Med. Sch.* **1973**, *51*:8–9.

32. Levander, O.A., Sutherland, B., Morris, V.C., and King, J.C.    *Amer. J. Clin. Nutr.* **1981**, *34*:2662–2669.

33. Robberecht, H., Van Grieken, R., Van Sprundel, M., Vanden Berghe, D., and Deelstra, H. *The Science of the Total Environment* **1983**, *26*:163–172.

34. NRC.    *Recommended Dietary Alowances, 10th edition*. National Academy of Sciences. Washington, D.C., 1989.

35. Chen, X., Yang, G., Chen, J., Chen, X., Wen, Z., and Ge, K.    *Biological Trace Element Research* **1980**, *2*:91–107.

36. Quarterman, J.    In: M. Kirchgessner (ed.) *Trace Element Metabolism in Man and Animals – 3. Proceedings of the 3rd International Symposium*. Weihenstephan, 1978.

# Chapter 12

# Placenta: Elemental Composition, and Potential as a Biomarker for Monitoring Pollutants to Assess Environmental Exposure

K. S. Subramanian[1] and G. V. Iyengar[2]

[1]Environmental Health Directorate, Postal Locator 0800B3, Health Canada, Tunney's Pasture, Ottawa, Ontario K1A 0L2, Canada
[2]Biomineral Sciences International Inc., 6202 Maiden Lane, Bethesda, MD 20817

This chapter presents a review of the available literature data on the elemental composition of placenta consistent with healthy fetal development, examine the influence of sampling and sample preparation on the results reported, explore possible correlation with maternal and cord blood, and assesses its usefulness as a biomarker for monitoring environmental exposure to metal pollutants.

Many metals are known to adversely affect human health even at relatively low concentrations than previously defined as *toxic levels* of exposure. In some cases it may simply be the effect of cumulative dose due to the chronic nature of the exposure aided by factors such as those arising from periodic industrial emissions, and acid rain and its effects on soil-plant-food chain. The developing fetus and the new born are particularly vulnerable when exposed to such situations (*1*). Therefore, placenta could play an important role in the assessment of fetal effects of toxic metals. Although the principal role of placenta is to nurture the fetus, the same processes that aid transport of nutrients can also act as pathways to toxic constituents due to chemical similarities with some of the nutrient metabolites, or simply as a result of passive diffusion.

However, in some cases, placenta acts as a barrier by preferentially concentrating specific toxicants and thereby reducing the exposure potential of the fetus at least to some degree (*2*). This feature makes placenta a potential biomarker from the point of monitoring the health of the fetus, and of the mother too to some extent. Of particular significance is the fact that unlike many other clinical specimens, placental samples are not difficult to obtain and therefore, present excellent opportunities to carry out epidemiological studies (*3*). General exposure related biomarkers that are routinely collected in human subjects are blood, urine and breath. In specific cases, scalp hair is also used but it is not free of objections due to the fact that it can be extraneously contaminated. On the other hand, placenta is still a less known analytical specimen among biological trace element researchers for a variety of reasons such as problems of sample collection and handling (see sampling section) and ascertaining representativeness. All this has contributed to the fact that only a handful of epidemiological studies have been carried out, and even these relate to the ability of placenta to act as a barrier to toxic metals such as As, Cd, Hg and Pb. The available information is discussed under biomonitoring section.

The purpose of this chapter is to evaluate the literature data on placental analysis for minor and trace elements with emphasis on the following points: (a) baseline elemental composition data based on a critical appraisal of the literature database; (b) epidemiological link to biomonitoring in order to address the maternal-fetal exposure issues; and (c)evaluation of the usefulness of placenta as a biomarker from a multidisciplinary perspective so that the information would be useful to not only the analytical chemist, but also the biologist, epidemiologist, physician, and the environmental, biomonitoring, clinical and biochemical communities. In doing so, we will also address such questions as details to be observed for development of appropriate lifestyle questionnaire to obtain relevant donor information, problems encountered in sampling placenta (including sample preservation), long-term storage and preservation strategy by adopting specimen banking approaches, and relevant guidance related to analytical procedures.

**The Placenta**

Placenta is a temporary organ found only in eutherian mammals and composed of cells from two different individuals. The embryonic tissues of the embryo and the endometrial tissues of the mother are involved in the formation of the placental complex. The placenta functions as a multi-faceted organ system in order to sustain the fetus through its developmental period to birth. Thus, the placenta provides fetal nutrition and transfer of oxygen; functions as the organ of excretion for waste products of the fetus and removal of carbon dioxide; and functions as a storage organ where during the first few months of pregnancy it accumulates proteins, calcium, and iron to be used in the later months for fetal growth.

According to the data available from the International Commission on Radiological Protection (4), the average weight of placenta at full term is close to 500 g without cord and membranes, and the volume is about 490 mL. Gross composition of placenta is: water 84.6%, protein 12%, fat 0.11% and ash 1 %. Specific gravity of the placenta is 0.995. Table I shows the physico-chemical characteristics of placenta. Table II gives recent data on the diameter and weight of placenta at various developmental stages of the embryo/fetus. The large surface area of the placenta which rapidly expands up to 11 square meters at term from about 5 square meters around 28 weeks, helps in the transportation of substances from maternal blood. The transfer can take place across the placenta in both directions. Oxygen and nutrients (e.g. water and electrolytes) are supplied from maternal blood to fetus in one direction and readily pass through

Table I.  Some Physicochemical Characteristics of Placenta

| Parameter | Av. Value | Parameter | Av. Value |
|-----------|-----------|-----------|-----------|
| weight | 420±8.5 g | villous area | 11±1.3 m² |
| volume | 488±99 mL | decidual area | 253±6.6 cm² |
| thickness | 1.6±0.1 cm | water | 84.6±0.2% |
| diameter | 15-20 cm | protein | 12.0±1.0% |
| sp. gravity | 0.995 | fat | 0.11±0.02% |
| blood vol | 250 mL | ash | 1.0±0.1% |

Table II.  Placental Diameter and Weight as a Function of Embryo Age

| Embryo Age (Weeks) | Placental Diameter (mm) | Placental Weight (g) |
|--------------------|-------------------------|----------------------|
| 6 | - | 6 |
| 10 | - | 26 |
| 14 | 70 | 65 |
| 18 | 95 | 115 |
| 22 | 120 | 18 |
| 26 | 145 | 250 |
| 30 | 170 | 315 |
| 34 | 195 | 390 |
| 38 | 220 | 470 |

Table III. Elemental Composition of Placenta: Pre-1975 Data

| Element | Method | $N^a$ | Av. Concn $(\mu g/g)^b$ |
|---|---|---|---|
| Al | AAS | 60 | 1.1±0.3 |
| As | NAA | 6 | 0.011±0.008 |
| Br | INAA | 7 | 3.7±0.8 |
| Ca | AAS | 118 | 1310±260 |
| Cd | AAS | 180 | 0.019±0.002 |
| Cl | NAA | 15 | 2170±310 |
| Co | NAA | 838 | 0.004±0.002 |
| Cr | AAS | 60 | 0.036±0.008 |
| Cu | AAS, NAA | 74 | 2.5±1.4 |
| Fe | AAS, NAA | 789 | 310±170 |
| Hg | AAS, NAA | 1076 | 0.034±0.025 |
| K | NAA | 16 | 1430±280 |
| Mn | AAS | 59 | 0.47±0.19 |
| Na | AAS | 16 | 1990±320 |
| Pb | AAS | 289 | 0.33±0.02 |
| Rb | NAA | 826 | 4.0±0.5 |
| Se | NAA | 822 | 0.3±0.1 |
| Zn | NAA | 811 | 12.0±2.3 |

SOURCE: Adapted from refs. 5 and 42.
AAS: Flame/furnace atomic absorption spectrometry;
NAA: Neutron acativation analysis.
[a]Number of samples analyzed.
[b]Expressed as wet weight of sample. Concentration values given in dry weight were converted to wet weight using the ratio: dry wt/wet wt = 6.

placenta to nurture the fetus. The elimination of carbon dioxide and other waste (e.g. urea, creatinine and bilirubin) products produced by the fetus proceed in the other direction.

Sometimes harmful substances also pass through placenta and damage the embryo. Several factors, such as life style, drug abuse, therapeutic excesses and exposure to bacterial and virus type of contamination and environmental chemicals are linked with fetal abnormalities. Toxic effects of lithium (heart anomalies) resulting from its therapeutic use as an antidepressant agent, the role of methyl mercury inducing mental retardation, cerebral atrophy and blindness, the action of lead in impairing central nervous system even at low concentrations, and several other metals and their compounds are good examples of impact of chemicals. Pregnant women are considered to be at risk as pollutants such as As, Cd, Hg and Pb are transferred from the maternal blood to the fetus through the placenta. It is stated that in the USA alone an estimated 3.7 million people live with a 1 mile radius of toxic waste dumps (5). Several adverse health effects have been documented in the literature as a result of the exposure of the fetus to noxious substances causing congenital malformations, including still births in extreme cases.

**Minor and Trace Elements in Placenta**

As a biomedical specimen, placenta has caught the attention of researchers who studied its elemental composition during the 60s and the 70s (4, 6). These pre-1975 data are compiled in Table III. However, due to several shortcomings that are mainly analytical in nature (as discussed below), the data generated were inadequate and poor in quality. Continued efforts in the 80s and 90s, have resulted in a somewhat better situation than before in that data for several elements are now available. Moreover, refined analytical approaches have been used to generate the information including use of reference materials to check the performance of the methods. These post-1975 data are shown in Table IV. For several elements there is still paucity of data and well controlled analyses are needed.

**Sampling and Sample Preparation of Placenta**

Sampling placenta for elemental composition studies is by no means an easy task as several aspects have to be considered at the time of sampling. Adequate care has to be taken to understand the impact of presampling factors (7) as well as the integrity of the sampled material to represent placental tissue. Placenta is a complex tissue consisting mainly of the chorionic villi, amnionic membranes and the cord detached at the point of entry to the placenta, and is highly vascularized by maternal and fetal blood vessels. Thus, often a piece of placenta dissected as a sample may simply be a heterogeneous mix of placental cells and decidual matter tainted with maternal and fetal blood. Metals are not uniformly distributed in these components of the placenta and therefore, raises the question of representativeness of the sampled material and the measured composition is

Table IV. Elemental Composition of Placenta: Post-1975 Data

| Element | Method(Ref) | $N^a$ | Av. Concn $(\mu g/g)^b$ | Ref |
|---|---|---|---|---|
| Al | NAA | 100 | 0.25±0.05 | 20, 21 |
| As | NAA, AAS | 118 | 0.008±0.004 | 5, 20 |
| B | NAA | 100 | 0.35±0.07 | 20, 21 |
| Br | NAA | 118 | 1.3±0.5 | 20, 43 |
| Cd | AAS | 171 | 0.005±0.002 | 5, 26, 32, 44 |
| Cl | NAA | 100 | 350±45 | 20, 21 |
| Co | NAA | 118 | 0.006±0.002 | 20, 43 |
| Cr | NAA | 100 | 0.004±0.003 | 20, 21 |
| Cu | AAS, ICP | 146 | 0.12±0.03 | 27, 44, 45 |
| Fe | AAS, NAA | 209 | 80±22 | 20, 27, 46 |
| Hg[total] | AAS | 26 | 0.02±0.01 | 44 |
| Hg[methyl] | AAS | 120 | 1.49±0.16 | 27, 32 |
| K | NAA | 100 | 1560±320 | 20, 21 |
| Mg | AAS, ICP | 177 | 95.6±6.0 | 44, 46, 47 |
| Mn | NAA | 100 | 0.14±0.09 | 20, 21 |
| Na | NAA | 100 | 370±80 | 20, 21 |
| Pb | AAS, ASV | 419 | 0.04±0.01 | 5, 9. 26, 27, 32, 44, 47, 48 |
| Rb | NAA | 118 | 1.2±0.7 | 20, 21 |
| Se | NAA, AAS, SF | 171 | 0.26±0.06 | 39, 43, 44, 49 |
| Zn | NAA, AAS, ICP | 279 | 6.3±4.3 | 20, 27, 43, 44, 46 |

AAS: Flame/furnace atomic absorption spectrometry. Arsenic and mercury were determined by hydride generation-AAS, and cold vapour-AAS, respectively; NAA: Neutron activation analysis; ICP: Inductively coupled plasma atomic emission spectrometry; AES: Atomic emission spectrometry; ASV: Anodic stripping voltammetry; SF: uv-visible spectrofluorometry.
[a]Number of samples analyzed.
[b]Expressed as wet weight of sample. Concentration values given in dry weight were converted to wet weight using the ratio: dry wt/wet wt = 6.

subject to considerable variation. For example, Manci and Blackburn (8) sampled four anatomic segments (one paracentral near the umbilical cord, two from the central part of the placenta and one from the periphery of the placenta), and showed variations for Ca, Cu and Zn in individual sample pieces. Also, analyses of very small samples run the risk of under-, or over-estimation, depending upon the segment sampled and the element in determined. Baghurst et al (9) have illustrated the problem of over-estimation of Pb in placenta based on results of a small sample of placental body containing membrane tissue.

At the first stage of sampling, it is advisable to collect the entire placenta even if it includes the umbilical cord and place it in clean polyethylene bags and freeze it at -20°C. Further preparatory work should be carried out in the laboratory. Under clean bench conditions, the sample can be thawed, cord detached at the point of entry to the placenta, and gently squeezed to remove excess blood and wiped with fibre-free tissue. If elements such as V, Mn and Cr are likely to be determined, blending the entire sample in a high purity stainless steel blender would be acceptable. The homogenous slurry can then be taken up for analysis depending upon the requirement of the particular analytical approach. The danger of contamination arising out of different circumstances relevant to trace element problems is discussed elsewhere (10, 11).

Homogenizing the entire sample may prove to be technically challenging and cumbersome if large numbers of samples have to be handled. However, for epidemiological studies, wherein representative samples are crucial and the analytical phase is likely to involve several analytes, handling the entire sample may be unavoidable. Further, use of metal-free accessories made of Teflon or high purity high density polyethylene, or use of specialized implements made of titanium will be necessary for the control of contamination.

An alternate method of procuring representative samples would require random samples from multiple sites, pooled, homogenized and assayed. Miller et al (12) recommend the following procedure: obtain multiple pieces from both peripheral and central lobules. Remove cubes of tissues from the tissue on the maternal side of the placenta (decidua basalis). Cubes obtained from the periphery should not contain marginal tissue since this is not placental by origin. Finally, cut the chorionic plate to sample the trophoblastic tissue. Collect a minimum of six such pieces from the different sections mentioned above. At this stage it is to be decided whether the sample should be frozen as it is or rinsed to remove trapped blood. Removal of blood (both maternal and fetal) is a tricky operation and involves washing out metals from the parent tissue. For example, cutting the tissue into 500 mg pieces and rinsing with buffered salt solutions removes maternal blood (but not fetal blood). However, one has to simultaneously evaluate the danger of contamination associated with excessive sample preparation processes. Therefore, it is a good practice to collect maternal and fetal blood samples, analyze them to evaluate the possible contributions (if significant) of specific metals, for correction of the results. In some cases (e.g. Cd) where the concentration of Cd is same in placenta and maternal blood (under normal living conditions), the correction required is minimal.

Long-term storage and preservation of placenta can be accomplished by adopting the basic procedures practised in specimen banking projects. Under this approach, samples are stored in specially fabricated liquid nitrogen-filled facilities. The concept of specimen banking is discussed briefly under the biomonitoring section.

Analytical procedures using nuclear and non-nuclear techniques for a variety of biomatrices have reached a reliable level as summarized in a recent WHO Report (13) wherein several methods have been evaluated for their applicability to bioenvironmental sample analysis.

As for documentation of the sample history, at the very minimum, details covering age, parity, time of expected delivery, medication, occupation, general characteristics of the environment surrounding the residence, source of drinking water, dietary and smoking habits, etc., should be recorded. Further, if the subjects are classified as non-smokers it should be clarified whether never smoked, not smoked during pregnancy or recently stopped smoking. Concerning diets, consumers of fish should be identified. Recording the month of parturition is helpful in identifying seasonal changes, if any, in the composition of placenta. For example, seasonal changes have been reported for placental levels of Na, K, Mg and Pb (14).

## Variations in Placental Composition

Several environmental factors induce changes in metal concentrations of blood and placenta. Industrial activities related to mining, metallurgy, smelting, emissions, and even tobacco smoke are all known to affect the concentration levels of elements such as As, Cd, Cr, Hg, Ni and Pb (5, 9, 16, 23, 26-29). However, good data are scarce for these and other elements including major elements such as Ca, and well designed epidemiological studies are required to fill the information gap. Data from different countries, especially for elements such as Pb show great differences, but definitive conclusions as to whether or not these differences are true geographic indications must wait until the quality of the analysis is beyond doubt. Thorough homogenization is the key and can be easily achieved by the use of brittle fracture technique (15). Further, with the exception of methyl-Hg, systematic information about speciated analytes for several elements are not available. This is extremely important to have a clear understanding of the toxic effects and risk assessment. All these concerns should be addressed in the future.

Compared with pre-1975 results (Table III), the concentration of Pb in placenta has declined by several folds as can be seen from the probable reference value of 40-50 ng/g arising out of evaluation of recent results (Table IV). At least two factors may contribute to this; one, improvements in analytical techniques (particularly control of contamination) and second, attempts to reduce environmental sources of Pb to comply with regulatory requirements.

## Placental Transfer of Cd, Pb and Hg

Placental transfer of several trace elements, in particular those that are detrimental to health have been investigated by many researchers. These studies have focused on both non-exposed (i.e. environment without any known pollutant exposure sources) as well as industry based conditions related to mining, smelting and intense metallurgical activities likely to affect the health of populations living nearby. Under both conditions, placental transfer to varying degrees of metals such as Cd, Hg and Pb, among others have been confirmed. Pb has been very extensively investigated for its impact on fetal effects and the ability of placenta to protect the fetus from this pollutant. However, the biochemical mechanism of the transfer process for many of these metals is poorly understood.

**Cadmium.** In a study from the former Yugoslavia, placentas collected under different environmental conditions have shown that those samples coming from industrial areas (predominantly smelting operations) showed a 50% increase in placental Cd (*16*). In another study investigating Cd levels in blood in mother-newborn pairs from the surroundings of Cu smelter in Sweden (*17*) and a control area reported that the Cd levels of the new-borns were about 70% of those in the mothers. Berlin et al (*18*) measured Cd in placenta from workers from a battery factory and found that the placental Cd levels were positively correlated with maternal blood Cd concentrations. Several studies have reported similar findings as summarized in various reports (*2, 19*). The most important source of environmental concern for Cd seems to be smoking as evidenced from higher Cd in blood in women who smoke than nonsmokers (*12, 20-22*).
      Literature data for Cd in maternal and fetal blood and placenta are not consistent. This problem is compounded by analytical difficulties associated with measuring very low concentrations of Cd in blood and related matrices, and the difficulty in meaningfully classifying the so-called normal and other types of exposure of the investigated subjects to environmental Cd. With the exception of samples clearly separated due to occupational exposure, control populations from rural and urban areas need to be carefully qualified (e.g. food and living habits, age, etc.) to be reference groups. Not withstanding any of these drawbacks, placenta Cd concentrations have been found to be positively correlated with maternal blood Cd concentrations, and there are clear indications that fetal blood Cd levels are below that of maternal blood Cd levels, thus showing that placenta acts as a partial barrier and lessens potential harmful effects to the fetus.

**Lead.** Several investigations have conclusively shown that Pb readily crosses the placental barrier (*23-25*). In an extensive study conducted in the former Yugoslavia, placenta samples collected under exposure conditions ranging from low, mid and high in Pb, the Pb levels in placenta (compared with no known additional sources of exposure) were found to be 75%, 150% and 450% higher, respectively (*16*). The effects of chronic exposure to Pb and the impact on its concentration in the umbilical cord and placental tissue (both body and membrane) have been studied (9). The results indicated modest increases in Pb concentrations in the placental body. Significantly, the Pb concentrations in membranes from late fetal deaths (2.73 mg/kg, dry) were very high when compared with those obtained from normal births (0.78 mg/kg, dry). In a Zambian study, the mean blood levels were very high being 412 and 370 ng/mL for mothers and infants, respectively (24). Surprisingly, the infants did not show any noticeable ill effects from this high Pb levels suggesting the presence of other protective factors. High intake of Pb through drinking water (26) and exposure to elevated Pb in the urban environment (27) have also been shown to moderately elevate Pb burdens in maternal blood, placenta as well as the blood Pb of the newborn (or cord blood) confirming efficient transfer of Pb through placenta. In a very recent Polish study carried out in a very heavily polluted area, following levels of Pb were observed: maternal blood, 72.5 ng/mL, 500 ng/g in placenta and 38 ng/mL (28). Other documented exposures to Pb, increases in placental Pb and concomitant stillbirth occurrences have been seen in cases studied in Stoke-on-Trent area in UK (29). The occupationally exposed women were employed in pottery, lithography and painting jobs. Other examples are documented in the literature (26, 30-32). Therefore, pregnant women must be prevented from exposure to Pb in drinking water as well as other emission sources.

Lead is transferred across the placenta by diffusion and is believed to occur at the very early stages of pregnancy. The duration of the gestation period, birth weight and length are inversely related to fetal cord blood Pb concentrations(2).

**Mercury.**  Hubermont et al (26) who investigated the placental transfer of Hg in women living in a rural area of Belgium have demonstrated that the Hg levels of blood samples from newborns (15.9 ng/mL) were higher than that found in maternal (14.5 ng/mL), and placental (9.7 ng/g) samples. In a Czechoslovakian study involving 50 subjects (two groups belonging to an industrial environment matched with controls), Truska et al (32) found practically no difference in the total-Hg concentrations in maternal blood erythrocytes and plasma, cord blood erythrocytes and plasma, and placenta. Dennis and Fehr (33) have investigated the relationship between Hg levels in maternal and cord bloods in paired samples from fish eating and non-fish eating women from north and southern Saskatchewan in Canada, respectively. Their results demonstrate that when maternal blood level is minimally influenced by Hg absorbed from fish, the cord blood levels are not significantly different from maternal blood levels, but that when maternal blood Hg levels are elevated, higher cord blood levels are found. In another study carried out by Nishima et al (34) on pregnant women from the city of Tokyo, both total-Hg and methyl-Hg found in cord blood (25±10 ng/g and 13±6 ng/g, respectively)were higher than those found in the maternal blood (14±6 ng/g and 6±3 ng/g). In a radiotracer study on rats investigating maternal-fetal transfer of organic and inorganic Hg via placenta and milk, it was found that organic Hg is more readily transferable across the placenta than inorganic Hg (35). However, both forms were transferred through milk at approximately the same efficiency.

The above findings suggest that Hg (both total and methylated forms) readily crosses the placental barrier. Also, subjects of child bearing age consuming fish frequently and into the pregnancy period run the risk of exposing the fetus to abnormal levels of Hg, particularly organic-Hg. There have also been studies reporting both organic and inorganic Hg in amniotic fluids from human studies carried out in Japan (*36*). The overall significance of these findings is yet to be understood.

On the other hand, Tsuchiya et al (*27*) determined total as well as methylated Hg in over 200 placental and related samples from Japanese women from the Nagoya city area, a typical urban environment characterized by high density traffic and large population. These results show that total Hg content in placenta (185 ± 452 ng/g) was significantly higher than in maternal (19 ± 36 ng/g) and cord (30 ± 62 ng/g) blood. On the other hand, methyl-Hg content of placenta (14 ± 8 ng/g) was not significantly different from that of maternal (9 ± 5 ng/g) or cord (14 ±9 ng/g) blood. Further, the umbilical cord content of both total and methyl-Hg showed the same tendency as the placenta. This finding indicates that placenta does act as a barrier. This may be applicable to high level exposure situations coupled with chemical forms of Hg in the exposed environment.

**Other Elements.**   Among other metals, *Mn* concentrations in maternal and fetal blood were lower than that in placenta indicating a role for placenta in holding back Mn. More significantly however, differences in *As* content of human placenta under exposed (5.88 ng/g) and normal (2.59 ng/g) environmental conditions have been demonstrated (*5*). It has been shown that placenta partially blocks the passage of As. Maternal and fetal blood As levels are not available from this study hence a firm conclusion awaits further information. In the case of *Se* a very recent investigation (*37*) reported that maternal Se concentrations were significantly greater than fetal concentrations, and that placental Se levels were four times that of fetal levels suggesting that placenta may act as a barrier or regulator of fetal Se transfer. Animal experiments conducted on calves and lambs (*38*) suggest a possible block of placental Se or an impaired Se metabolism at the placental level as a causative factor in some prenatal muscular defects. Investigations of Shariff et al (*39*) on sheep supports the concept of a bidirectional transfer of Se across the sheep placenta in utero. They have demonstrated that Selenium-75 f rom the fetus was transferred to the ewe indicating that placenta is permeable to Se on both sides. Data on placental transfer for elements such as Cr, Mn, Ni and V are scarce and warrant investigations.

### Placenta as a Biomarker (Biomonitoring)

The potential of placenta, or any other specimen for that matter, as a biomarker depends upon the possibility to associate the presence of a chemical (in the form of metals, nonmetals or any other pollutant sources) to a specific, and preferably a measurable biological parameter (i.e. a metabolite or some kind of a biological effect). Having identified such a link the next task is to use it as a monitoring tool. Such procedures are carried out as either real-time monitoring (RTM) or as long-term monitoring (LTM) operations.

The RTM is a means of frequently checking short-term changes of pollutant profiles using routine clinical specimens such as blood and urine (and occasionally also breast milk and saliva). The samples collected for RTM are analyzed as soon as access to laboratory is available. Usually, these samples are not stored for a long time. On the other hand, samples for LTM should be reliable indictors of long-term body burden of the chemicals identified in them. Examples are hair (with some limitations), adipose tissue (for organic pollutants) and liver for both organic and inorganic pollutants). Also, samples obtained for LTM are an integral part of a long-term preservation scheme. Such a storage facility, when technically improved, transforms itself into what is now known as a specimen bank, which facilitates the preservation of specimens for decades if necessary, and enables deferred chemical characterization.

The criteria governing the acceptability of samples for the clinical diagnosis of deficiency or toxicity, for the biological monitoring of environmental pollutants (i.e. exposure), nutritional surveillance and monitoring, and for forensic investigations differ markedly (22). An indication of the suitability of various human tissues including placenta, and body fluids for use in the biomonitoring of a number of trace elements and in assessing exposure to toxic elements is presented in Tables V and VI. The feasibility of the selective monitoring of essential and other trace elements through the analysis of whole blood and its components or of hair, urine, faeces or milk is discussed elsewhere (13). By combining this information with what is known about placental physiology (e.g. as a barrier for selected trace element transport) as discussed earlier, several useful approaches can be developed for practical use.

As in many cases, making decisions based on trace analytical results whether it is to assess health effects of the presence of a pollutant or for judging the suitability of a biological specimen as a biomarker that involves some kind of analytical measurements to be performed on it, depends on how well the biological meaning of the specimen is understood and how consistent the measurements are. The placenta is no exception to this situation and if any, it presents a situation that is complicated by its physiology, susceptibility to great variations during sampling and finally the analysis itself, all of which can be truly challenging in some cases. Trace metal determinations are particularly prone to all of the above parameters.

It is an essential requirement that the maternal and fetal compartments collected for investigation be accurately defined to facilitate meaningful interpretation of analytical data on such specimens. Further, considering that during the term of pregnancy women undergo extensive changes in body composition resulting in the redistribution of chemical constituents, any additional burdens from environmental and other sources are superimposed on

Table V.  Suitability of Human Clinical Specimens for Biological Monitoring Programs

| Element | Specimen | Exposure | Risk |
|---------|----------|----------|------|
| Al | U | Moderate | Unknown |
| Al | B | Strong | Strong |
| As(inorg.) | U, P(?) | Moderate | Weak |
| As(org.) | U | Moderate | Unknown |
| Cd | B, U, P | Strong | Strong |
| Co | U | Weak | Unknown |
| Cr | U | Moderate | Weak |
| Hg(inorg.) | B, U, P(?) | Moderate | Moderate |
| Hg(org.) | B, H, P(?) | Strong | Strong |
| Mn(inorg.) | B, U | Unknown | Unknown |
| Mn(org.) | U | Moderate | Unknown |
| Ni | B, U | Weak | Unknown |
| Pb(inorg.) | B, P(?) | Strong | Strong |
| Pb(org.) | U, P(?) | Weak | Weak |
| Sb | B | Weak | Unknown |
| Se | B, U | Moderate | Moderate |
| Sn | B, U | Unknown | Unknown |
| V | B, U | Weak | Unknown |

SOURCE: Adapted from **Ref. 22** (chapter 4).
B: Whole blood, serum or plasma; H: Hair; P: Placenta;
U: Urine.

Table VI. Human Clinical Specimens Suitable for Assessing Exposure to Some Non-Essentail Elements

| Tissue | As | Cd | Hg(inorg) | Hg(org) | Pb |
|--------|-----|-----|-----------|---------|-----|
| Blood | X | X | X | X | X |
| Bone | | | | | X |
| Brain | | X | | X | |
| Faeces | | X | | | X |
| Hair | X | | | X | X[a] |
| Kidney | | X | X | | X |
| Liver | | X | | X | X |
| Placenta | ?[b] | X | ? | ? | X |
| Tooth | | | | | X |
| Urine | X | X | X | X | X |

SOURCE: Adapted from **Ref. 22** (chapter 4).
[a]Sampling and preparatory steps critical.
[b]See text.

the changes already taking place. This combination of events becomes an important parameter in data interpretation and can present a difficult situation. To understand the role of placenta as a biomarker, it is useful to compare it with amniotic fluid which is a recognized specimen in determining congenital abnormalities. This process is repeated if needed at different interval or if reconfirmation of the diagnosis is needed. Thus, amniotic fluid is used as an established diagnostic specimen. In the case of placenta, samples are usually obtained at time of parturition, a one-time event. Hence, each pregnancy has to be looked upon as a RTM process since the affected species is exposed to the placental source of pollutants only during the course of that particular pregnancy. As far as the fetus is concerned, exposure to placental sources of pollutants is an *in utero* event. Hence, the usefulness of placenta as a biomarker comes into picture in an unusual way: an assessment of the presence of a pollutant in placenta as a means to reduce the pollutant burden of the maternal environment; this, while beneficial for future pregnancies, may not necessarily help the infant, since it is the mammary barrier that is the deciding factor at this stage; in cases where placenta acts even as a partial barrier (e.g. Cd), it suggests a decrease of the harmful effects, if any. However, in the case of organo-metalic compounds such as methyl-Hg, although the levels measured tend to be similar in maternal and fetal specimens due to a lack of barrier-effect, the information is still very useful as a forewarning since neurological effects of methyl-Hg are likely to show up later in the adult life (1). Preventive measures to reduce the post-natal exposure to methyl-Hg is warranted under these conditions. A similar case can be made for Pb also.

**Arsenic.** The best evidence for placenta as a biomarker for As has come from the studies of Diaz-Barriga et al (5). Placental samples collected from an industrial belt known for smelting and related metallurgical operations, 2.3 times greater As levels were measured in samples from industrial areas than those collected from rural districts, some 300 km away. Measurement of As was also undertaken in water, soil, air and household dust samples, and were shown to contain higher levels in the industrial environment than those from rural source confirming a generally high exposure, confirming the link between exposure and elevated As levels. The As levels were also high in scalp hair, pubic hair and urine in the exposed subjects (Table V). Considering that hair is susceptible to external contamination (although a good indicator for As exposure)  and placental As correlates with that of urine (10.3 times greater in industrial areas compared with controls), the combination of urine (as a RTM) and placenta, as an intermediate-to-long-term indicator of As exposure, provides a practical tool to monitor pregnant women.

**Cadmium.** Based on the well designed study of Hubermont et al (26), at term the Cd concentration of placenta reaches a 10-fold high average value in comparison with that found in maternal blood, providing the most convincing evidence of the barrier role played by placenta in preventing Cd from reaching the fetal tissues. Another compelling evidence is presented in a very recent study by Baranowska (*28*). In this investigation samples of maternal and neonatal blood as well as placenta collected from the intensely industrial Upper Silesian region of Poland (98% of all the coal mined, 100 % of the Zn and Pb ore processed, 53 % of steel production, 35% coke consumed and 29 % of energy generated) were analyzed and the following results obtained: Cd in venous blood, cord blood and placenta = 4.9, 1.1 and 110 ng/g, respectively. Thus, high concentrations of Cd were found in placentas and in maternal blood, but not in the neonatal samples, clearly indicating the role of placenta as a barrier in sequestering excess Cd from the blood and minimizing transfer to the fetus.

The review by Miller et al (*12*) indicates that there is sufficient evidence to support variation of Cd levels in placenta to reflect the geographic differences (8 ng/g in the Netherlands to 176 ng/g in the Ruhr Valley in Germany) and related environmental sources. Further, a distinct correlation has been established between smokers and nonsmokers (*17*) designating smoking as the most important environmental exposure in relation to Cd. For other metals such as Cu, Pb, Hg and Zn no such correlation has been established. The best evidence for placenta as a biomarker for Cd has come from the studies of Diaz-Barriga et al (*5*). Placental samples collected from an industrial belt known for smelting and related metallurgical operations, 8.1 times greater Cd levels were measured in samples from industrial areas than those collected from rural districts, some 300 km away. Measurement of Cd was also undertaken in water, soil, air and house hold dust samples, and were shown to contain higher levels in the industrial environment than those from rural source confirming the link between excessive exposure and elevated Cd levels. Among other samples, hair turned out to be unsuitable while urine showed 8.8 times greater Cd levels when compared with samples from control areas. Thus placenta has been shown to block Cd efficiently. Similar to the case of As, the combination of urine (as a RTM) and placenta as an intermediate-to-long-term indicator of Cd exposure, provides a practical tool to monitor pregnant women. Thieme et al (*40*) have used placenta as a monitor of Cd in the urban areas of Munich. Ward et al (*21*) have reported highly significant negative relationships between placental Cd and Pb levels and birth weight. However, Cd accumulation in placenta is not unequivocal. Truska et al (*32*) have reported that in their study of placental samples from industrial exposure areas, no Cd uptake was found.

**Mercury.**   Monitoring for Hg through placenta appears to be still evolving. The animal experiment demonstrating that the ratio of methyl-Hg concentrations in the placenta to that of the blood of the newborn being close to 1 indicates that the placental value can serve as an indicator (*35*). Maternal and fetal blood are equally effective. Monitoring the maternal and fetal blood for Hg, especially in women consuming large quantities of fish, has been shown to be effective. Although placenta was not studied in this case, when maternal blood Hg levels are elevated, higher cord blood levels are found (*33*). Further, fetal levels have been shown to rise more sharply than corresponding maternal levels. Similar results have also been reported by Baglan et al (*30*). On the other hand, based on the study of Hubermont et al (*26*), placental barrier does not exist for Hg. In a Japanese study using amniotic fluid as a monitoring specimen it was demonstrated that Hg was present at detectable levels in almost all of the samples, and that the level of inorganic Hg was consistently higher than that of organic Hg by a factor of 3 (36). These investigators also reported that Hg level changed according to gestational age for both forms of Hg, with the highest value being observed in the 7th month of gestation. The reasons for the increase of Hg during the 7th month of pregnancy are still not clear. Of interest is the observation that most of the fetal Hg was in the inorganic form, contrary to the supposition that transplacental route favours organic forms of Hg. Also, there is no evidence of biotransformation of organic Hg in the fetus to inorganic Hg (36).

**Lead.**    In the case of Pb also, evidence for placenta as a possible biomarker for Pb has come from the studies of Diaz-Barriga et al (5). Placental samples collected from an industrial belt known for smelting and related metallurgical operations, 2.2 times greater Pb levels were measured in samples from industrial areas than those collected from rural districts, some 300 km away. Measurement of Pb was also undertaken in water soil, air and house hold dust samples, and were shown to contain higher levels in the industrial environment than those from rural source. With the exception of water which showed no difference in the two sources, the remaining samples confirmed a generally high exposure to Pb, providing a link between exposure and elevated Pb levels. The Pb levels were also high in samples of pubic hair (but not so in scalp hair) but not significantly different in urine specimens of neither of the populations. In another investigation (28) samples of maternal and neonatal blood as well as placenta collected from the intensely industrial Upper Silesian region of Poland (98% of all the coal mined, 100 % of the Zn and Pb ore processed, 53 % of steel production, 35% coke consumed and 29 % of energy generated) were analyzed and the following results obtained: Pb in maternal blood, cord blood and placenta were 72.5, 38.3 and 500 ng/g, respectively. Thus, high concentration of Pb was found in placentas, but the levels observed in maternal blood sample are only twice that of cord blood samples indicating efficient transfer of Pb to the fetus. Although high levels of Pb in the environment are also reflected in placenta, in terms of protecting the fetus the barrier factor appears to be insufficient. Based on the study of Hubermont et al (26) investigating the influence of low and high (11.8 and 247 $\mu$g/L) Pb in drinking water, a significant difference in the Pb levels in maternal and cord blood and in placenta was found in the two groups of subjects. Although the differences in placental and blood Pb levels were small, the maternal and fetal blood Pb levels were correlated with water Pb. The similarity of the influence of water Pb on both maternal and fetal blood levels indicates a rapid transfer of Pb from mother to fetus, and masks the influence of placenta as a barrier, if any. Therefore, for monitoring purposes during pregnancy, water analysis in combination with maternal blood levels provides a good tool and emphasizes the need to prevent undue Pb exposure during pregnancy (26). Considering that even pubic hair is susceptible to some form of contamination (although a good indicator for Pb exposure), placental

Pb appears to have some value as an intermediate-to-long-term indicator of Pb exposure and appears to be useful tool to monitor pregnant women. Thieme et al (*40*) have used placenta to monitor for Pb in the urban areas of Munich. Finally, Baghurst et al (*9*) have observed that there is considerable inter-individual variation in the efficiency of transfer of Pb from maternal blood to placental tissue.

**Other Elements.** Systematic information through human studies for many other elements in context of placenta are lacking and offer many research opportunities. Se is a good example. Many human studies, on an epidemiological scale have been carried out in non-placental specimens such as blood, erythrocytes, urine and hair (*13*). Also, metals such as Cr, Mn, Mo, Ni and V are scarcely (or not at all) researched.

**Concluding Remarks**

Reliable results for trace elements in placenta are still very scarce and well designed epidemiological scale studies are needed to establish baseline values and the changes arising from exposure to specific industrial and other environmental conditions.

   Representative sampling of placenta for routine use is still a debated issue. If whole sample is not collected and homogenized, then sampling should be from multiple sites. Improved sample handling techniques (e.g. grinding under cryogenic conditions) need to be developed to minimize the occurrence of extraneous contamination.

   As a biomedical sample, its easy accessibility makes placenta a very attractive specimen to study environmental impact of several elements considered to be harmful to health. However, its effectiveness as a biomonitoring specimen is enhanced by supplementing the information by analyzing additional specimens such as blood and urine (e.g. for Pb and As).

   Based on the available information, monitoring for *Hg* through placenta appears to be still evolving, hence no definitive statement can be made. Hair is a better specimen especially for monitoring methyl Hg. The determination of placental *Pb* appears to have some value as an intermediate-long-term indicator of Pb exposure and appears to be a useful tool to monitor pregnant women. However, maternal blood appears to be a more reliable indicator that can be used to predict the status of Pb in the fetus. In combination with urine (as a RTM), placenta is a practical tool to monitor pregnant women for *As* exposure. Hair continues to be an effective and practical indicator specimen for As. Consistently high concentrations of *Cd* found in placentas and in maternal blood, but not in the neonatal samples, clearly confirms the role of placenta as a barrier in sequestering excess Cd from the blood and minimizing transfer of this harmful element to the fetus. A distinct correlation has been established between smokers and nonsmokers, designating smoking as the most important environmental exposure in relation to Cd.

## Literature Cited

(1)     *Similarities and Differences between Children and Adults. Implications for Risk Assessment*; Guzelian P. S.; Henry, C. J.; Olin, S. S., Eds.; ILSI Press: Washington DC, 1992.

(2)     Goyer, R. A. In *Toxicology of Metals*, Goyer R. A.; Cherian, M. G., Springer: New York, NY, 1995, pp 1-17.

(3)     Human Tissue Monitoring and Banking. *Environ. Health Persp.* **1995**, *103 (suppl. 3)*, 1-112.

(4)     Snyder, W. S.; Cook, M. J.; Nasset, E. S.; Karhausen, L. R.; Howells, G. P.; Tipton, I. H. *Report of the Task Group on Reference Man, ICRP-23*, Pergamon, NY, 1975.

(5)     Diaz-Barriga, F.; Carrizeles, L.; Calderon, J.; Batres, L.; Yanez, L.; Tabor, M. W.; Castelo, J. In *Biomonitors and Biomarkers as Indicators of Environmental Change;* Butterworth, F. M., Ed.; Plenum Press, NY, 1995, pp 139-148.

(6)     Iyengar, G. V.; Kollmer, W. E.; Bowen, H. J., *The Elemental Composition of Human Tissues and Body Fluids - A Compilation of Values for Adults;* Verlag Chemie: Weinheim, NY, 1978.

(7)     Iyengar, G. V. *Anal. Chem.* **1982**, *54*, 554A-558A.

(8)     Manci, E. .A.; Blackburn W. R. *Placenta.* **1987**, *8*, 497-502.

(9)     Baghurst, P. A.; Robertson, E. F.; Oldfield, R. K.; King, B. M.; McMichael A. J.; Vimpani, G. V.; Wigg N. R. *Environ. Health Persp.* **1991**, *90*, 315-320.

(10)    Iyengar, G. V.; Iyengar, V. In *Quantitative Trace Analysis of Biological Materials;* McKenzie, H. A.; Smythe, L. E. Eds.; Elsevier: Amsterdam, 1988, pp 401-417.

(11)    Iyengar, G.V.; Iyengar, V. In *Handbook of Metal-Ligand Interactions in Biological Fluids, Bioinorganic Chemistry*; Berthon, G., Ed,; Marcel Dekker: NY, 1995, Vol. 2; pp 1127-1137.

(12)    Miller, R. K.; Mattison, D. R.; Plowchalk, D. In *Biological Monitoring of Toxic Metals*; Clarkson, T. W.; Friberg, L.; Nordberg, G. F.; Sager, P. R., Eds.; Plenum, NY, 1988, pp 567-602.

(13)    Food and Agricultural Organization of the United Nations.; International Atomic Energy Agency. *Trace Elements in Human Nutrition and Health;* WHO, Geneva, 1996.

(14)    Dawson, E. B.; Croft H. A.; Clark, R. R.; McGanity, W. J. *Am. J. Obstet.. Gynecol.* **1968**, *102*, 354-61.

(15)    Iyengar, G. V.; Kasperek, K. *J. Radioanal. Chem.* **1977**, *39*, 301-315.

(16)    Loiacono, N. J.; Graziano, J. H.; Kline, J. K.; Popovac, D.; Ahmedi, X.; Gashi, E.; Mehmeti, A.; Rajovic, B. *Arch. Environ. Health.* **1992**, *47,*, 250-255.

(17)    Lagerkvist, B. J.; Nordberg, G. F.; Soderberg, H. A.; Ekesrydh, S.; Englyst, V.; Gustavsson, M.; Gustavsson, N. O.; Wiklund, D. E. In *Cadmium in the Human Environment: Toxicity and Carcinogenicity;* Nordberg, G. F., Ed.; IARC, Lyon, 1992, pp 287-291.

(18) Berlin, M.; Blanks, R.; Catton, M.; Kazantzis, G.; Mottet, N. K.; Samiullah, Y. *IARC Sci. Publ.* **1992,** *118,* 257-262.

(19) Karp, B.; Robertson, A. F. In *Cadmium in the Environment, Part II Health Effects*; Nriagu, J. O., Ed.; Wiley, NY, 1981, pp 729-742.

(20) Ward, N. I.; MacMohan, T. D.; Mason, J. A. *J. Radioanal. Nucl. Chem.* **1987a,** *113,* 501-514.

(21) Ward, N. I.; Watson, R.; Bryce-Smith, D. *Internat. J. Biosocial Res.* **1987b,** 63-81.

(22) Iyengar, G. V. *Elemental Analysis of Biological Systems;* CRC Press, Boca Raton, FL. 1989; Vol. 1.

(23) Baumann, A. *Arch. Gynakol.* **1933,** *153,* 584-592.

(24) Clark, A. R. L. *Postgrad. Med. J.* **1977,** *53,* 674-678.

(25) Miller, C. D.; Buck, W. B.; Hembrough, F. B.; Cunnigham, W. L. *Vet. Human Toxicol.* **1982,** *24,* 163-166.

(26) Hubermont, G.; Buchet, J. P.; Roels, H.; Lauwerys, R. *Internat. Arch. Occupat. Environ. Health.* **1978,** *41,* 117-124.

(27) Tsuchiya, H.; Mitani K.; Kodama K.; Nakata T. Arch. Environ. Health. **1984,** *39,* 11-17.

(28) Baranowska, I. *Occupat. Environ. Med.* **1995,** *52,* 229-232.

(29) Khera, A. K.; Wibberley, D. G.; Dathan, J. G. *Br. J. Ind. Med.* **1980,** *37,* 394-396.

(30) Baglan, R. J.; Brill, A. B.; Schulert, D.; Wilson, D.; Larsen, K.; Dyer, N.; Mansour, M.; Schaffner, W.; Hoffman, L.; Davies, J. *Environ. Res.* **1974,** *8,* 64-70.

(31) Takaes, Von S.; Barkani, L.; Tatar, A.; Hardonyi, A. *Zentralbl. Gynakol.* **1984,** *106,* 1204-1214.

(32) Truska, P.; Rosival, L.; Balazova, G.; Hinst, J.; Rippel, A.; Palusova, O.; Grunt, J. *J. Hyg. Epidemiol. Microbiol. Immunol.* **1989,** *33,* 141-147.

(33) Dennis, C. A. R.; Fehr, F. *Sci. Total Environ.* **1975,** *3,* 275-277.

(34) Nishima, T.; Ikeda, S.; Tada, T.; Yagyu, H.; Mizoguchi, I. *Ann. Rep. Tokyo Metr. Res. Lab.* **1977,** *28,* 215-220.

(35) Mansour, M. M.; Dyer, N. C.; Hoffman, L. H.; Schulert, A. R.; Brill, A. B. *Environ. Res.* **1973,** *6,* 479-484.

(36) Suzuki, T.; Takemoto, T.; Shishido, S.; Kani, K. *Scand. J. Work Environ. Health.* **1977,** *3,* 32-35.

(37) Lee, A. M.; Huel, G.; Godin, J.; Hellier, G.; Sahuquillo, J.; Moreau, T.; Blot, P. *Sci. Total Environ.* **1995,** *159,* 119-127.

(38) Bostedt, H.; Schramel, P. *Biol. Trace Elem. Res.* **1990,** *24,* 163-171.

(39) Shariff, M. A.; Krishnamurti, C. R.; Schaefer, A. L.; Heindze, A.M. *Can. J. Animal Sci.* **1984,** *64,* 252-54.

(40) Thieme, R.; Schramel, P.; Keiler, G. *Gerburtshilfe Frauenheilrd.* **1986,** vol. *46,* 180-184.

(41) Kaufmann, P.; Scheffen, I. In *Fetal and Neonatal Physiology;* Polin, R; Fox, W., Ed.; Sanders, Philadelphia, 1992, p 48.

(42) Snyder, W. S.; Cook, M. J.; Nasset, E. S.; Karhausen, L. R.; Howells, G. P.; Tipton, I. H. *Report of the Task Group on Reference Man,* International Commission on Radiological Protection No. 23, Pergamon, New York, NY, 1975.

(43)    Alexiou, D.; Grimanis, A. P.; Papaevangelou, G.; Koumantakis, E.;
        Papadatos, C. *Pediat. Res.* **1977,** *11,* 646-648.
(44)    Schramel, P.; Lill, P.; Hasse, S.; Klose, B.-J. *Biol. Trace Elem. Res.*
        **1988,** *16,* 67-75.
(45)    Riemschneider, R.; Martins, A. F. *Gerburtsch. U. Franhelik.* **1978,**
        *38,* 371-375.
(46)    Riemschneider, R.; Martins, A. F. *Gerburtsch. U. Fraunhelik.* **1978,**
        *38,* 971-977.
(47)    Radomanski, I.; Sikorski, R. *J. Perinat. Med.* **1992,** *20,*
(48)    Romero, R.; Grandillo, J. A.; Navarro, B.; Rodriguez-Iturbe.;
        Pappaterra, J.; Pirela, G. *J. Trace Elem. Electrolyte. Health Dis.* **1990,**
        *4,* 241-243.
(49)    Korpela, H.; Loueniva, R.; Yrjanheikki, E.; Kauppila, A. *Internat. J.
        Vitam. Nutr. Res.* **1984,** *54,* 257-261.

Chapter 13

# Cadmium, Lead, Mercury, Nickel, and Cesium-137 Concentrations in Blood, Urine, or Placenta from Mothers and Newborns Living in Arctic Areas of Russia and Norway

J. O. Odland[1], N. Romanova[2], G. Sand[2], Y. Thomassen[2],
B. Salbu[3], Eiliv Lund[1], and E. Nieboer[1,4]

[1]Institute of Community Medicine, University of Tromso,
N–9037 Tromso, Norway
[2]National Institute of Occupational Health, Pb8149 Dep,
0033 Oslo, Norway
[3]Laboratory for Analytical Chemistry, Agricultural University
of Norway, Ås, Norway
[4]Department of Biochemistry and Occupational Health Program,
McMaster University, 1200 Main Street West, Hamilton,
Ontario L8N 3Z5, Canada

Cadmium, lead, mercury and nickel concentrations were measured in blood and urine samples collected pre- or postpartum of pregnant women living in arctic regions of Norway and Russia. Levels of these metals were also determined in cord blood and of nickel in the first urine void of newborn babies. For cadmium, lead and mercury, the concentrations found were within baseline reference intervals, even for Russian communities considered to be heavily polluted. Urinary nickel levels were significantly higher in the Russian communities (p<0.0001), even in the absence of a nickel refinery as a point source. Proximity of communities to the Chernobyl fallout zone appears to account for the concentration distribution pattern observed for Cesium-137 in 10 placentas collected in each of 11 northern communities.

Considerable attention is being paid to the industrial pollution in the regions adjoining the Norwegian/Russian border. There have been investigations of the pollution of air, water, soil and biota, but very little on human health (1-3). However, the Norwegian-Russian Health Group (NRHG) has recently initiated human health studies. In the popular media, stories have been featured about enhanced perinatal mortality and congenital malformations, as well as about reduced life expectancy and malnutrition. In Russian metal refineries, such as in the Kola Peninsula, female workers are likely to be exposed to high levels of

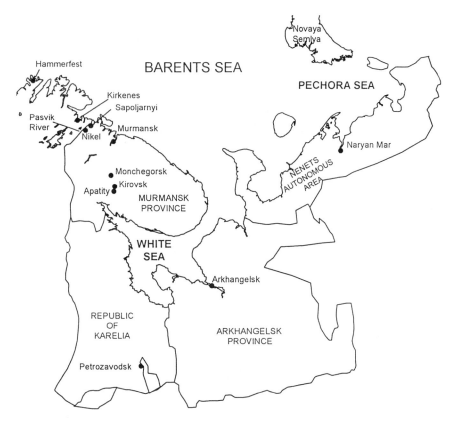

Figure 1. Schematic map of the Russian Barents Region. Courtesy of Harvey Goodwin, Akvaplan-niva, Tromsø, Norway.

toxic compounds during pregnancy, and this is suggested as a contributing factor to these adverse health effects (4,5). A major goal of the NRHG is to investigate possible correlations between metal exposure and toxic outcomes experienced by mothers, foetuses and children during pregnancy or the perinatal period. The first step is to assess exposure and this is the focus of the present paper. Preliminary data are presented for cadmium, lead, mercury, nickel and cesium-137 ($^{137}$Cs) in blood, urine or placenta from pregnant women collected before and/or after delivery. The results obtained are part of the Russian and Norwegian contributions to the Arctic Monitoring and Assessment Programme (AMAP). This organization involves scientists from the eight circumpolar countries.

The principal Russian geographic sites in our study were in the Kola Peninsula, namely the towns of Nikel, Sapoljarnyi, Monchegorsk, Kirovsk,

Apatity, Murmansk and Arkhangelsk. Their location is shown in Figure 1. For the [137]Cs work, additional sites were at Gomel in Belo-Russia and Kiev in the Ukraine; respectively, they are located in the vicinity of the Chernobyl fallout zone or away from it south of the reactor site. Nikel, Sapoljarnyi and Monchegorsk are cities with large nickel refineries. Kirovsk and Apatity feature apatite production. Murmansk is the biggest city in the north of Russia, with approximately 450,000 inhabitants. There is, however, no major industry. Arkhangelsk is almost of the same size as Murmansk, with 5 big pulp and paper plants in the surrounding area.

The nickel plants in Nikel and Sapoljarnyi are located, respectively, 10 and 40 km from the Norwegian-Russian border. Here, and around Monchegorsk, the surrounding forest has been severely damaged largely due to sulphur dioxide emissions. For Nikel and Sapoljarnyi, the dead forest areas are ca. 3 km from the Pasvik border-river. The local lateritic ore deposits in the Kola Peninsula are not substantial, and from 1978 on an increasing fraction of sulphide-based ore has been imported annually from Norilsk in Siberia. The production in Nikel is mainly nickel matte; in Sapoljarnyi there is some mining and milling, ore separation/purification and roasting; while in Monchegorsk nickel is refined electrolytically, by the Mond process and by pyrometallurgical methods (*4,5*). The technology is very old, and releases of metals and sulphur dioxide to the environment are substantial. There are now plans for renovation of the oldest plant in Nikel, but the timeframe of this project is uncertain.

Monitoring of the air pollution on the Norwegian side of the border started in Sør-Varanger in 1974 and on the Russian side in 1985. The Joint Norwegian-Russian Commission of Environmental Cooperation was established in 1988, with NRHG as a sub-group established in 1991. In 1989 the total $SO_2$-contamination from the Nikel and Sapoljarnyi nickel-plants was 272,000 t, 3 times the total $SO_2$-pollution in Norway the same year (*1-3*). Concomitantly, a total amount of 1,100 t of metals were emitted, including 510 t of nickel, 310 t of copper and 18 t of cobalt. Air measurements at Norwegian border stations have exceeded 3,000 $\mu g/m^3$ $SO_2$. On the Russian side, $SO_2$ concentrations exceeding 500 $\mu g/m^3$ are followed by a decrease in production. Nickel and copper concentrations in rivers and lakes of the area have been directly correlated to the distance from the pollution source. Analyses of parenchymatous organs from elk and caribou in areas on the Norwegian side of the border have showed increased values of nickel, lead, copper and cadmium, but these are not regarded to be of human health concern (*6*). In adjoining Finnish Lappland, enhanced levels of cadmium have also been reported in the blood of reindeer herders living close to the Russian border (*7*). Based on this information, it is important to look at human biological material from individuals residing in these areas and to compare the observed concentrations with background levels for reference populations not exposed to known point sources of metal pollutants.

Over the past 30 years, there has also been considerable concern about human health effects associated with radioactive fallout in arctic areas of Russia and Norway (*8*). The major reasons for public sensitivity about this issue are: the nuclear testing programme on Novaya Semlya in the 1950s and 1960s; the Chernobyl accident in 1986 with contamination of large areas of the Ukraine, Belo-Russia, Russia, Sweden and Norway; the potential releases from old nuclear power stations in the Kola Peninsula and from civilian and military installations in the Kola Peninsula and Arkhangelsk Region (8-10). In Narjan Mar, located in the Arkhangelsk Region close to Novaya Semlya, unverified medical data claim that there are very high incidences of spontaneous abortions, congenital malformations and cancer (11). Measurements of beta and gamma radiation have been reported in wet or dry precipitation, sediments, flora and fauna (8,9). In terms of human exposure, limited whole-body counting has been done, although direct sampling of tissue has not been reported. Because $^{137}$Cs is an important anthropogenic radionuclide and its dietary intake can be significant, its concentration in human placentas collected from a number of northern communities are determined in the present study.

**Subjects**

Personal contacts with colleagues in the different delivery departments were established, and all procedures and protocols were provided in Norwegian, English and Russian. The Norwegian reference towns were Kirkenes, Hammerfest and Bergen. Kirkenes is near the Russian-Norwegian border. Hammerfest is a coastal town of Finnmark and the northern-most town in the world. Bergen is the second biggest city in the south-west of Norway. For the placenta work, samples were also collected at Sømna in southern Nordland, Namdal in northern Trøndelag and Valdres in the central south of Norway; all these locations were significantly contaminated by fallout from the Chernobyl accident. In addition, placentas were obtained from aboriginal individuals (Sami people) living in central Finnmark.

The goal was to collect information, maternal and cord blood, maternal urine and the first-voided urine of the new-borns from 50 consecutive delivery patients in each location; the number of placentas was limited to 10. The registration of participants and sampling were performed in the following time periods: Kirkenes, Hammerfest, Bergen and Nikel during April-June 1991; Murmansk, Kirovsk, Apatity and Monchegorsk in the spring 1992, Arkhangelsk in the spring of 1993; and a final sampling period in Kirkenes, Hammerfest, Bergen and Nikel from November 1993 to June 1994. The whole placentas from Kiev, Gomel, Nikel, Valdres, Namdal, Sømna, Kirkenes (Sør-Varanger) and Hammerfest were collected in the autumn of 1993, while those from the Finnmark Sami group were collected in the autumn of 1994. The patients were asked to join the study by means of completing a consent form. No Norwegian patients refused to join the study. On the Russian side, the patients attitude fluctuated from the positive to negative and then once again to positive at the end of the sampling period, probably because of concurrent social and economic changes in Russia.

## Methods

Fifty consecutive patients at the delivery departments of the local hospitals were asked to complete a written questionnaire addressing the following particulars: age, parity, ethnic background, places of residence exceeding 6 months, schooling, occupation, smoking habits, alcohol, medication, serious diseases, dietary habits, especially related to food intake. The following information was collected about the births: Naegele term, date of birth, length of baby, weight of baby, weight of placenta, APGAR score, congenital malformations, gestational age, and individual comments by the doctor or midwife. The completion of the informed consent form and collection of anamnestic information were done before the delivery process started, to minimize stress. Identical information was obtained for the patients donating their placenta for $^{137}$Cs-counting. Collection of the specimens of blood, urine and placenta was performed without interruption of the delivery situation. The appropriate volume of cord blood was transferred directly into uncapped syringes tested to be free of metals (Sarstedt 02.264.020 10 ml Monovette, AH 23510) after cutting the umbilical cord with a plastic or titanium knife. Samples of antecubital vein blood were taken from the mothers on the first standard sampling day after the delivery (normally two or three days post partum) after installing a Viggo Spectramed Venflon (1.4 mm, 17 Gauge L 45 mm) in the vena cubitii. The blood was drawn directly into the same syringe type as used for the collection of cord blood; it was immediately stored at -20°C at the local hospital.

The urine of mothers was sampled at two stages: the first time at week 20 in the pregnancy (only in Kirkenes and Nikel) and the second time to coincide with the blood sampling. Maternal urine was sampled directly into a cup for transfer to containers (NØD-0438 CERBO Norge A/S, volume 20 mL); both the cup and the container tested free of nickel. The sampling of urine from the neonate was a challenge in most cases. Immediately post-partum, a uridome (Hollister Norge 126-0004) was plastered to the childs genitalia externa until the first void was produced. For ethical reasons, we had to stop the sampling if the uridome irritated the very sensitive skin of the newborns. The whole placentas were collected directly into a small plastic bag immediately after delivery and all concern about the newborn child was resolved. They were then frozen at -20°C. Within 3 months, the blood and placental materials were transported frozen to Norway for storage in a -70°C freezer. A number of samples had to be rejected because the prescribed protocols for collection, storage and shipping were not followed.

Cadmium and lead in whole blood, as well as urinary nickel, were determined by electrothermal atomic absorption spectrometry (*12*) employing a Perkin Elmer Model 5100PC Zeeman-based atomic absorption spectrometer with L'vov platform atomization. Urine was analyzed without pre-treatment, while whole-blood samples were first digested in hot nitric acid. The concentrations measured in whole blood and urine quality control materials (Nycomed, Norway) were in accordance with the values recommended by the manufacturer (± 5%). The detection limits (DLs) were: 1.0 nmol/L, cadmium; 0.01 μmol/L, lead; and 10 nmol/L, nickel. The mercury analyses were performed at Le Centre de toxicologie du Quebec, Canada. Mercury in whole blood was determined by cold vapor atomic absorption spectrometry, using a Pharmacia Model 100 Mercury Monitor. A detection limit of 5 nmol/L was achieved, using a sample volume of 500 µL. Run-to-run coefficient of variation of 6% was observed at the 60 nmol/L level. Whole placentas were counted for their [137]Cs activity without further treatment. The measurement was carried out employing a Canberra germanium detector (efficiency 20%, resolution 2 keV). The samples were placed in 1-L containers with a surface area of 20 cm$^2$. The counting times were 7 or 16 h, corresponding to DLs of 1.3 Bq/kg and 0.84 Bq/kg (wet weight). The quality control protocol regularly used in the Laboratory for Analytical Chemistry, Agricultural University of Norway, was strictly followed.

   In the statistical analyses, the Russian and Norwegian communities were grouped as two separate populations. Significance was tested using the z-statistic two-tailed test for groups of different size. Individual groups were tested against each other similarly. For statistical purposes, concentrations below DL were arbitrarily assigned the value of ½ DL.

   This study was approved by The Regional Ethical Committee, University of Tromsø, Norway, the Norwegian National Scientific Data Control (Datatilsynet) and the Regional Health Administrations of Murmansk and Arkhangelsk, and the local health authorities at the sites in Belo-Russia and Ukraina.

**Results**

The mean ±SD concentrations for cadmium, lead, mercury, nickel and [137]Cs in maternal and cord blood for the Russian and Norwegian communities are compiled in Table I. In all cases there were some outliers, and thus the sample distribution frequencies observed were not normal.

**Cadmium.** The comparison in Table I indicates that the cadmium concentrations in mothers blood are marginally lower in the Russian group ($p < 0.05$); both sets of mean data are well below the baseline reference value of <18 nmol/L. This comparison could not be made for the cord blood levels, since many of the values in both groups were below the DL of 1.0 nmol/L. Visual comparison of the cadmium concentrations in maternal blood is provided in Figure 2 for a number of the towns. As expected, the smokers (10% of the total) had the highest blood cadmium concentrations (median of 18.5 nmol/L). Two mothers each in the Norwegian and Russian group smoked more than 10 cigarettes/day; none smoked more than 20 cigarettes daily.

**Lead.** It is clear from the data in Table I, that the mean lead levels in mothers and neonates in Russia are significantly higher than in Norway ($p < 0.0001$); both sets of data conform to the expected baseline reference value. The relationship between maternal-blood and cord-blood lead is depicted in Figure 3.

**Mercury.** The mean maternal-blood mercury levels are marginally higher in the Norwegian population ($p < 0.01$; see Table I), and for both groups correspond to

Table I. A Comparison of Cadmium, Lead, Mercury, Nickel and [137]Cs in Whole Blood (WB), Cord Blood (CB), Urine of Mothers (UM), First-void Urine of Neonates (UC) or Placenta (P) in Selected Norwegian and Russian Communities with Accepted Baseline Reference Values (*13-17*)

| Element (reference interval; unit) | Fluid | Concentration ($X\pm D$) | | | | |
|---|---|---|---|---|---|---|
| | | Russian Communities | N | Norwegian Communities | N | P |
| Cadmium (<18 nmol/L; non-smokers) | WB CB | $3.4\pm1.5$ $\leq1.0^a$ | $206^b$ $73^c$ | $4.5\pm5.5$ $\leq1.0^a$ | $120^b$ $69^c$ | <0.05 |
| Lead (<0.20 μmol/L) | WB CB | $0.16\pm0.08$ $0.10\pm0.05$ | $206^b$ $73^c$ | $0.06\pm0.03$ $0.06\pm0.03$ | $120^b$ $69^c$ | <0.0001 <0.0001 |
| Mercury$^d$ (<12 nmol/L) | WB | $10\pm5$ | $60^e$ | $12\pm4$ | $25^e$ | <0.01 |
| Nickel (8.5-104 nmol/L) | UM UC | $153\pm165$ $43\pm53$ | $98^f$ $64^g$ | $17\pm12$ $10\pm3$ | $142^f$ $77^g$ | <0.0001 <0.0001 |
| [137]Cs (Bq/kg; DL 1.3 Bq/kg) | P | $1.9\pm1.4$ (range 1.3-7.9) | $41^h$ | $5.5\pm11.0$ (range 1.3-83.2) | $74^h$ | |

a  For statistical purposes, concentrations <DL = 1.0 were assigned the value of 0.5 nmol/L, corresponding to ½ DL; DL, detection limit.

b  Communities include: Bergen, Hammerfest and Kirkenes (Norway); Murmansk, Monchegorsk, Apatity, Arkhangelsk and Nikel (Russia).

c  Communities include: Kirkenes and Bergen (Norway); Nikel and Arkhangelsk (Russia).

d  Brune et al (16) report ranges of 0.2-12 nmol/L for non-fish eaters, 4-21 nmol/L for eating <2 fish meals/week, and 7-40 nmol/L for >2 fish meals weekly; one fish meal corresponds to approximately 200 g.

e  Communities include: Hammerfest (Norway); Nikel and Arkhangelsk (Russia).

f  Communities include: Kirkenes and Bergen (Norway); Nikel and Arkhangelsk (Russia), samples from middle of pregnancy and post partum, values are unadjusted.

g  Communities include: Kirkenes and Bergen (Norway); Nikel and Arkhangelsk (Russia); values are unadjusted.

h  Communities include: Sami-population of Finnmark, Kirkenes, Hammerfest, Sømna, Namdal, Bergen and Valdres (Norway); Arkhangelsk and Nikel (Russia), Gomel (Belo-Russia) and Kiev (Ukraine).

## Maternal-Blood Cadmium

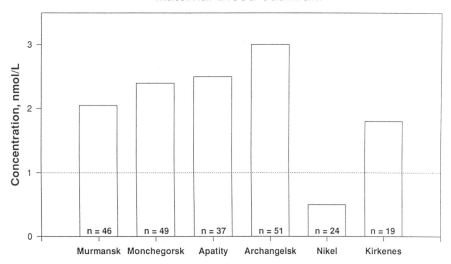

Figure 2. Median post-partum whole-blood concentrations of cadmium in females living in Norwegian and Russian arctic communities. Levels for the town of Nikel mothers were significantly lower relative to the other communities ($p<0.02$). The bar denotes SD and the dotted line the detection limit of 1.0 nmol/L.

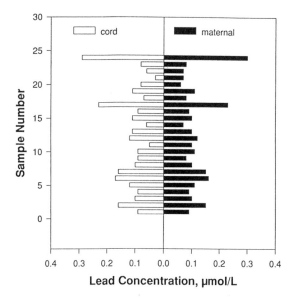

Figure 3. A comparison of individual lead levels in matched maternal whole-blood and cord-blood samples in the Russian community of Nikel ($r=0.96$; $p<0.02$; $N=24$).

reference values for moderate to no fish consumption. When broken down by community, the following mean values are found (in nmol/L): 9 (Arkhangelsk, N=37); 11 (Nikel, N=23); 12 (Hammerfest; N=25), with the difference between Hammerfest and Arkhangelsk reaching significance (p<0.0001).

**Nickel.** Both the neonatal and maternal mean urinary nickel levels are considerably higher among the Russian communities (p<0.0001; see Table I). Those reported for the Norwegian group fall within the baseline reference interval of 9-100 nmol/L (*23,24*), while the Russian values on average lie outside of it. There appears to be no apparent relationship between the maternal and corresponding neonatal urinary nickel concentrations (r=0.11, p>0.5, N=224). In Figure 4, the urinary nickel concentrations for mothers are broken down by communities. The values for the Russian towns of Nikel and Archangelsk are significantly higher than those for the Norwegian centres (p<0.0001) and different from each other (p<0.01). In terms of the neonatal first voids, there was no statistical differences between individual communities, whether in Russia or Norway.

**Cesium-137.** The Russian results show a more homogenous pattern than the Norwegian material, with ranges 1.3-7.9 Bq/kg and 1.3-83.2 Bq/kg, respectively. In terms of specific communities, we find no detectable values in Arkhangelsk, only 2 low values in Nikel and only 1 low value in Kiev. The remainder are below the DL of 1.3 Bq/kg. In Gomel we find detectable values in 6 out of 7 placentas, range 1.3-7.9 Bq/kg; in Bergen we find no detectable values; all values were detectable in Sømna, range 1.4-7.3 Bq/kg, and Valdres (range 2.8-15.9 Bq/kg). All samples from 10 Finnmark Sami women had detectable values, ranging 1.3-14.6 Bq/kg. For Kirkenes and Hammerfest, which are near the Russian border, a more mixed picture is seen, with detectable values in 2 out of 10 and 5 out of 8 placentas, respectively (ranges 1.3-3.9 and 1.3-6.1 Bq/kg). For Namdal in Trøndelag, the results were variable, with 6 out of 11 below the DL and 2 high values of 46.0 and 83.2 Bq/kg.

**Discussion**

**Cadmium.** The main environmental sources of cadmium are atmospheric deposition, agricultural application of fertilizers and sewage sludge, and solid wastes (*13*). A major source is tobacco smoke. Potential occupational exposure occurs in smelting and refining, alloy and battery production, pigment production and use (including plastics production), and welding (also cadmium/silver soldering). Exposure is mainly to cadmium metal, its oxide or water-soluble salts. Cadmium is an environmental contaminant of concern because of its long biological half-life (more than 10 y in human liver and kidney, compared to 80 d in the blood compartment) (*13*). It is relatively poorly absorbed; the daily Western dietary intake is less than 50 µg/day. Its main toxic effect is on the kidney, producing tubular dysfunction characterized by enhanced excretion of renal tubular tissue enzymes, low-molecular-mass proteins, and other essential metabolites such as calcium and phosphate. Mild kidney dysfunction has been

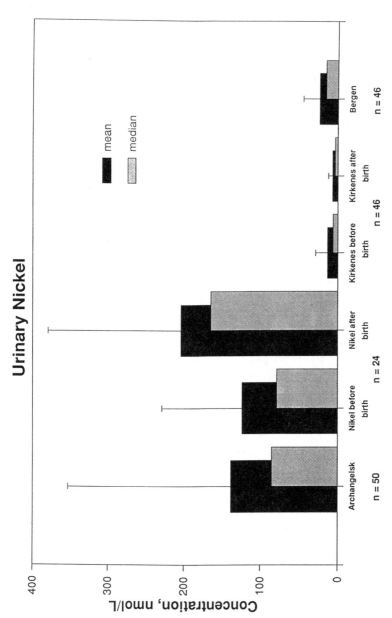

Figure 4. Mean and median urinary nickel concentrations in pregnant females and/or new mothers in Norwegian and Russian communities; the bar denotes SD. The results for Arkhangelsk and Bergen correspond to post-partum urine collections.

associated with dietary intake of environmentally contaminated food and heavy smoking (*13*).

The cadmium values in blood in both the Russian and Norwegian populations of study are very low, indicating insignificant environmental exposure. There is, however, a significant difference between the non-smoking mothers and the mothers smoking more than 10 cigarettes per day (p<0.0001). Interestingly, some of the mothers smoking the latter amount have children with blood concentrations below the DL, suggesting a possible threshold for the passage of cadmium through the human placenta. There are differences between different mammals, with free passage of cadmium through the Wistar rat placenta, while the hamster placenta is totally blocked for this metal (*18,19*). It seems that the human placenta serves as a selective barrier to cadmium with an average attenuation of 40-50% (*13,20,21*). Because many of the cord-blood cadmium concentrations were below the DL, it may be that individual variability related to physiological phenomena such as metallothionein induction in the placenta is operative.

**Lead.** In countries permitting the use of lead in gasoline, the automobile is the major environmental source of lead. The refining of lead and its industrial use (e.g., in battery manufacture) are typical point sources (*22*). Other non-occupational sources are drinking water (from lead or lead-soldered pipes) and paints in older homes. The latter is still the most important source of lead for children in the U.S.A. (*22*). Gun-powder fumes and contamination of game shot with leaded ammunition are others (*23*). Even though lead is a systemic poison (*24,25*), the main environmental concern is the impact on the unborn and children. Perinatal development (physical and neurological) is known, and young children have cognitive and behaviourial difficulties. The level of medical concern in children is 0.48 µmol/L, which is well below any action level for adults.

Both population groups have mean lead concentrations below the baseline reference value of <0.20 µmol/L. The levels are, however, significantly higher in the Russian populations (p<0.0001). Russian cars are still using leaded gasoline, but the traffic density is not as high as in western cities, even in a town like Monchegorsk with a population of about 60 000 inhabitants. If we compare the blood values of the Russian women of today with western measurements 20 years ago (*22*), we find much lower concentrations in the Russian women. The total Norwegian results correspond to some of the lowest concentrations observed in the literature (*24,25*). Improvement in preventing contamination during collection, sampling, handling and storage of the specimens might be a reason for this finding. The other, of course, is that the aftermath of the automobile lead emissions is diminishing. The data of Figure 3 point out the very easy transport and equilibrium of lead through the placental barrier, even at very low concentrations (*26*).

**Mercury.** Natural sources of mercury are volcanic activity and oceans (because of its volatility); and weathering of the earth crust, which is enhanced by acidification and flooding (*27,28*). The world-wide oceanic output of elemental mercury is estimated at about 30-40% of the annual Hg-emissions (it is the result of microbial activity); long-range transport has been established. In-lake synthesis of methyl-mercury by microbes responds to depositional mercury fluxes and can result in enhanced accumulation in fish. Typical anthropogenic sources are: mining, smelting, pulp and paper industry, industrial and laboratory manometers, alkyl-mercury fungicides; and medical uses (earlier used as diuretics, amalgams in dentistry and broken thermometers) (27,28). In terms of environmental health, the consequence of consuming fish contaminated by methyl mercury is the main concern (29,30). This compound is also secreted into breast milk. When methyl mercury is taken in, whole blood and hair mercury are good exposure indices. Some neurological/developmental risk to the foetus may occur for maternal blood levels >100 nmol/L, with the risk being more certain for levels $\geq$ 500 nmol/L (15, also see 29,30).

It is clear that the levels observed in Hammerfest, Nikel and Arkhangelsk do not pose any health risk. Consistent with the previously observed correlation between fish intake and blood mercury levels (16), the mean for the fishing town of Hammerfest was the highest. The very low levels for Arkhangelsk suggest that methyl mercury is not released from or used in the pulp and paper mills situated near this town. An early tradition in the pulp and paper industry in North America was its use as a fungicide (27).

**Nickel.** Unlike for cadmium and mercury, there is little global transport of nickel. Point sources such as nickel mining/refining plants constitute the primary environmental concern. In the occupational setting, lung and nasal cancers are the primary health effects associated with exposure to both particulate forms (mainly oxide and sulphides) and aerosols of dissolved nickel (as chloride or sulphate) (31-34). Nickel contact dermatitis is a public health concern and in the West occurs most commonly in women (7-10% in females and 1-2% in males). Contact with poor quality jewelry is the most established cause. Work-related nickel dermatitis also occurs, as well as nickel-related asthma with either an immediate or delayed response, or both. The reproductive/developmental effects of nickel have not been firmly established in humans, although some concern exists about apparent increases in spontaneous abortions and structural malformations (especially cardiovascular and musculo-skeletal) in newborn babies whose mothers were employed in a Russian nickel refinery (4,5). Follow-up studies are in progress. The evidence of teratogenic and developmental effects of nickel in animal studies is difficult to interpret with confidence (31,32).

The reason for the higher mean urinary nickel levels observed in mothers and their newborns in the Russian communities is not obvious. When the observed concentrations are corrected using urinary creatinine levels (not shown), the difference between the Russian and the Norwegian communities is reduced considerably, but nevertheless remains significant (p<0.001). As discussed by Nieboer et al (17), specific gravity corrections might have been more suitable, but unfortunately these were not measured. The creatinine normalized data does, however, suggest some additional environmental contribution for the residents of Nikel, compared to Arkhangelsk. This is not unexpected and has been observed for the nickel mining/refining town of Sudbury, Canada (17,35). For random, but corrected urine samples from Sudbury residents, urinary nickel levels averaged

54±48 nmol/L (range 7-170 nmol/L), compared to 34±34 nmol/L (range 12-90 nmol/L) for residents of Hartford, CT, USA, a city without an obvious nickel point source (*17,35*). Clearly the values observed for Kirkenes and Bergen are lower than Hartford, CT, presumably reflecting less industrialization and urbanization. Interestingly, the average nickel levels in tap water reported by Hopfer et al. (*35*) were 7 nmol/L in Hartford and 1860 nmol/L in Sudbury. The industrial emissions of nickel in Sudbury are clearly reflected in the nickel levels in the water supply source. Drinking water as a source of nickel may well have to be explored to resolve the relatively high levels observed in the Arkhangelsk community. Leaching of nickel from stainless steel items is also known (*32*). Although the average Western diet contains about 150 µg/day of nickel, there are certain food items that are relatively rich in this metal, such as chocolate (*31*). Special dietary sources in the Russian communities might therefore also be investigated. Interestingly, Sunde et al. (*36*) have reported urinary nickel measurements for inhabitants of Sør-Varanger community near the Norwegian-Russian border in April 1990. All participants were over 40 years of age, with equal distribution of sex (N=22 from the district of Pasvik, N=24 from the district of Bugøynes). The mean value in Pasvik, nearest to Nikel, was 22 nmol/L, while that in the reference group in Bugøynes was 17 nmol/L. Since these concentrations are from a group of adult citizens living in the same community as our women and children were drawn from in the Kirkenes group, it is not surprising that the latter exhibited comparable concentrations.

Since the measurement of toxic metals in first-void urine specimens of newborns have apparently not been attempted before, no comparison with literature values is possible. The finding of significant levels of nickel in these samples attest to the free-flow of nickel across the placental barrier.

Compared to urine nickel levels of nickel workers, which typically range 100-1700 nmol/L, the exposure to nickel in the Russian communities studied might be termed to be mild to moderate. Although the health consequences of this are not believed to be of concern, the observations reported should be subjected to closer scrutiny. The use of 24-hour urine samples are advised in future studies in order to reduce the natural variation due to dilution effects (*17*). In addition, non-pregnant females should be studied, since pregnancy and child birth involve complex physiological process and trauma and these may influence urinary nickel levels (*37*).

**Cesium-137.** This radioisotope is one of the most significant fission products (6-7% of the fission yield). $^{137}$Cs has a radioactive half-life ($T_{1/2}$) of 30.2 y and its beta-particle decay (average energy of 0.19 MeV) is accompanied by a gamma ray of modest energy (0.66 MeV). Dietary transfer is generally from grain products, meat and milk. The lichen-caribou-human food chain constitutes a major pathway. The whole body retention $T_{1/2}$ varies with age and physiology: 19 (infants), 57 (children), 49 (pregnant women), 84 (women), and 105 days (men) (*10*). The concentration ratio maternal tissue:placenta:fetal tissue = 1:1:1. This is the basis for using the determination of $^{137}$Cs in placenta for dose estimation purposes (*38*).

The preliminary results in our study are somewhat surprising. $^{137}$Cs levels found in the Norwegian samples have a wider range than the Russian material, with 2 real high outliers in Namdal (46.3 Bq/kg and 83.2 Bq/kg, respectively). These two individuals revealed in their answers to the questionnaire that a substantial amount of their total diet was of local, natural origin, such as reindeer meat and berries. Namdal is situated in an area that was heavily affected by the Chernobyl fallout and presumably a connection between $^{137}$Cs and its accumulation in the food chain is indicated. The placentas from other Norwegian areas subject to Chernobyl fallout, namely Sømna and Valdres, also exhibit considerable activity (X=4.6 Bq/kg, SD 4.2, range 1.4-7.3, and X=5.9 Bq/kg, SD 5.2, range 2.8-15.9, respectively; detectable levels occurred in all samples). The values for Gomel are also of comparable magnitude (X=4.5 Bq/kg, SD 4.2, range 2.3-7.9, with one below DL), which is consistent with its location relative to Chernobyl. By contrast, no $^{137}$Cs could be detected in the Bergen samples. Bergen is located well outside the Chernobyl zone. Although the Sami population of Finnmark live in an area not influenced by the Chernobyl accident, the placental $^{137}$Cs levels (X=7.1 Bq/kg, SD 5.2, range 0.9-14.5) are comparable to those observed for individuals living in areas which were affected. Most likely this inconsistency reflects the combined effects of significant consumption of local traditional foods by the Sami group and the persistence of the influence of global fallout and perhaps the proximity to the Novaya Semlya testing area. The Russian populations studied, namely at Nikel and Arkhangelsk, and the Ukrainian city of Kiev just south of Chernobyl show placenta $^{137}$Cs-activities at the DL or below. The wind direction immediately after the accident protected Kiev from contamination, while Gomel was directly in its path. Arkhangelsk is south of Novaja Semlja, the centre of the Russian atomic bomb testing programme. Traditional food diets are not common in Nikel or Arkhangelsk.

**Concluding Remarks**

This article constitutes a preliminary overview of the status in arctic regions of Russia and Norway of the environmental exposure by humans to toxic elements. Concentrations in blood, urine or placenta from new mothers and in cord blood and urine from their newborn babies served as indices of exposure. The levels found were within the baseline reference values or intervals in case of cadmium, lead and mercury in all the populations examined, including a group from the Russian town of Nikel which is regarded as seriously polluted. The urine nickel levels were significantly higher in the Russian populations, but were independent of the nearness of a nickel refinery. Other sources need to be explored. The $^{137}$Cs levels in placenta seem to correlate with the residence in areas affected by the Chernobyl fallout. Consumption of traditional foods also appears to be a determining factor in the accumulation of this radionuclide.

The interpretations given will, of course, be re-evaluated when a multivariate analysis incorporating the questionnaire information is completed.

## Acknowledgments

This work has been supported by the University of Tromsø, Steering Group of Medical Research in Finnmark and Nordland, and the Royal Norwegian Department of Foreign Affairs, East-European Secretariate. The authors wish to thank the staff at the obstetric departments of Bergen, Kirkenes, Hammerfest, Namdal, Lillehammer, Brønnøysund, Nikel, Murmansk, Monchegorsk, Kirovsk, Apatiti, Arkhangelsk, Gomel and Kiev for the excellent cooperation in the administration of the questionnaire and collection of specimens. Acknowledgment is extended to Dr. Knut Dalaker, Dr. Natalya Glukhovetz and Dr. Jean-Philippe Weber for kind support in different phases of the project.

## Literature Cited

1. Sivertsen, B.; Schjoldager, J. *Luftforurensninger i Finnmark fylke;* NILU OR 75/91; Norwegian Institute of Air Research: Lillestrøm, Norway, 1991.
2. Sivertsen, B.; Makarova,T.; Hagen, L. O.; Baklanov, A. A. *Air Pollution in the Border Areas of Norway and Russia, Summary Report 1990-91;* NILU OR 8/92; Norwegian Institute of Air Research: Lillestrøm, Norway, 1992.
3. Hagen, L. O.; Sivertsen, B. Overvåking av luft- og nedbørkvalitet i grenseområdene i Norge og Russland, Oktober 1991-Mars 1992; NILU OR 82/92; Norwegian Institute of Air Research: Lillestrøm, Norway, 1992.
4. Norseth, T. *Sc. Total Environ.* **1994**, *148*, 103-108.
5. Chashschin, V. P.; Artunina, G. P.; Norseth, T. *Sc. Total Environ.* **1994**, *148*, 287-291.
6. Sivertsen, T.; Daae, H. L.; Godal, A.; Sand, G. *Opptak av tungmetaller i dyr i Sør-Varanger;* DN-notat 1991-15, Fagrapport 22, Naturens tålegrenser; Direktoratet for naturforvaltning: Trondheim, Norway, 1991.
7. Naya, S.; Korpela, H.; Pyy, L.; Hassi, J. *Lancet* **1991**, *338*,1593.
8. Paakkola, O. *Radioactivity in the Arctic Region - A Status Report. Protection of the Arctic Environment;* State of the Environment Reports, Finland; Special Report prepared for the Arctic Monitoring and Assessment Programme, 1990.
9. Moberg, L.; Reizenstein, P. *Nord Med* **1993**, *4*, 117-120.
10. WHO. *Environmental Health Criteria, Vol 25: Selected Radionuclides.* World Health Organization: Geneva, Switzerland, 1983.
11. Tkatchev, A. V.; Dobrodeeva, L. K.; Isaev, A. I.; Podjakova, T. S. *Senfølger av kjernefysiske prøvesprengninger på øygruppen Novaya Semlya i perioden 1955 til 1962;* Rapport etter programmet "Liv"; University of Tromsø, 1994. 12. Welz, B.; Schlemmer, G.; Mudakavi, J. R. *J. Anal. At. Spectrom.* **1992**, *7*, 1257-1271.
13. WHO. *Environmental Health Criteria, Vol 134: Cadmium;* World Health Organization; Geneva, Switzerland, 1992.
14. Minoia, C.; Sabbioni, E.; Apostoli, P.; Pietra, R.; Pozzoli, L.; Gàllorini, M.; Nicolaou, G.; Alessio, L.; Capodaglio, E. *Sc.Total Environ.* **1990**, *95*, 89-105.
15. Health Canada/Ontario Ministry of Environment. *Health and Environment: A Handbook for Health Professionals;* Health Canada; Ottawa, Canada, 1995.

16. Brune, D.; Nordberg, G. F.; Vesterberg, O.; Gerhardsson, L.; Wester, P. O. *Sc.Total Environ.* **1991**, *100*, 235-282.

17. Nieboer, E.; Sanford, W. E.; Stace, B. C. In *Nickel and Human Health: Current Perspectives;* Nieboer, E.; Nriagu, J. O., Eds; John Wiley: New York, 1992; pp.49-68.

18. Tsuchiya, H.; Shima, S.; Kurita, H.; Ito, T.; Kato, Y.; Kato, Y.; Tachikawa, S. *Bull. Environ. Contam.Toxicol.* **1987**, *38*, 580-587.

19. Hanlon, D. P.; Ferm, V. H. *Experientia,* **1989**, *45*, 108-110.

20. Lauwerys, R.; Buchet, J. P.; Roels, H.; Hubermont, G. *Environ Res.* **1978**, *15*, 278-289.

21. Roels, H.; Hubermont, G.; Buchet, J. P.; Lauwerys, R. *Environ Res.* **1978**, *16*, 236-247.

22. CDC. *Preventing Lead Poisoning in Young Children;* Centres for Disease Control, Public Health Service, U.S.Department of Health and Human Services: Atlanta, GA, 1991.

23. Frank, A. *Sc. Total Environ.* **1986**, *54*, 275-281.

24. Environ Health Persp, **1990**, *89*, 3-144; special issue.

25. Environ Health Persp, **1991**, *91*, 3-86; special issue.

26. Hubermont, G.; Buchet, J.P.; Roels, H.; Lauwerys, R. *Int Arch Occup Environ Health,* **1978**, *41*, 117-124.

27. WHO. *Environmental Health Criteria, Vol 101: Methylmercury;* World Health Organization; Geneva, Switzerland, 1990.

28. WHO. *Environmental Health Criteria, Vol. 118: Inorganic Mercury;* World Health Organization; Geneva, Switzerland,1991.

29. Hansen, J. C.; Tarp, U.; Bohm, J. *Arch. Environ. Health,* **1990**, *45*, 355-358.

30. Foldspang, A.; Hansen, J. C. *Am. J. Epidemiol.* **1990**, *132*, 310-317.

31. WHO. *Environmental Health Criteria, Vol. 108: Nickel;* World Health Organization, Geneva, Switzerland, 1991.

32. Nieboer, E.; Fletcher, G. G. In *Handbook of Metal-Ligand Interactions in Biological Fluids, Bioinorganic Medicine;* Berthon, G. Ed; Marcel Dekker, New York, 1995; pp. 412-417, 709-715, 1014-1019.

33. Doll, R. *Scand. J. Work Environ. Health,* **1990**, *16*, 1-82.

34. IARC. *Evaluation of Carcinogenic Risks to Humans, Vol. 49: Chromium, Nickel and Welding;* International Agency for Research on Cancer, WHO: Geneva, Switzerland, 1990.

35. Hopfer, S. M.; Fay, W. P.; Sunderman, Jr., F. W. *Ann. Clin. Lab. Sci.* **1989**, *19*, 161-167.

36. Sunde, H. G.; Alexander, J. *Tidsskr Nor Lægeforen* **1992**, *112*, 2384-6.

37. Nomoto, S.; Hirabayashi, T.; Fukuda, T. In *Chemical Toxicology and Clinical Chemistry of Metals;* Brown, S. S.; Savory, J., Eds; Academic Press: New York, USA, 1983; pp. 351-352.

38. Stather, J. W.; Harrison, J. D.; Kendall, G. M. *Radiat. Prot. Dosim.* **1992** , *41*, 111-118.

Chapter 14

# Total Mercury and Methylmercury Levels in Scalp Hair and Blood of Pregnant Women Residents of Fishing Villages in the Eighth Region of Chile

Carlos G. Bruhn[1], Aldo A. Rodriguez[1], Carlos A. Barrios[2], Victor H. Jaramillo[2], José Becerra[2], Nuri T. Gras[3], Ernesto Nuñez[4], and Olga C. Reyes[5]

[1]Departamento de Análisis Instrumental, Facultad de Farmacia, Universidad de Concepción, Casilla 237, Concepción–3, Chile
[2]Departamento de Farmacia, Facultad de Farmacia, Universidad de Concepción, Casilla 237, Concepción–3, Chile
[3]Comisión Chilena de Energia Nuclear, Centro de Estudios Nucleares La Reina, Laboratorio de Análisis por Activación, Santiago, Chile
[4]Departamento de Pediatria, Facultad de Medicina, Universidad de Concepción, Casilla 237, Concepción–3, Chile
[5]Secretaria Regional Ministerial de Salud Octava Región, Concepción–3, Chile

The extent of environmental exposure to mercury through the diet (*i.e.*, by consumption of fish and seafood), of a population group having higher than average fish and seafood consumption was assessed between 1991 and 1994 in pregnant women (PW) and nursing women (NW) residing in selected fishing villages (FVs) along the coastal zone of the Eighth Region of Chile. The control population was an equivalent group with negligible seafood consumption. The samples were scalp hair collected from the occipital region, and blood obtained by venipuncture. Total mercury (Hg-T) levels were determined by cold vapor atomic absorption spectrometry (CVAAS) in all samples, and in 20% of them by instrumental neutron activation analysis (INAA) as a reference method. Methylmercury (Me-Hg) was determined by gas chromatography with electron capture detector (GC-ECD) in hair samples of the 1994 study, after its selective isolation from human hair by volatilization in a microdiffusion cell. Significant efforts were made to validate the analytical methodology for Hg-T and Me-Hg determination in human hair, and for Hg-T determination in blood. Multiple comparison tests confirmed significant differences in Hg-T content between PW and NW in each FV and in the control group, with higher concentrations in at least five FVs which were selected for further in depth studies. A new survey was performed in the selected five FVs and in the control group in 1994, including sample collection of scalp hair (for Hg-T and Me-Hg) and blood (for Hg-T) of new PW. These results showed significantly higher mean concentrations for Hg-T in scalp hair ($x = 2.44 \pm 1.30$ mg/kg); range: 0.76 - 6.26 mg/kg; n = 40) and in blood ($x = 9.04 \pm 5.05$ μg/kg; range: 1.5 - 24.1 μg/kg; n = 39) with respect to the control group (scalp hair: $x = 0.40 \pm 0.16$ mg/kg; range: 0.20 - 0.69 mg/kg; n = 9; blood: $x = 2.7 \pm 1.3$ μg/kg; range: 0.2 - 4.8 μg/kg; n = 9). Also, high correlation was observed between Hg-T concentrations in scalp hair and in blood within the study group.

Environmental exposure to mercury has raised worldwide concern due to the element's high toxicity and ubiquitous nature, despite the fact that the environmental levels of this element are generally low. This is due to the ability of Hg to bioaccumulate in some organisms, and to the wide distribution of this element in both terrestrial and aquatic plants and animals (1). The presence of Hg as a contaminant at relatively high concentrations in commercial seafood, especially fish, constitutes the main human dietary source of this element (2). The Hg contamination risk in humans is based on its accumulation and biotransformation in fish which convert inorganic-Hg to organic forms of higher toxicity. For example, Me-Hg constitutes between 70 - 90% of the net mercury content (3) in the edible tissue of fish. Health effects of mercury are seen mainly in the nervous system, and among the first symptoms observed in adults are paresthesia, numbness and blurred vision. Toxic effects of Me-Hg include alterations in the nervous and locomotor system, mainly a reduction in mental abilities which may produce maniac and depressive signs followed by tremors, athaxia, seizure and choreoathetoid movements (4-7). For example, 7.4% of people who ate fish 3 times per week, resulting in a mean mercury concentration in hair of 11.9 mg/kg (0.6 - 21.8 mg/kg), presented neurological symptoms (8,9). However, the effects of exposure to mercury as Me-Hg in prenatal and early postnatal life may become quite significant, producing severe disorders in the development of the central nervous system (mental retardation and cerebral palsy) despite the fact that the mothers of these children showed no, or only mild effects (10) Clinical and epidemiological studies indicate that embryos and fetuses are more sensitive to the toxic effects of Me-Hg than adults (11), and that there are harmful effects in early childhood development and in the mental ability of children whose mother ingested 3 to 4 times the tolerable recommended weekly intake (PTWI) set by WHO and FAO during pregnancy (12). Recent studies in children aged between 12 and 30 months and exposed to mercury prenatally in populations with high fish consumption (13), showed a positive correlation with abnormalities in the muscle tone and in reflex acts. The maximum mercury concentration in maternal hair during pregnancy was 23.9 mg/kg. Also, retardation was reported in the development of children 4 years old, whose mothers had mean mercury concentrations in hair of between 6 - 86 mg/kg (14) during pregnancy.

The Eighth Region of Chile is one of the most industrialized areas of the country, including steel production, petroleum refining, petrochemical plants, chloro-alkali plants, cellulose plants and paper mills, fishmeal plants, coal mines, forest products industries, textile and fertilizer production. Since 1977 (15-17) mercury contamination has been assessed in the Bio-Bio river and in the nearby coastal waters of the Arauco Gulf and San Vicente Bay (Figure 1). Mercury in this area is of anthropogenic nature, and is present at relatively high levels in water, sediments and marine organisms. Relatively high levels of mercury were measured in coastal waters of

Concepcion Bay (mean concentration of 1.5 μg/L), in sediments of the Lenga Estuary (72.9 mg/kg, due to past discharges from a chloro-alkali plant) and in shellfish (*Aulacomya ater*, 0.24 mg/kg fresh-weight) from the Arauco Gulf. (*17*). Recent studies on sediments of Concepcion and San Vicente Bays, and Arauco Gulf showed mean mercury concentrations of 0.16 mg/kg, 0.38 mg/kg and 0.25 mg/kg respectively (*18*). The concentrations found in San Vicente Bay and Arauco Gulf were higher than those in Concepcion Bay, which was consistent with the mercury inflows received through the Lenga estuary and Bio-Bio river, respectively. Also, mercury was determined in sediments (0.36 mg/kg), in suspended particulate matter (23 μg/kg), and in several samples of crustaceans ("langostino colorado", *Pleuroncodes monodon*) and fish

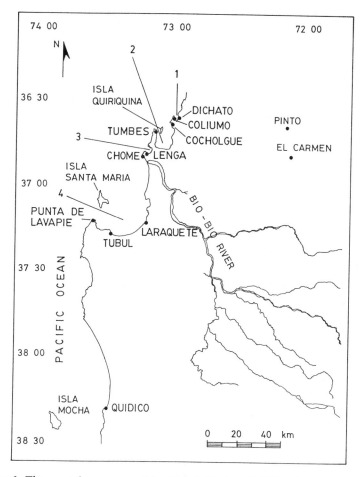

Figure 1. The coastal zone map of the Eighth Region of Chile. Fishing villages in the coastal group: Dichato, Coliumo, Cocholgue, Tumbes, Lenga, Chome, Isla Santa María, Laraquete, Tubul, Punta Lavapié and Quidico. Interior group: Pinto, El Carmen. (Adapted from ref. 27.).

("congrio negro", *Genypterus maculatus*) obtained in the Arauco Gulf between February 1992 and November 1993 (*19*). The mean total mercury concentration in "langostino colorado" was 0.31 mg/kg and in "congrio negro" was 0.34 mg/kg and the mean in the edible muscle portion was 0.33 mg/kg. These species belong to a food chain which was used as a dietary source by fishermen and their families. A dietary survey performed among 117 family groups corresponding to 779 residents in towns close to this coastal area, indicated an estimated mean annual fish consumption of 15.4 kg. The main species consumed were *Trachurus murphyi* (31%), *Merluccius gayi* (21%), and to lesser amount, *Genypterus maculatus* (3.3%). In samples of canned *Trachurus murphyi* the mean total mercury concentration was 0.26 ± 0.12 mg/kg (dry weight, n = 9)(Navarrete, G., Universidad de Concepción, unpublished data). To the best of our knowledge, no data exist on mercury and methylmercury levels in humans living in the coastal zone although they are exposed to this contamination. Our group started this IAEA (International Atomic Energy Agency) research contract in 1991 aimed at assessing mercury exposure in selected population groups for whom fish and seafood constitute the major source of this exposure.

In this study scalp hair was chosen as sample material because Hg-T and Me-Hg concentrations are higher in this tissue than in other organs and body fluids (approximately 250 times compared to levels in blood)(*1,3*). Once mercury enters the hair, its concentration remains invariable and its longitudinal distribution provides a measure of the Me-Hg levels in blood (*3*). Hair sampling and storage are simpler than the methods for other biological materials (*8,20*), and the rate of hair growth reflects the biomedical and environmental history of a human being under exposure for weeks and months. Scalp hair grows an average of 10 - 15 mm per month, samples can be taken painlessly and mercury concentration in its proximal end corresponds to the metal stored in hair and reflects the body burden of mercury during its anagen phase. In particular, the hair segment up to 5 mm from the scalp provides information about the last 2 - 3 weeks before sample collection (*21*). However, hair is susceptible of exogenous contamination from the environment, dust, dirt, perspiration, cosmetics (special shampoos, waving and dying products) and pharmaceutical formulations (*22*). Nevertheless, for epidemiological studies and under certain conditions regarding sample collection and cleaning, hair is a good indicator of environmental exposure to mercury through the diet (*23*).

Due to its relatively long mean biological half-life (39 - 70 days)(*3*), Hg accumulates in the body, and hair is one of the main storage compartments. Although the daily intake of mercury by the general population is estimated between 1 - 20 µg (*1*), among fishermen and their families this figure may increase significantly depending on the amount of fish consumed and on its mercury concentration. A long-term intake of Me-Hg (between 3 and 7 µg/kg of body weight) leads to an accumulation of 50 - 125 mg/kg in hair and 20 - 50 µg/mL in blood (*1,24*). Lesser amounts, corresponding to the provisional tolerable weekly intake (PTWI) established by WHO and FAO (0.2 mg/week of mercury as Me-Hg), lead to mercury levels in hair of 5 - 6 mg/kg (*24*). The relation between mercury exposure (as Me-Hg) and its accumulation in hair may be established based on a one compartment "metabolic model" (*1*). For a constant daily intake of 10 µg Me-Hg by an adult of 70 kg body weight, one half of the final concentration is reached rapidly on average in about 70 days. The maximum mercury level in hair of 2.5 mg/kg is reached in a period of 5 biological mean lifetimes (*ca.* 350 days)(*1*). Even in persons with low fish consumption (10 - 20 g per day), mercury present in fish may contribute significantly to intake as Me-Hg. Thus, the consumption of 200 g of fish containing 500 mg of mercury per kg (*i.e.*, in fishes of the highest trophic levels), may result in the intake of 100 μg of mercury (mainly Me-Hg), making one half of the PTWI established by WHO (*25*).

## Selection of the study group

A study was performed during 1991-1993 in 153 healthy pregnant women (PW, n = 129) and nursing women (NW, n = 24) not occupationally exposed to mercury, aged 13 - 40 and residing in 11 fishing villages along the coastal zone of the Eighth Region of Chile (Figure 1), to establish Hg-T levels in scalp hair of a selected human group consuming at least one fish meal per week (coastal group). The control group was selected alike and consisted of a group of 26 healthy PW (n = 22) and NW (n = 4) of similar age range, social status, not occupationally exposed to mercury, residing in rural towns within the interior of this region (Figure 1, Pinto and El Carmen), and with negligible fish or shellfish consumption (interior group). The following criteria were applied in the selection of the sample population: a) only women of fertile age, b) with recent maternal control (*i.e.*, up-to-date prenatal or postnatal (only in 1991) control) of the state health service c) residing in the village and d) not gainfully employed outside their homes. Both groups belonged to the same socioeconomic stratum, as all women were beneficiaries of the state child-maternal health program. The PW and NW were recruited at the health stations of these fishing villages, within the state child-maternal health program, and the response rate was 95% of the PW and NW selected at random when attending the health stations. At this stage, only Hg-T was analyzed in hair because our efforts concentrated mainly in establishing a potential high risk group within this study group in which Hg-T and Me-Hg would be determined. Both groups were selected during January 1991 and March 1993. However, NW were included only in 1991 and not thereafter because the focal point of this study were PW as was agreed within the CRP supporting it. The working hypothesis was "the high risk population selected had mercury levels in hair significantly higher than the normal population, due to their dietary habits and due to the proximity of their communities to an ecosystem receiving wastes containing mercury".

A questionnaire (*26*) (improved in 1992) was administered to all participating women. The questions included: age, period of residency in the fishing village (or rural town), work and activities, weight, height, nutritional history, consumption of fish (*i.e.*, hake, jurel, conger, corvina, cojinova, mackarel, sardine, skate, robalo, blanquillo, cabrilla, angler, flounder, tuna, dogfish, among others) and seafood, number of fish meals per day and per week (only surveyed during 1992 and 1993), the quantity of fish per meal and the origin and mode of preparation, the source of drinking water, habits (use of cosmetic products for hair waving or dyeing, special shampoos, alcohol consumption and smoking), number of conceptions, morbidity during the previous fortnight and the expected delivery dates. A scalp hair sample (2 g) was obtained from each woman during the survey.

## Sampling and pretreatment

The scalp hair samples were collected and analyzed according to the approved protocol within the IAEA Co-ordinated Research Programme (CRP) on Assessment of Environmental Exposure to Mercury in Selected Human Populations as Studied by Nuclear and Other Techniques (*12*). The hair was cut from the proximal ends into segments of 2 - 5 mm using stainless steel scissors, transferred to a Pyrex bottle, washed with acetone for 10 min and dried at room temperature. At this point, instead of continuing the washing sequence using the small hair segments, the dry sample was powdered in a PTFE homogenizer (*i.e.*, using a microdismembrator) after freezing in liquid nitrogen, and the washing sequence was resumed afterwards with water and acetone, according to the protocol. The powdered sample was dried, protected from dust and draughts and stored in precleaned polyethylene bag at room temperature.

Blood (10 mL) was collected by venipuncture with disposable syringes and a stainless steel needle (Terumo), transferred into a precleaned polypropylene tube (Sarsted), mixed slowly for a few minutes with an anticoagulant ($Na_2EDTA$) and was stored at 4°C until analysis. Several tubes loaded with the anticoagulant (18.6 mg $Na_2EDTA$) and 10 mL ultrapure water were checked for Hg-T content by CVAAS and the levels were always below the limit of detection (0.13 ng, $3\sigma_{Blank}$)(*27*).

All glassware used in the determination of Hg-T was carefully cleaned as detailed elsewhere (*27*). The polypropylene tubes used for blood storage were cleaned with 20% (v/v) hydrochloric acid for 48 h, followed by several washings with ultrapure water and drying at 35 °C in an oven. These tubes were loaded with 0.5 mL of 0.1 M $Na_2EDTA$, dried, capped and stored in polyethylene clean bags in a clean hood until use. All glassware used in the determination of Me-Hg was precleaned with diluted neutral detergent (1% sol.), washed copiously with deionized water and stored in 10% HCl until used.

## Analytical methods

Hg-T was determined by CVAAS in all samples and by INAA only in 20% of the scalp hair samples for periodic external quality control. Me-Hg was determined by GC-ECD. Selenium was determined in some samples by INAA.

**CVAAS.** Human hair was digested with concentrated $HNO_3$ and blood in a $HNO_3$ : $HClO_4$ mixture, both in sealed Pyrex ampoules (*28,29*). Other biological materials were digested in concentrated $HNO_3$ in PTFE bombs or in $HNO_3$ : $HClO_4$ mixture in sealed Pyrex ampoules, as described elsewhere (*30*). The acid mixture was used to enhance oxidation of organomercury compounds in such highly organic matrices. Aliquots of blood (0.5 mL) measured with a micropipette and weighed accurately in an analytical balance were acid digested under pressure in sealed Pyrex ampoule by addition of 2.5 mL of a mixture of (3 : 1) $HNO_3$ : $HClO_4$. After standing overnight at normal pressure the acid - blood mixture in the ampoule was sealed and placed in a thermal block at 90 °C for 2 h. The sample was completely digested as indicated by the light - yellow color of the solution. The ampoule was removed from the thermal block, cooled to room temperature, washed with ultrapure water, cooled with liquid nitrogen (*ca.* 5

min), notched with a diamond knife and opened. The digest was quantitatively transferred to a 10-mL volumetric flask and diluted with ultrapure water. The CVAAS procedure with amalgamation of mercury in a gold/platinum grid has been described previously (*27,30*). Using this CVAAS procedure, Hg-T was determined in several reference materials (RMs), in two blind hair samples sent for analytical quality control exercise (AQCE) (IAEA-085 (Elevated Level Human Hair) and IAEA-086 (Natural Level Human Hair) Intercomparison Study of Me-Hg and Hg-T), in several blood samples of the Interlaboratory Comparison Programme of the Centre de Toxicologie du Québec (CTQ), in 203 scalp hair samples of the coastal and interior groups (collected between 1991 and 1994) and in 50 blood samples of pregnant women of the coastal and interior groups collected in 1994. The detection limits ($3\sigma_{Blank}$) for Hg-T were 0.033 mg/kg in scalp hair (*27*) and 0.5 µg/kg in blood.

**INAA.** The analytical methodology for Hg-T (measured at the $^{203}$Hg photopeak at 279 keV) and Se has been described elsewhere (*27*). The following RMs were used as solid standards: Citrus Leaves (NIST SRM 1572), Human Hair (BCR-397), Oyster Tissue (NIST, SRM 1566), and the IAEA RMs: Trace Metals in Tuna Fish (IAEA-350), Fish Tissue Lyophilized (MA-B-3/OC) and Shrimp Tissue Lyophilized (MA-A-3/TM). These were prepared and irradiated in the same form as the hair samples. In parallel, standard solutions of Hg (5.0 µg/mL) and Se (5.0 µg/mL) were encapsulated in quartz (Vitreosil) and surrounded by an iron ring used as neutron flux monitor. Samples (20% of the scalp hair samples of the study and control groups) and standards were irradiated in a nuclear reactor (RECH-1, Comisión Chilena de Energía Nuclear) for 24 hr with a neutron flux oscillating between 0.9 - 2.0 x $10^{13}$ n/sec x cm$^{2}$. After an appropriate decay period (*ca.* 10 days), the vials were cooled with liquid nitrogen, opened and transferred to polyethylene containers for measurement by γ-spectrometry. Samples were counted for 1 hr on a γ-spectrometer composed by a HPGe detector with a 35% relative efficiency for $^{60}$Co gamma emission of 1332 keV, resolution of 1.8 keV (FWHM); a Canberra model 1510 electronic system with amplifier, high voltage power supply and analogue-to-digital converter; a Canberra model 1407 pulse generator; and a Canberra S-100 multichannel card installed in a compatible PC (386, 8Mb RAM) for data collection. The software for data processment and calculation of concentrations (APACEN) was designed by the informatic group of the Nuclear Research Centre of La Reina (Santiago, Chile). A pneumatic sample changer, controlled through an IBM/PC compatible computer, was used for the analysis. No losses of Hg-T and Se were detected as a result of the irradiation process.

**GC-ECD.** A procedure based on selective isolation of Me-Hg from human hair by volatilization in a micro-diffusion cell (Convey dish) was adopted in order to isolate and extract Me-Hg in biological materials (*31*). It provides a much cleaner extract resulting in simpler chromatograms, requires less sample, gives more concentrated toluene extracts, permits large sample throughput, requires fewer reagents, and is cost-effective.

In the micro-diffusion approach, the cysteine impregnated filter paper containing the Me-HgCN (trapped on it following volatilization) was acidified with 4M KBr and 4M $H_2SO_4$ (precleaned by extraction with toluene). The Me-Hg (as Me-HgBr) was extracted into toluene (1 mL x 2) under mechanical shaking for 10 min. This was followed by a 10-min centrifugation for phase separation. After transferring the toluene phase into a separate glass vial, the toluene was dehydrated with a few anhydrous $Na_2SO_4$ crystals. A Varian 2700 gas chromatograph equipped with $^{63}Ni$ electron capture detector was used. Prior to sample analysis, the chromatographic glass column (1.8 m long, 2 mm i.d., packed with 5% diethylene glycol succinate (DEGS-PS) on 100/120 Supelcoport) was pre-conditioned (*32*); first, for 24 h with a saturated solution of $HgCl_2$ in toluene. Following this, a 1 μL injection of a 10 μg/mL Me-Hg solution was applied into the column and was left for another 12 h before use. For actual sample analysis, the separation conditions used were: Column temperature, 160°C; Injector temperature, 180°C; Detector temperature, 220°C; gas flow, 50 mL/min (ultrapure $N_2$); injection volume, 2 μL.

After cross-checking our Me-HgCl working stock solution with a standard provided by the IAEA Marine Environmental Laboratory (MEL) in Monaco, and upon confirmation of the Me-Hg and ethylmercury (Et-Hg) chromatographic peaks with standard solutions provided by MEL, recovery results obtained for Me-Hg in aqueous solutions (at 0.50, 0.10, 0.05 and 0.025 μg/mL) by the micro-diffusion technique yielded between 86 ± 3 % (at 0.025 μg/mL ) and 92 ± 3 % (at 0.5 μg/mL). These recoveries were reproducible and considered reasonable for pure Me-HgCl solutions. However, recoveries were better when Me-Hg was analyzed in real samples as is shown in the RMs that were analyzed (see below). The absolute detection limit for Me-Hg was estimated as 16 pg (*32*) and corresponded to 0.06 mg/kg in scalp hair.

**Quality assurance**

A number of approaches were followed for both *internal* and *external* analytical quality controls (AQC) in our studies of Hg-T and Me-Hg concentrations in scalp hair and Hg-T in blood. *Internal* AQC was performed for Hg-T through the periodic analysis of either NIES RM N°5 (Human Hair, 4.40 ± 0.40 mg/kg) or a in-house human hair sample. The latter is segmented but not powdered hair, and was available in this laboratory in small amount (*ca.* 3 g), with an established Hg-T value by CVAAS (1.10 ± 0.05 mg/kg, 95% confidence interval based on 16 independent determinations). This sample was used as internal control to assess intra-laboratory variability. Its homogeneity was established at the 100 mg level by comparing the relative standard deaviation (RSD (%)) obtained in independent determinations of Hg-T in 8 subsamples, with the RSD (%) found when independent sub-samples of RM NIES N° 5 were analyzed in a similar way. The AQC tests showed 100% and 92.3% of results within mean ± 2 SD in NIES RM N°5 and in-house human hair sample, respectively. The importance of analysing a very homogeneous powdered human hair sample (*e.g.*, NIES N°5) rather than segmented hair was realized in this study, revealing that the Hg-T determination by CVAAS in powdered hair was within statistically acceptable limits.

The results of the analyses of Hg-T and Me-Hg in human hair RMs which are certified for Hg-T and have reference values for Me-Hg, and of Hg-T in the CTQ blood samples are given in Table I. The Hg-T values were found to be satisfactory with the exception of low recovery in one blood sample which had a relatively high Hg value. No significant differences with respect to target values were observed by paired *t*-tests

TABLE I. Hg-T LEVELS AND Me-Hg IN HUMAN HAIR REFERENCE MATERIALS AND HG-T IN BLOOD SAMPLES OF THE INTER LABORATORY COMPARISON PROGRAM[a]

| | Hg-T (mg/kg)[b] | | Me-Hg (mg/kg as Hg) | |
|---|---|---|---|---|
| SAMPLE | CVAAS | CERT./TARGET | GC-ECD | REFERENCE |
| NIES N° 5 | 4.52 ± 0.18(13) | 4.40 ± 0.40[c] | 3.37 ± 0.164(6) | 3.3 - 3.4[e] |
| Human Hair | | 4.50 ± 0.50[d] | | |
| | | | | |
| BCR-CRM-397 | 11.63 ± 0.22(5) | 12.3 ± 0.60 | 0.715 ± 0.034(5) | (0.646 ± 0.110)[e] |
| Human Hair | 12.69 ± 0.48(6)[f] | | | |
| | | | | |
| GBW 09101 | 2.12 ± 0.10 (5) | 2.16 ± 0.21 | 1.89 ± 0.252(3) | 0.90 ± 0.04[d] |
| HH Human Hair | | | | 1.54 ± 0.04[d] |
| | | | | |
| IAEA-085 Human hair elevat. level | 26.03 ± 2.13(5) | 22.9 ± 3.4 (P.R.)[g] | 24.69 ± 2.35 (8) | 21.6 ± 4.2 (P.R.) |
| | | | | |
| IAEA-086 Human hair natural level | 0.636 ± 0.045(10) 0.526 ± 0.043(12)[h] | 0.574 ± 0.15(P.R.) | 0.362 ± 0.041(8) | 0.291 ±0.092 (P.R.) |

[a]Centre de Toxicologie du Québec (Canada);  [b]Paired *t*-Test, *tcalc.*= 0.899 (*d.f.* = 6, p = 0.403). [c]certificate value; [d]Horvat M. et al, IAEA, NAHRES-13,Vienna (1992). [e]Horvat M. *et al, Water, Air and Soil Poll.* 56, 95 (1991). [f]Results July 1994. [g]Preliminar results. [h]Acid digestion in PTFE bomb.

| | Hg-T (nmol/l)[a] | | |
|---|---|---|---|
| BLOOD SAMPLE | CVAAS | TARGET[b] | ACCEPTABLE RANGE |
| M-9314 | 45.07 ± 8.86(3) (9.1 ± 1.8 ng/ml) 35.4 - 52.8 | 48 | 33 - 63 |
| M-9315 | 130.1 ± 10.8(3) (26.7 ± 2.6 ng/ml) 118.3 - 139.2 | 180 | 140 - 220 |
| M-9407 | 134.8 ± 4.9 (3) (27.0 ± 1.0 ng/ml) 129.8 - 139.5 | 140 | 107 - 173 |
| M-9408 | 74.7 ± 11.4(3) (15.0 ± 2.3 ng/ml) 61.8 - 83.6 | 75 | 55 - 95 |
| M-9409 | 43.6 ± 2.6(3) (8.7 ± 0.5 ng/ml) 40.6 - 45.1 | 44 | 30 - 58 |

[a]Paired *t*-Test, *tcalc.*= 1.226 (*d.f.* = 4, p = 0.2875). [b]Centre de Toxicologie du Québec (CTQ), Interlaboratory Comparison Programme, Québec, Report N° 5 (1993) and N°3 (1994).

for Hg-T in the results obtained in both sample matrices ($t_{calc.}$ = 0.899, $d.f.$ = 6, $P$ = 0.403 (Hair); $t_{calc.}$ = 1.226, $d.f.$ (degrees of freedom) = 4, $P$ = 0.2875 (Blood)). Results for Me-Hg show good agreement with recommended values in NIES RM N°5, BCR CRM-397, and with the preliminary means reported for human hairs IAEA-085 (Elevated level, spiked with Me-Hg) and IAEA-086 (Natural level, unspiked), the latter two corresponding to the AQCE and sixth intercomparison exercise of one of the external AQC approaches used. Recoveries for Me-Hg were higher than in aqueous standards (96 to 124 %), possibly because Me-Hg is very sensitive to light exposure and in acid medium it may decompose to some extent during the digestion procedure, whereas in real samples the volatilization efficiency of Me-Hg from the sample is increased (Horvat, M., IAEA-Marine Environment Studies Laboratory, personal communication, 1995.). The preliminary results obtained in the sixth intercomparison study were somewhat higher, in particular for Hg-T and Me-Hg in Sample N° 085. Due to the high Hg-T concentration present in this sample, it was necessary to significantly dilute the sample solution before analysis (5-fold) and to use CVAAS without the amalgamation unit; hence, determinate errors could not be discarded in this procedure, in particular regarding blank corrections and incomplete destruction of the organic matrix by acid digestion at low temperature in sealed Pyrex ampoules. The same intercomparison exercise included the determination of Hg-T and Me-Hg in RM IAEA-350 (tuna fish homogenate): Hg-T = 4.90 ± 0.35 mg/kg (n = 6) and Me-Hg = 3.38 ± 0.18 mg/kg (n = 6) and BCR-CRM-397 (human hair): Hg-T = 12.69 ± 0.48 mg/kg (n = 6) and Me-Hg = 0.715 ± 0.034 mg/kg (n = 6). These results were quite consistent with the certified or recommended values. Also, Se was determined in IAEA-085 and 086 samples and the results were 1.11 ± 0.12 (mg/kg)(IAEA-085) and 1.37 ± 0.13 (mg/kg)(IAEA-086)(n = 6).

*External* AQC was approached in two ways; first, by comparing the results obtained for Hg-T by CVAAS and INAA in several biological and environmental SRMs and CRMs including two RMs of human hair; and second, through intercomparison exercises. In comparing the results of the analyses of various RMs by CVAAS and INAA using the $t$-test, there were no significant differences ($P$ < 0.05) in the results of both methods, which showed a correlation coefficient of 0.993 (*30,32*).

Using the second approach seven intercomparison exercises were accomplished. Three of them (the first, fifth and seventh exercises) were periodic AQC using INAA as reference method, and consisted in the analysis of 20% of the scalp hair samples collected at three different times and analyzed by CVAAS. In the first exercise, regression analysis of 21 hair samples of the coastal and interior groups collected in 1991 and 1992 showed a significant correlation between results obtained by CVAAS and INAA for all the women studied ($r^2$ = 0.88, $P$ < 0.0001); no significant differences were established by paired $t$-test (mean difference = 0.178, std. error= 0.155, $t_{calc.}$ = 1.152, $P$= 0.2628)(*32*). The fifth exercise with 23 scalp hair samples of the same groups collected in 1992 and early 1993 also showed good correlation ($r^2$ = 0.889, $P$ < 0.0001) without significant differences by $t$-test (mean difference = 0.206, $t_{calc.}$ = 1.490). Finally, a new comparison (seventh exercise) with 19 hair samples collected in 1994 again showed excellent correlation between CVAAS and INAA results (slope = 0.937; $r^2$ = 0.986, $P$ < 0.0001) with no significant differences by $t$-test (mean difference = 0.096, std. error = 0.275, $t_{calc.}$ = 1.442, $P$ = 0.1666).

The third exercise was the delivery of human hair samples collected and homogenized in our laboratory from the study and control groups, to other CRP participants for analysis of Hg-T. Table II shows a comparison of our results and of the data received from six laboratories. Some samples corresponded to segmented (not powdered) hair because in 1991 and early 1992 there was no microdismembrator available in our laboratory. Considering the different analytical techniques used, these results are fairly consistent, particularly at very low concentrations. However, at very high Hg concentrations (> 9.0 mg/kg), which are very unusual in our study population group, even with a powdered hair sample (coded RR-) prepared in our laboratory, significant differences were obtained between our results (by CVAAS) and results obtained by NAA using different analytical approaches by the participants. The sample RR- was also analyzed by our reference method (INAA) and the result (9.05 ± 0.27 mg/kg, n=2) was quite consistent with the CVAAS result and with the NAA result obtained by one collaborator. Unfortunately, the available sample amount was very small and it was not possible to repeat the analysis to test the sample homogeneity.

The fourth exercise was the intercomparison of Hg-T and Me-Hg in three biological reference materials provided by the IAEA Marine Environmental Studies Laboratory (IAEA-MESL): Trace Metals in Tuna Fish IAEA-350, Shrimp Tissue Homogenate (MA(S)-MED-86/TM), Shrimp Tissue Lyophilized (MA-A-3/TM) and Fish Tissue Lyophilized (MA-B-3/OC). The results are shown in Table III. The Table also shows the data on IAEA Samples N° 1 and 2 of the second exercise, that were reanalysed for Me-Hg. The second exercise was an intercalibration procedure based on the analysis of Hg-T (by CVAAS and INAA) and Me-Hg (GC-ECD) in two "blind" biological RMs provided by the IAEA to all participants in this CRP. In general, reasonably good agreement was obtained with the certified and reference values for Hg-T and Me-Hg in RMs IAEA-350 (bottles Nos. 220 and 346), and MA-B-3/OC. Relatively high Hg-T and Me-Hg results were obtained for RM MA(S)-MED-86/TM and MA-A-3/TM, and also for Me-Hg in IAEA Samples N° 1 and 2. Three of these samples have relatively low Me-Hg and high Et-Hg contents and both chromatographic peaks were not well resolved under these conditions due to peak overlap. Studies are in progress to further purify the toluene extract of Me-Hg to solve this problem. Also, the results of Se obtained by INAA (Table III) were quite comparable with the certified or recommended values in RMs IAEA-350 (bottles Nos. 220, 346 and 400), in MA-B-3/OC and IAEA Sample N° 1.

**Results of the study group**

The results for Hg-T in scalp hair are summarized in Table IV and combine the data obtained in 1991 (*27*), 1992 and 1993, following the same methodology. All the samples were above the limit of quantitation of Hg-T (0.11 mg/kg)(*27*). The Student t-test was an appropriate statistic for comparison of the study and interior group because the data was not too large and both groups were well defined. The arithmetic mean for the coastal group (1.79 ± 1.50 mg/kg) was significantly higher than the mean for the interior group (0.42 ± 0.15 mg/kg)($t_{calc.}$ = 4.643; *d.f.* = 177; $t_{0.05;177}$ = 1.645)( *P* < 0.01). The mean concentration found in our study group agrees with the results reported by Airey (*8*) in scalp hair of female populations of 13 countries who ate fish

TABLE II. COMPARISON OF INTERLABORATORY RESULTS FOR Hg-T AND Se IN HUMAN HAIR SAMPLES[a] AS PART OF EXTERNAL QUALITY CONTROL

| SAMPLE # | Hg-T (mg/kg) OUR RESULTS | Hg-T (mg/kg) OTHERS[b] | Se (mg/kg) OUR RESULT | Se (mg/kg) OTHERS |
|---|---|---|---|---|
| 2 - 91 | 3.40 ± 0.25 (2) I[c] | 2.50 ± 0.25 (2) [CH] | 0.78 ± 0.11 (2) | - |
|  | 4.00 ± 0.09 (2) II |  | - | - |
| 2 - 92 | 2.31 ± 0.21 | 4.09 ± 0.3 [IT] | - | 0.45 ± 0.08 |
| 14 - 92 | 223.3 ± 21.6 (4) I | 160.0 ± 3.9 (4) [CH] | < 0.19 (2) | - |
| 18 - 92 | 1.19 ± 0.05 | 2.58 ± 0.04 [CZ] | - | - |
| 31 - 92 | 0.60 ± 0.06 (2) I | 0.71 ± 0.07 (2) [BR] | - | - |
| 36 - 92 | 0.70 ± 0.01 (2) I | 1.7 ± 0.4 [VI] | - | - |
| 42 - 92 | 1.80 ± 0.15 (2) I | 1.07 ± 0.1[IT] | - | 0.66 ± 0.07 [IT] |
| 62 - 92 | 1.38 ± 0.14 (2) I | 1.4 ± 0.1 [BR] | 0.49 ± 0.08 (2) | - |
|  |  | 2.28 ± 0.02 [CZ] | - | - |
| 67 - 92 | 1.92 ± 0.09 (2) I | 2.3 ± 0.5 [VI] | - | - |
| 68 - 92 | 1.19 ± 0.01 (2) I | 1.35 ± 0.05 [BR] | - | - |
| 87 - 92 | 0.52 ± 0.01 (2) I | 0.42 ± 0.08 [IT] | - | 0.45 ± 0.09 [IT] |
| 99 - 91 | 4.37 ± 0.21 (2) I | 4.0 ± 0.4 [IT] | < 0.24 (2) | 0.08 ± 0.03 [IT] |
|  | 4.75 ± 0.09 (2) II |  | - | 0.68 ± 0.07 [IT] |
| 101 - 92 | 0.56 ± 0.01 (3) I | 0.42 ± 0.06 [IT] | 0.51 ± 0.07 (2) | 0.68 ± 0.07 [IT] |
|  | 0.50 ± 0.02 (2) II |  | - | - |
| 102 - 92 | 0.35 ± 0.01 (2) I | 0.25 ± 0.06 [BR] | 0.47 ± 0.09 (2) | - |
|  | 0.36 ± 0.03 (2) II |  | - | - |
| 111 - 92 | 2.16 ± 0.03 (2) I | 3.48 ± 0.11 [CZ] | - | - |
| 117 - 92 | 1.01 ± 0.02 (2) I | 0.86 ± 0.13 [BR] | - | - |
| 257 - 91 | 0.44 ± 0.04 (2) I | 0.49 ± 0.01 (2) [CH] | - | - |
| RR -2 | 9.27 ± 1.09 (3) I | 8.27 ± 0.8 0 [IT] | < 0.57 | 0.56 ± 0.07 [IT] |
| RR-5 | 9.27 ± 1.09 (3) I | 11.1 ± 0.14 [CZ] | < 0.57 | - |
| RR-7 | 9.27 ± 1.09 (3) I | 5.9 ± 0.6[VI] | < 0.57 | - |
| RR-8 | 9.27 ± 1.09 (3) I | 9.40 ± 0.16 [BR] | < 0.57 | - |

[a]Samples of pregnant and nursing women obtained and prepared in this laboratory; [b][BR] = Brazil; [CH] = China; [CZ] = Czech; [IT] = Italy, [VI] = Vietnam. [c]I = CVAAS; II = INAA.

TABLE III. RESULTS FOR TOTAL MERCURY, METHYL MERCURY AND SELENIUM OBTAINED IN SOME IAEA REFERENCE MATERIALS

| REFERENCE MATERIAL | Bott. N° | Hg-T (mg/kg) | | | MeHg (as Hg) (mg/kg) | | Se (mg/kg) | |
|---|---|---|---|---|---|---|---|---|
| | | CVAAS | INAA | CERTIFIED | GC-ECD | REFERENCE | INAA | CERTIFIED |
| IAEA-350 (Trace metals in tuna fish) | 220 | 4.89 ± 0.213(12) (4.48 - 5.53) / 4.84 ± 0.267(6)A[a] (4.55 - 5.23) / 4.95 ± 0.438(6)B[a] (4.48 - 5.75) | 4.35 ± 0.383(4) | 4.99 ± 0.26 | 3.38 ± 0.180(6) (3.14 - 3.54) | 3.65 ± 0.31[b] | 5.98 ± 0.35(2) | 5.57 / 4.42 - 5.97 |
| IAEA-350 (Trace metals in tuna fish) | 346 | 5.13 ± 1.323(4)B (4.33 - 6.02) | 5.20 ± 0.229(2) | 4.99 ± 0.26 | 3.83 ± 0.189(6) (3.58 - 4.12) | 3.65 ± 0.31[b] | 5.92 ± 0.77(3) | 5.57 / 4.42-5.97 |
| IAEA-350 (Trace metals in tuna fish) | 400 | 5.27 ± 0.973(4)B (4.93 - 5.70) | | 4.99 ± 0.26 | N.D. | 3.65 ± 0.31[b] | 6.02 ± 0.15(2) | 5.57 / 4.42-5.97 |
| MA(S)-MED-86/TM (Shrimp tissue homog.) | 131 | 2.50 ± 0.372(4) (2.37 - 2.76) | 0.976 ± 0.077(2) | 1.79 ± 0.24 (Prov.val.) | 0.110 ± 0.015(6) (0.093 - 0.129) | 0.08 - 0.09[c] | 3.17 ± 0.04(2) | 2.02 / ±0.41 |
| MA-A-3/TM (Shrimp tissue lyophilized) | 171 | 2.06 ± 0.177(7) (1.80 - 2.36) | 1.87 ± 0.052(2) | 1.79 ± 0.24 (Prov.val.) | N.D. | 0.08 - 0.09[c] | | |
| MA-B-3/OC (Fish tissue lyophylized) | 194 | 0.513 ± 0.048(7) (0.447 - 0.603) | 0.560 ± 0.098(2) | 0.510 ± 0.070 | 0.398 ± 0.052(3) (0.378 - 0.419) | 0.411-0.503[c] | 1.61 ± 0.07(4) | 1.46 / 1.35-1.70 |
| IAEA Sample # 1 (IAEA-MA-A-1/TM) (Dried Copepoda) | | 0.280 ±0.028(6)[d] | 0.264 ± 0.019(2) | 0.28 ± 0.01 | 0.205 ± 0.050(3)[e] (0184 - 0.224) | 0.018 ± 0.002[b] | 3.42 ± 0.11(2) | 3.03 / 2.43-3.39 |
| IAEA Sample # 2 (IAEA-MA-A-2/TM) (Fish Homogenate) | | 0.479 ±0.053(8)[d] | 0.477 ± 0.011(2) | 0.47 ± 0.02 | 0.397 ± 0.048(3)[e] (0.380 - 0.418) | 0.300 - 0.318[b] | 1.44 ± 0.18(2) | N.A. |

[a] A = Acid digestion in PTFE bomb; B = Acid digestion in sealed Pyrex ampoules. [b] Result of intercalibration IAEA/RUN TUNA 350 (Mee L.D. et al, IAEA/AL-052, IAEA/ MEL-53. Monaco (1992)). [c] Horvat M. et al, IAEA, NAHRES-13, Vienna (1992). [d] Bruhn C. et al, IAEA, NAHRES-13. Vienna (1992). [e] Sample of the first analytical AQCE reanalyzed for Mc-Hg. N.A. = Reference value not available.

between once a month and once every week ($x$ = 2.1 ± 1.6 mg/kg, range = 0.3 - 8.0 mg/kg, n = 102). Significant differences were found between the arithmetic means for PW in the coastal (1.73 ± 1.47 mg/kg, n = 129) and interior group (0.44 ± 0.16 mg/kg, n = 22)($t_{calc.}$ = 4.131; $d.f.$ = 149; $t_{0.05;149}$ = 1.967)( $P$ < 0.01), as well as for NW (coastal: 2.12 ± 1.67 mg/kg, n = 24 and interior: 0.35 ± 0.10 mg/kg, n = 4)($t_{calc.}$ = 2.089; $d.f.$ = 26; $t_{0.05;26}$ = 2.056)( $P$ < 0.01), but no significant difference was found between the arithmetic means of PW and NW in the coastal group ($t_{calc.}$ = 1.672; $d.f.$ = 151; $t_{0.05;151}$ = 1.960)( $P$ < 0.01) and interior group ($t_{calc.}$ = 1.082; $d.f.$ = 24; $t_{0.05;24}$ = 2.064)( $P$ < 0.01). These mercury levels determined in hair are considered normal for populations with dietary habits based mainly on seafood products. Hence, it is not expected that they will adversely affect the health of PW and NW living in the coastal zone, although it could not be ruled out that they and their children could suffer sub-clinical effects. Available evidence indicates that neurobehavioral dysfunction in children may occur if the mercury concentration in hair is > 6 µg/g ( 30 nmol/L). This value corresponds to a blood mercury concentration of *ca.* 24 µg/L ( 120 nmol/L)(*33*). Also, the levels of Hg-T in scalp hair of the interior group compared relatively well with those of control groups consuming negligible and small amounts of fish as reported by Dermelj *et al* (*22*). In studies performed in the Mediterranean, the Hg-T concentrations in scalp hair of groups of fish consumers were invariably higher than those found in groups with low or negligible fish consumption (*22*). Although the underground or well water consumed by the population of the fishing villages is another potential source of mercury, the regional water authority never noted concentrations beyond the maximum acceptable mercury level established for drinking water in Chile (0.001 mg/L) in these water sources. Therefore, mercury was not measured in water consumed by these communities.

Table V shows the arithmetic means obtained for Hg-T in scalp hair of PW and NW in 11 coastal fishing villages. The variance analysis (ANOVA) comparison for this group of results with the mean of the interior group confirmed a statistically significant difference ($F$ = 12.24, $P$ < 0.05). In agreement with our expectations, the Tuckey test (honestly significant difference test, HSD) indicated no statistically significant differences ($P$ < 0.05, $q_{0.05; 12; 167}$ = 4.62; HSD = 1.029) between the mean concentrations in scalp hair of the PW and NW of the interior group and PW and NW of the fishing villages of Dichato ($x$ = 0.75 mg/kg) and Quidico ($x$ = 0.60 mg/kg), located near the northern and southern limits, of this coastal zone (Figure 1). A statistically significant difference was found ($P$ < 0.05; $t_{calc.}$ = 6.218, $d.f.$ = 152) by comparison of the mean concentration in scalp hair of women of these two fishing villages ($x$ = 0.698 ± 0.379 mg/kg, n = 42) with the mean of women of the other 9 fishing villages of the coastal group ($x$ = 2.220 ± 1.566 mg/kg, n = 112). When the Hg-T levels obtained within the coastal group in the 1992-1993 period were examined with respect to the number of fish meal/week (Table VI), it was found that the mean fish consumption for Dichato and Quidico (weighed mean = 2.7 meals/week) was lower than the mean for the other 9 fishing villages (weighed mean = 5.7 meals/week). The relatively higher concentration of Hg-T in the scalp hair of the women living within the polluted coastal area can be attributed to consumption of fresh fish, shellfish and other edible sea products contaminated by this metal. Moreover, fish consumption in Lenga

TABLE IV. CONCENTRATION OF TOTAL MERCURY (mg/kg) IN SCALP HAIR OF PREGNANT (PW) AND NURSING (NW) WOMEN (COASTAL AND INTERIOR GROUP) OF THE EIGHTH REGION OF CHILE, 1991 - 1993.

| GROUP | n | $\bar{x}$ | SD | RANGE |
|---|---|---|---|---|
| COASTAL | | | | |
| PW | 129 | 1.73 | 1.47 | 0.14 - 9.72 |
| NW | 24 | 2.12 | 1.67 | 0.16 - 6.62 |
| PW + NW* | 153 | 1.79 | 1.50 | 0.14 - 9.72 |
| | | | | |
| INTERIOR | | | | |
| PW | 22 | 0.44 | 0.16 | 0.25 - 0.79 |
| NW | 4 | 0.35 | 0.10 | 0.20 - 0.44 |
| PW + NW* | 26 | 0.42 | 0.15 | 0.20 - 0.79 |

* $t_{calc.} = 4{,}643$; $d.f. = 177$; $t_{0.05;177} = 1{,}645$

TABLE V. CONCENTRATION OF TOTAL MERCURY (mg/kg) IN SCALP HAIR OF PREGNANT AND NURSING WOMEN. DISTRIBUTION BY FISHING VILLAGES EIGHTH REGION OF CHILE, 1991 - 1993

| FISHING VILLAGE | n | $\bar{x}$ | SD | RANGE |
|---|---|---|---|---|
| DICHATO | 27 | 0.75 | 0.38 | 0.14 - 1.80 |
| COLIUMO | 9 | 1.47 | 1.22 | 0.43 - 3.81 |
| COCHOLGUE | 14 | 2.07 | 0.76 | 0.77 - 2.97 |
| TUMBES | 15 | 2.97 | 2.17 | 1.19 - 9.72 |
| CHOME | 4 | 3.54 | 0.70 | 2.71 - 4.40 |
| LENGA | 6 | 2.63 | 0.43 | 2.04 - 3.15 |
| I. STA. MARIA | 10 | 3.23 | 2.27 | 1.27 - 9.12 |
| LARAQUETE | 30 | 1.48 | 0.76 | 0.48 - 3.09 |
| TUBUL | 12 | 1.53 | 1.16 | 0.76 - 4.93 |
| PTA. LAVAPIE | 11 | 3.07 | 2.09 | 0.61 - 7.11 |
| QUIDICO | 15 | 0.60 | 0.38 | 0.18 - 1.30 |

$F_{calc.} = 12{,}24$ ($F_{0.95;11;167} = 1{,}84$)

and Chome located within the most polluted zone (see Figure 1; weighed mean = 1.2 meals/week) was lower than for Dichato and Quidico, but Hg-T concentration in scalp hair was significantly higher in the former two. Inhabitants of Lenga do eat less fish because they are informed about the industrial mercury wastes released into the Lenga Estuary.

Table VII shows the relationship between Hg-T concentration in scalp hair and the consumption of fish or seafood in general (fish, shellfish and marine algae) of PW from the coastal group based on the 1992 and 1993 survey data. By *t*-test, statistically

TABLE VI. CONCENTRATION OF TOTAL MERCURY (mg/kg) IN SCALP HAIR HAIR OF PREGNANT WOMEN. DISTRIBUTION AND MEAN NUMBER OF FISH FISH MEALS PER WEEK BY FISHING VILLAGES. 1992 - 1993

| FISHING VILLAGE | n | $\bar{x}$ | SD | RANGE | # FISH MEALS/WEEK |
|---|---|---|---|---|---|
| DICHATO | 14 | 0.69 | 0.36 | 0.14 - 1.51 | 3.3 |
| COLIUMO | 5 | 0.94 | 0.56 | 0.45 - 1.75 | 7.4 |
| COCHOLGUE | 7 | 1.51 | 0.65 | 0.77 - 2.59 | 8.3 |
| TUMBES | 10 | 2.68 | 2.54 | 1.19 - 9.72 | 6.9 |
| CHOME | 1 | 4.40 | — | — | 1.0 |
| LENGA | 3 | 2.69 | 0.38 | 2.31 - 3.06 | 1.3 |
| I. STA. MARIA | 10 | 3.23 | 2.27 | 1.27 - 9.12 | 7.1 |
| LARAQUETE | 15 | 1.23 | 0.87 | 0.48 - 3.09 | 2.4 |
| TUBUL | 8 | 1.68 | 1.36 | 0.76 - 4.93 | 5.1 |
| PTA. LAVAPIE | 7 | 2.17 | 0.73 | 0.79 - 3.07 | 8.4 |
| QUIDICO | 12 | 0.52 | 0.35 | 0.18 - 1.30 | 1.9 |

significant differences ($P < 0.05$) were confirmed between PW consuming fish one or two times per week and those consuming one or more times per day ($t_{calc.} = -2.551$, $d.f. = 67$, $P < 0.05$). No significant differences were detected between PW consuming fish one or two times per week and those consuming fish three to four times and five to six times per week. However, in PW consuming one or two meals per week based on fish, shellfish or marine algae, the Hg-T concentration in scalp hair was significantly lower than in PW consuming five or six meals per week ($t_{calc.} = -2.421$, $d.f. = 37$, $P < 0.05$) and seven or more meals per week ($t_{calc.} = -3.554$, $d.f. = 58$, $P < 0.05$) of these foods. As the accuracy of the answers to the questionnaire depended on the memory and veracity of the PW surveyed and since no record was kept of the daily food consumed, some answers could be biased. By simple regression between Hg-T concentrations ($Y$) and fish, shellfish and marine algae consumption ($X$, meals/week), the equation obtained was: $Y = 0.072X + 1.105$; $r = 0.279$; $P = 0.007$), indicating a slight positive correlation between both parameters. These results confirm the influence of mercury intake in the diet.

No significant differences were established when the Hg-T content in scalp hair was related to the type of food preparation (raw, cooked or fried) and to women's age, at least when PW of the coastal group were divided into two sub-groups (Table VIII). Nevertheless, the $t$-test showed significant differences between Hg-T concentrations in scalp hair of women living in the same village for less than 20 years ($x = 1.380 \pm 1.082$ mg/kg, n = 80) and those living there for 20 years or more ($x = 2.375 \pm 1.843$ mg/kg, n = 64). Assuming no dietary exposure to mercury from sources other than fish and seafood, this observation is consistent with the greater diet exposure to mercury to be expected for the subgroup with a longer residence period.

These results prove that the main source of mercury in the coastal group was fish and seafood consumed in their diets. However, none of the subjects examined had

TABLE VII. EFFECT OF FISH AND SEAFOOD CONSUMPTION ON TOTAL MERCURY CONCENTRATION IN SCALP HAIR PREGNANT WOMEN. COMPARISON OF CONSUMPTION FREQUENCIES IN THE COASTAL GROUP. EIGHTH REGION OF CHILE. 1992 - 1993

| FEEDING: | ONLY FISH | | | | | FISH-SHELLFISH-ALGAE | | | | |
|---|---|---|---|---|---|---|---|---|---|---|
| Freq. consumption (meals/week) | N° Indiv. | Mercury (mg/kg) $\bar{x}$ | SD | $t_{calc}$ [a] | d.f. [b] | N° Indiv. | Mercury (mg/kg) $\bar{x}$ | SD | $t_{calc}$ [a] | d.f. [b] |
| (1 - 2) | 45 | 1.262 | 1.570 | - | - | 26 | 1.057 | 0.933 | - | - |
| (3 - 4) | 13 | 1.605 | 0.893 | -0.750 | 56 | 20 | 1.393 | 2.028 | -0.749 | 44 |
| (5 - 6) | 11 | 1.534 | 0.707 | -0.557 | 54 | 13 | 1.778 | 0.745 | -2.421 | 37 |
| 7 | 24 | 2.182 | 1.102 | -2.551 | 67 | 34 | 2.012 | 1.1 | -3.554 | 58 |

[a] Student's *t*-test.   [b] Degrees of freedom.

TABLE VIII. EFFECT OF AGE AND RESIDENCE PERIOD IN THE COASTAL GROUP. PREGNANT AND NURSING WOMEN. EIGHTH REGION OF CHILE. 1991 - 1993

a. EFFECT OF AGE.

| Age range (years) | n | Hg-T (mg/kg) $\bar{x}$ | SD |
|---|---|---|---|
| < 25 | 81 | 1.669 | 1.465 |
| ≥ 25 | 60 | 1.997 | 1.647 |

$t_{calc} = -1.246$; $d.f. = 139$; $t_{0,05;139} = 1.960$

b. EFFECT OF THE LENGTH OF RESIDENCE.

| Residence Per. (years) | n | Hg-T (mg/kg) $\bar{x}$ | SD |
|---|---|---|---|
| < 20 | 80 | 1.380 | 1.082 |
| ≥ 20 | 64 | 2.375 | 1.843 |

$t_{calc} = -4.038$; $d.f. = 142$; $t_{0,05;142} = 1.960$

hair mercury concentrations corresponding to the onset of toxicological symptoms (*34,35*) as was confirmed by the morbidity data in the questionnaire.

**High risk group (1994 Study)**

A new survey was performed in order to confirm the influence of dietary mercury intake on the Hg-T and Me-Hg levels in scalp hair and blood. Also studied were the Me-Hg/Hg-T ratios in scalp hair, possible correlations between Hg-T and Me-Hg content in hair, Hg-T in hair and blood, Me-Hg in hair and Hg-T in blood, as well as Hg-T and Me-Hg content in hair in relation to the frequency of fish and seafood consumption in 1994, the age, and the residency period in the same village or town. The study was conducted in 40 new PW not occupationally exposed to mercury, aged 17 - 42, residing in 5 fishing villages (Tumbes, Lenga, Chome, Isla Santa Maria and Punta Lavapié) located within the more polluted area of the coastal zone (Figure 1), and consuming at least one fish meal per week (coastal group). For comparison, 9 new PW, aged 14 -35, residing in the town of El Carmen, located inland in the same region far from the coastal area, and with negligible or no fish or seafood consumption were considered to represent the interior group. PW of both groups were concurrently surveyed, and samples of scalp hair and blood were obtained from them. These fishing villages presented the largest differences in Hg-T content with respect to the interior group surveyed in 1991 - 1993. Their pooled arithmetic mean ($3.06 \pm 1.97$ mg/kg, n = 46) was almost 70% higher than the mean of the coastal group, and therefore they were considered at high risk due to dietary exposure to mercury. Unfortunately only one sample was obtained in Chome and no statistical analysis was possible for this village.

The results obtained for scalp hair samples of the coastal and interior groups, and of the former arranged by fishing villages, are summarized in Table IX (Hg-T and Me-Hg) including the age range, the arithmetic and geometric means, the standard deviation and the concentration ranges. The Hg-T results in blood are summarized and arranged by fishing villages in Table X. The arithmetic mean of Hg-T in scalp hair of the coastal group ($2.44 \pm 1.30$ mg/kg) was significantly higher than the mean obtained for the interior group ($0.40 \pm 0.16$ mg/kg)($t_{calc.} = 4.658$; $d.f. = 47$; $t_{0.05;47} = 1.683$)( $P <$ 0.05), almost 35% higher than the mean obtained in the coastal group for the 1991-1993 period, but 20% lower than the pooled mean ($3.06 \pm 1.97$ mg/kg, n = 46) obtained in that period for the same five fishing villages. The arithmetic mean of Hg-T in blood from the coastal group ($9.04 \pm 5.05$ μg/kg) was also significantly higher than the mean obtained for the interior group ($2.73 \pm 1.27$ μg/kg)($t_{calc.} = 3.690$; $d.f. = 46$; $t_{0.05;46} = 1.680$)($P < 0.05$). Although the determination of Me-Hg has not yet been completed in all the hair samples, the preliminary arithmetic mean of Me-Hg (as Hg) in scalp hair was significantly higher for the coastal group ($2.34 \pm 0.98$ mg/kg) than the corresponding mean for the interior group ($0.30 \pm 0.08$ mg/kg)($t_{calc.} = 5.850$; $d.f. = 31$; $t_{0.05;31} = 2.041$)($P < 0.05$). The Hg-T(coastal)/Hg-T(interior) ratios in scalp hair and in blood were 6.1 and 3.3, respectively and the Me-Hg (coastal)/Me-Hg(interior) ratio in scalp hair was relatively higher (7.8). These results indicate that a good control group was selected. The ANOVA test to compare the mean result of each fishing village with the mean of the interior group confirmed a statistically significant difference both in hair

TABLE IX. TOTAL MERCURY (Hg-T) AND METHYLMERCURY (Me-Hg) IN SCALP HAIR OF PREGNANT WOMEN. COASTAL AND INTERIOR GROUPS SURVEYED IN 1994.

| GROUP | AGE RANGE | | Hg-T (mg/kg) | | | | |
|---|---|---|---|---|---|---|---|
| | (years) | n | $\overline{x}$ [a] | SD | $\overline{x}_G$ [b] | MEDIAN | RANGE |
| COASTAL | 17 - 42 | 40 | 2.44 | 1.30 | 2.16 | 2.13 | 0.76 - 6.26 |
| INTERIOR | 14 - 35 | 9 | 0.40 | 0.16 | 0.37 | 0.36 | 0.20 - 0.69 |
| | | | Me-Hg (as Hg, mg/kg) | | | | |
| COASTAL | 17 - 42 | 25 | 2.34 | 0.98 | 2.19 | 1.96 | 1.30 - 4.93 |
| INTERIOR | 14 - 35 | 8 | 0.30 | 0.08 | 0.29 | 0.31 | 0.16 - 0.43 |
| FISHING VILLAGE[c] | | | Hg-T (mg/kg) | | | | |
| LENGA | 20 - 30 | 5 | 4.41 | 1.62 | 4.14 | 4.88 | 2.17 - 6.26 |
| TUMBES | 19 - 33 | 7 | 1.57 | 0.54 | 1.49 | 1.44 | 0.91 - 2.30 |
| CHOME | 26 | 1 | 3.22 | - - - | 3.22 | 3.22 | - - - - |
| P. LAVAPIE | 18 - 35 | 12 | 1.99 | 0.80 | 1.88 | 1.68 | 1.27 - 3.99 |
| ISLA S. MARIA | 17 - 42 | 15 | 2.45 | 1.15 | 2.20 | 2.60 | 0.76 - 5.00 |
| | | | Me-Hg (as Hg, mg/kg) | | | | |
| LENGA | 20 - 30 | 4 | 2.93 | 0.34 | 2.92 | 2.90 | 2.62 - 3.31 |
| TUMBES | 19 - 33 | 2 | 2.20 | 0.61 | 2.16 | 2.20 | 1.77 - 2.63 |
| CHOME | 26 | 1 | 3.02 | - - - | 3.02 | 3.02 | - - - - |
| P. LAVAPIE | 18 - 35 | 7 | 1.77 | 0.37 | 1.74 | 1.65 | 1.48 - 2.50 |
| ISLA S.MARIA | 17 - 42 | 12 | 2.19 | 0.85 | 2.07 | 1.84 | 1.30 - 4.09 |

[a] $\overline{x}$ = Arithmetic mean; [b] $\overline{x}_G$ = Geometric mean. [c]Coastal group.

and blood ($F_{0.05;4;42} = 2.59$; $P < 0.05$). By multiple comparison test (honestly significant difference, HSD), a statistically significant difference ($\alpha < 0.05$) was established for Hg-T in scalp hair between the interior group and the fishing villages of Lenga, Punta Lavapie and Isla Santa Maria, and in blood between the former and the fishing villages of Tumbes, Lenga and Punta Lavapie. Also, by paired $t$-test a significant difference was confirmed for Hg-T in scalp hair of Tumbes ($t_{calc.} = 6.287$ ; $d.f. = 14$; $t_{0.05;14} = 2.145$)($P < 0.05$) and for Hg-T in blood of Isla Santa Maria ($t_{calc.} = 2.786$ ; $d.f. = 22$; $t_{0.05;22} = 2.074$)($P < 0.05$). The Me-Hg/Hg-T mean ratio in scalp hair of the coastal group ($0.820 \pm 0.159$, range = 0.373 - 1.008) was almost 20% higher than the same ratio in the interior group ($0.686 \pm 0.145$, range = 0.487 - 0.958), which was consistent with a higher dietary mercury intake of the former group. Nevertheless, by $t$-test a statistically significant difference was confirmed only at the level of $P < 0.10$ ($t_{calc.} = 2.004$ ; $d.f. = 30$; $t_{0.10;30} = 1.697$). The Me-Hg/Hg-T ratio found in the present study is consistent with the 75 - 89 % of total mercury present as Me-Hg in scalp hair of adults having high fish consumption, as reported by other participants in this CRP (*36,37*).

To examine more effectively the Hg-T concentration distributions in scalp hair and blood they are presented in the form of box plots in Figure 2. A quick insight is possible into the median values and concentration ranges in each fishing village and in

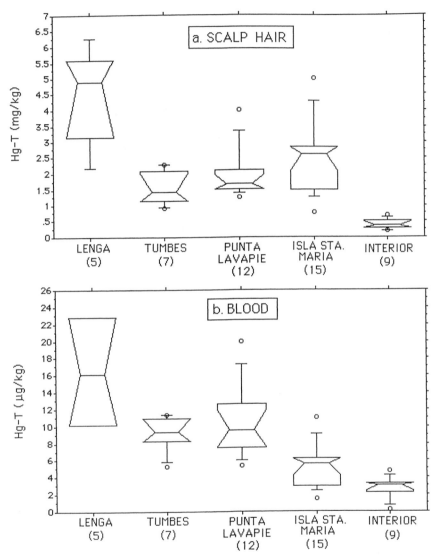

Figure 2. Distribution of Hg-T concentrations obtained in 1994 in four fishing villages (coastal group) and the interior group: a) scalp hair and b) blood.

the interior group. The same distribution pattern is observed for both the median and the range in scalp hair and blood of PW of Lenga, Tumbes, Punta Lavapie and the interior group (control). As expected, the latter shows a very narrow range in both sample matrices and very low median values. However, PW of Isla Santa Maria present surprisingly lower concentrations and narrower distribution in blood than the other fishing villages, in spite of greater fish and seafood consumption than women of the other fishing villages (see below). When the Hg-T (hair)/Hg-T(blood) ratios of the medians were calculated for each fishing village and the control group (Table IX), the figure obtained for Isla Santa Maria was 464, significantly higher than those found for the other villages (*e.g.*, 155 - 302) and for the interior group (*e.g.*, 120), and quite different from the estimated 200-250-fold higher mercury concentration in hair compared to blood as described in the literature (*1,3*). Possibly, these women were not eating as much fish and seafood as they indicated in the survey, because at the time of sampling (September 1994) there was fishing ban to preserve certain fish species and shellfish commonly eaten by these people, and which are being caught extensively also by the fish industries. Hence, the Hg-T levels determined in scalp hair are consistent with the dietary intake during the last two months when these women were eating fish and seafood daily, and the corresponding relatively low Hg-T levels found in blood probably are due to the more recent low mercury intake as already explained.

When the Hg-T and Me-Hg concentrations in scalp hair of the coastal group are compared by linear regression, a fairly high correlation coefficient was obtained (slope = 1.153, intercept = 0.275; r = 0.854, $P < 0.05$)(Figure 3a), whereas the correlation between Hg-T concentrations in hair and blood was weaker (slope = 0.124, intercept = 1.26; r = 0.50, $P < 0.05$)(Figure 3b). The latter result did not differ significantly from those of previous studies in which no significant correlation was observed for Hg-T (*38*). However, when the comparison between Hg-T in hair and blood was made with the whole data of both groups, the correlation was comparatively higher (r = 0.628, $P < 0.05$), and when this was done within the coastal group excluding the data of Isla Santa Maria, the correlation was slightly better (r = 0.637, n = 23, $P < 0.05$). The linear regression correlation of the concentration of Me-Hg in scalp hair and of Hg-T in blood of the coastal group was poor (r = 0.245, $P = 0.2478$). Also, when the concentrations of Hg-T in scalp hair and in blood were compared by linear regression with the women's age, the correlations was again poor (r = 0.155 and 0.104, $P = 0.3381$ and 0.5269, respectively).

Table XI shows the relationship between Hg-T and Me-Hg concentrations in scalp hair and Hg-T in blood, and the consumption of fish and seafood (fish, shellfish and marine algae) of PW from the coastal group for 1994. By *t*-test, statistically significant differences ($P < 0.05$) were confirmed for Hg-T and Me-Hg levels in hair of the coastal group between two sub-groups of PW: A, consuming among one and four meals per week based on fish, shellfish or marine algae, and B, consuming more than 4 meals per week. However, no significant differences were detected for Hg-T in blood between the same sub-groups. Each sub-group showed significant differences by *t*-test ($P < 0.05$) with respect to the control group. These data corroborated the previous inference that Hg content in scalp hair increased with the number of meals per week based on fish, shellfish or marine algae.

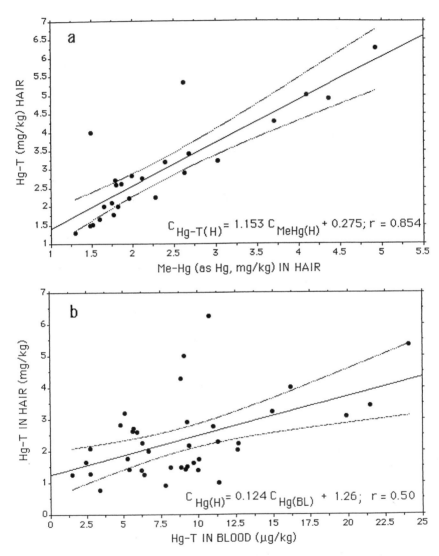

Figure 3. Correlations between a) Hg-T and Me-Hg concentrations obtained in scalp hair; and b) Hg-T concentrations obtained in hair and blood. 1994 Period.

**TABLE X. TOTAL MERCURY (Hg-T) IN BLOOD OF PREGNANT WOMEN SURVEYED IN 1994**

| GROUP | AGE RANGE (years) | n | Hg-T (µg/kg) x̄ | SD | x̄_g | MEDIAN | RANGE | Hg-T(H)/Hg-t(BL) RATIO[a] |
|---|---|---|---|---|---|---|---|---|
| COASTAL | 17 - 42 | 39 | 9.04 | 5.05 | 7.72 | 8.90 | 1.5 - 24.1 | 239 |
| INTERIOR | 14 - 35 | 9 | 2.73 | 1.27 | 2.13 | 3.03 | 0.20 - 4.77 | 120 |
| **FISHING VILLAGE[b]** | | | | | | | | |
| LENGA | 20 - 30 | 4 | 16.45 | 7.43 | 15.14 | 16.15 | 9.4 - 24.1 | 302 |
| TUMBES | 19 - 33 | 7 | 9.19 | 2.12 | 8.94 | 9.30 | 5.3 - 11.4 | 155 |
| CHOME | 26 | 1 | 14.90 | (---) | 14.90 | 14.90 | (---) | 216 |
| P. LAVAPIE | 17 - 35 | 12 | 10.51 | 4.24 | 9.81 | 9.50 | 5.4 - 19.9 | 177 |
| ISLA S. MARIA | 16 - 42 | 15 | 5.42 | 2.70 | 4.75 | 5.60 | 1.5 - 11.0 | 462 |

[a]Medians ratio: (H = hair, BL = blood). [b]Coastal group.

**TABLE XI. EFFECT OF FISH AND SEAFOOD CONSUMPTION ON THE LEVELS OF Hg-T AND Me-Hg IN SCALP HAIR AND Hg-T IN BLOOD OF PREGNANT WOMEN BASED ON CONSUMPTION FREQUENCIES IN THE COASTAL GROUP (1994 SURVEY)**

| FREQUENCY OF CONSUMPTION (meals / week) | n | Me-Hg (as Hg, mg/kg) x̄ | SD | t_calc. | d.f.[a] | n | Hg-T (mg/kg) x̄ | SD | t_calc. | d.f. |
|---|---|---|---|---|---|---|---|---|---|---|
| | | | | | | | SCALP HAIR | | | |
| (1 - 4) (A) | 9 | 1,752 | 0.324 | | | 23 | 1,849 | 0.744 | | |
| > 4 (B) | 16 | 2,482 | 0.799 | -2,603 | 23 | 19 | 3,025 | 1,518 | -3,136 | 40 |
| INTERIOR[b] | 8 | 0.303 | 0.084 | | | 9 | 0.395 | 0.160 | | |
| | | BLOOD, Hg-T (mg/kg) | | | | | | | | |
| (1 - 4) (A) | 21 | 9,262 | 3,993 | | | | | | | |
| > 4 (B) | 18 | 8,772 | 6,173 | 0.298 | 37 | | | | | |
| INTERIOR[b] | 10 | 2,714 | 1,195 | | | | | | | |

[a]d.f.=degrees of freedom; [b]By t - test, statistically significant differences were confirmed throughout for sub-groups A and B respect to the interior group (P < 0.05).

Conclusions

1. The arithmetic mean concentrations of Hg-T and Me-Hg found in scalp hair of PW in the coastal group were consistently higher and statistically different ($P < 0.05$) than the means obtained for the interior group. Moreover, the latter had low levels throughout this study and the concentration ranges were quite narrow, showing low dispersion of these results. As expected, the results obtained in this group denoted the rather scarce consumption of fish and fish products due to their inland location far from the coast.

2. By application of multiple comparison tests (ANOVA and DSH) to the Hg-T concentration means obtained in each FV with respect to the interior group, significant differences ($P < 0.05$) were confirmed in nine FVs in the 1991-93 period, all located within the more polluted coastal zone. The results obtained in two FVs (Dichato and Quidico), at the northern and southern limits respectively of this zone, were consistently low as expected, and did not differ significantly with respect to the interior group, because they are located beyond the critical zone. Instead, they differed significantly ($P < 0,05$) with the mean of the other 9 FVs. The relatively higher concentration of Hg-T in the scalp hair of the women living within the polluted coastal area could be attributed to consumption of fresh fish, shellfish and other edible sea products contaminated by Hg.

3. Within the coastal group and based on the Hg-T levels in scalp hair, statistically significant differences ($P < 0.05$) were confirmed between PW consuming fish one or two times per week and those consuming one or more times per day. No significant differences were detected between PW consuming fish one or two times per week and those consuming fish three to four times and five to six times per week. However, in PW consuming one or two meals per week of fish, shellfish or marine algae, the Hg-T concentration in scalp hair was significantly lower than in PW consuming five or six meals per week, and seven or more meals per week of these products. Consequently, the Hg content in scalp hair increased with the number of meals of fish, shellfish or marine algae per week. These results confirmed the influence of dietary mercury intake on hair levels.

4. Within PW and NW who ate fish and seafood at least one time per week, significant differences were observed between Hg-T concentrations in scalp hair of women living in the same village for less than 20 years and those living for 20 years or more in the same place. Assuming no dietary exposure to Hg from sources other than fish and seafood, this observation was consistent with the greater exposure to Hg to be expected for the subgroup with a longer residence period.

5. Five FVs (Lenga, Tumbes, Chome, Isla Santa María and Punta Lavapié) whose PW presented the highest Hg-T content in hair (*i.e.*, a pooled arithmetic Hg-T mean of 3.06 ± 1.97 mg/kg, n = 46, almost 70% higher than the coastal mean) were selected as high risk group with regard to dietary exposure to Hg. New PW were studied in this coastal group and in the interior group. Statistically significant differences were obtained for Hg-T and Me-Hg in scalp hair, for Hg-T in blood, and for the Me-Hg/Hg-T ratios in scalp hair of the coastal and interior groups. The higher Me-Hg/Hg-T ratio in the coastal group confirmed the relatively higher concentration of Me-Hg present in scalp

hair of adults having high fish consumption. Multiple comparison test confirmed significant differences for Hg-T in scalp hair and blood, at least between three FVs and the interior group.

6. A high correlation was observed between Hg-T and Me-Hg concentrations in scalp hair of the coastal group (r = 0.854, $P$ < 0.05), and a somewhat weaker correlation for Hg-T in scalp hair and blood (r = 0.50, $P$ < 0.05) of the same group. Poor correlations were observed when the concentration of Me-Hg in scalp hair of the coastal group was compared with Hg-T in blood, and when Hg-T in scalp hair and in blood were compared with the women's age.

7. Statistically significant differences ($P$ < 0.05) were confirmed for Hg-T and Me-Hg levels in hair of the coastal group between two sub-groups of PW, one consuming 1 - 4 meals per week of fish, shellfish or marine algae, and another consuming more than 4 meals per week. However, no significant differences were detected for Hg-T in blood between the same sub-groups. These data corroborated previous inference that Hg content in scalp hair increases with the number of meals per week of fish, shellfish or marine algae.

The concentrations obtained for Hg-T and Me-Hg in scalp hair of PW, and of Hg-T in blood of PW, do not suggest that the selected "high risk group" is truly at a high risk of adverse health effects due to Hg, because these concentrations are within the normal levels found in populations with fish consumption between once a month and once every week. However, the appearance of sub-clinical effects in the fetuses due to low dose dietary exposure of the mother to Hg and Me-Hg should not be disregarded. Neurobehavioral dysfunction in children may occur if the mercury concentration in hair is > 6 µg/g ( 30 nmol/L), corresponding to a blood mercury concentration of *ca.* 24 µg/L ( 120 nmol/L)(*33*).

8  Contribution from drinking water and other environmental sources to the total Hg intake of these population groups was minimal.

## Acknowledgments

The support to this work by the International Atomic Energy Agency (IAEA) through the Research Contract N° 6331/R1/R2/R3/RB and by the Dirección de Investigación of the Universidad de Concepción is highly recognized.

## Literature Cited

1. World Health Organization, *Environmental Health Criteria 1. Mercury;* WHO: Geneva; Switzerland, 1976.
2. Airey D., *Environ. Health Persp.* **1983**, *52*, 303-316.
3. World Health Organization. International Programme on Chemical Safety (IPCS/UNEP/ILO/ WHO). *Environmental Health Criteria 101. Methylmercury* WHO: Geneva, Switzerland, 1990.
4. Clarkson T.W., *Environ. Toxicol. and Chem.* **1990**, *9*, 957-961.
5. *Medical Toxicology;* Ellenhorn J., Barceloux D., Eds.; Elsevier; New York, NY, 1988; 1048-1052.
6. *Toxicology, the basic science of poisons;* Doull J., Klaassen C., Eds.; Macmillan: New York, NY, 1980; 423.

7. *Toxicología Clínica;* Marruecos L., Nogu, S., Nolla J., Eds.; Springer Verlag Ibérica: Barcelona, Spain, 1993; 283.

8. Airey D., *Sci. Tot. Environ.* **1983**, *31*, 157-180.

9. Paccagnella B., Prati L., and Bigoni A., *L'Igiene Moderna* **1973**, *66*, 480-504.

10. Takeuchi T., *Pediatrician* **1977**, *6*, 69-87.

11. Elhassani S.B., *J. of Toxicol.-Clin. Toxicol.* **1983**, *19*, 875-906.

12. International Atomic Energy Agency. *Co-ordinated research programme on assessment of environmental exposure to mercury in selected human populations as studied by nuclear and other techniques;* NAHRES-7; IAEA: Vienna, Austria, 1991; 1-14.

13. McKeown-Eyssen G.E., Ruedy J., Neims A., *Am. J. Epidemiol.* **1983**, *118*, 470-479.

14. Kjellstrom T., Kennedy P., Wallis S., et al. *Physical and mental development of children with prenatal exposure to mercury from fish. Stage 2. Interviews and psychological tests at age 6.* National Swedish Environmental Protection Board Report N° 3642. Solna, Sweden, 1989.

15. Hoffman W., *Distribution of mercury as contaminant in water, sediments and organisms of the Lenga Estuary* ; Memoria de título (Thesis); Universidad de Concepción: Concepción, Chile, 1978.

16. Barrios C., Rodríguez A.. In: *Actas Primer Congreso Iberoamericano de Toxicología,* National Institute of Toxicology (Sede Sevilla), Ed.; National Institute of Toxicology: Sevilla, Spain, 1982; 661- 665.

17. Secretaría Regional Ministerial de Planificación Octava Región Chile (SERPLAC). *Assessment of the degree of contamination of the littoral of the Eighth Region of Chile (Third Report).* Universidad de Concepción: Concepción, Chile, 1980; 35-45.

18. Salamanca M.A., Chuecas L. and Carrasco F., *Gayana Misc.* **1988**, *9*, 3-16.

19. González-Muñoz F.E.M., *Heavy metals in the trophic chain, organic matter contained in sediment-langostino colorado-congrio negro in the Arauco Gulf, Chile;* Doctoral Thesis; Universidad de Concepción: Concepción, Chile, 1994.

20. Phelps R.W., Clarkson T.W., Kershaw T.G., and Wheatley B., *Arch. of Environ. Health* **1980**, *35*, 161-168.

21. Bencze K, *Fresenius J. Anal. Chem.* **1990**, *338*, 58-61.

22. Dermelj M., Horvat M., Byrne A.R., et al. *Chemosphere* **1987**, *16*, 877-886.

23. U.S. Environmental Protection Agency. *Biological monitoring of toxic trace metals. Vol. 1 Biological monitoring and surveillance;* EPA-600/3-80-089; EPA: Las Vegas, Nevada, 1980.

24. International Atomic Energy Agency. *Co-ordinated research programme on assessment of environmental exposure to mercury in selected human populations as studied by nuclear and other techniques;* NAHRES-13; IAEA: Vienna, Austria, 1992; 1-13.

25. World Health Organization. International Programme on Chemical Safety (IPCS/UNEP/ILO/WHO). *Environmental Health Criteria 86. Mercury-Environmental Aspects;* WHO: Geneva, Switzerland, 1989.

26. Interamerican Group for Research in Environmental Epidemiology. *Internat. J. of Epidem.* **1990**, *19*, 1091-1099.

27. Bruhn C.G., Rodriguez A.A., Barrios C., et al. *J. Trace Elem. Electrolytes Health Disease* **1994**, *8*, 79-86.
28. Horvat M., Zvonaric T., Stegnar P., *Vestn. Slov. Kem. Drus.* **1986**, *33*, 475-487.
29. Horvat M., Lupsina V., and Pihlar B., *Anal. Chim. Acta* **1991**, *243*, 71-79.
30. Bruhn C.G., Rodriguez A.A., Barrios C., et al. *Journal of Analytical Atomic Spectrometry,* **1994**; *9*, 535-541.
31. Horvat M., Prosenc A., Smrke J., Liang L., In *Co-ordinated research programme on assessment of environmental exposure to mercury in selected human populations as studied by nuclear and other techniques*; IAEA, Ed; NAHRES-13; IAEA: Vienna, Austria, 1992; 65-81.
32. Bruhn C.G., et al, In *Co-ordinated research programme on assessment of environmental exposure to mercury in selected human populations as studied by nuclear and other techniques*; IAEA, Ed; NAHRES-13; IAEA: Vienna, Austria, 1992; 33-49.
33. Grandjean, P., Weihe, P., and Nielsen, J.B., *Clin. Chem.*, **1994**, *40*, 1395-1400.
34. Magos L., and Webb M., *CRC Crit. Rev. Toxicol.* **1980**, *8*, 1-42.
35. Kurland L.T., Faro S.N. and Siedler H., *World Neurol.*, **1960**, *1*, 370-390.
36. Vasconcellos M.B.A., et al, In *Co-ordinated research programme on assessment of environmental exposure to mercury in selected human populations as studied by nuclear and other techniques*; IAEA, Ed; NAHRES-27; IAEA: Vienna, Austria, 1995; 19-34.
37. Ingrao G, Belloni P, Santaroni GP, In *Co-ordinated research programme on assessment of environmental exposure to mercury in selected human populations as studied by nuclear and other techniques*; IAEA, Ed; NAHRES-27; IAEA: Vienna, Austria, 1995; 87-100.
38. Horvat M., Prosenc A., Smrke J., et al, In *Trace Elements in Health and Disease*, Aitio A., Aro A., Järvisalo J., and Vainio H., Eds.; Royal Soc. of Chemistry: London, UK, 1991.

# Chapter 15

# Personal Exposure to Indoor Nitrogen Dioxide

Toshihiro Kawamoto[1], Koji Matsuno[1], Keiichi Arashidani[2], and Yasushi Kodama[1]

[1]Department of Environmental Health, and [2]School of Nursing and Medical Technology, University of Occupational and Environmental Health, P.O. Orio, Yahata nishi ku, Kitakyushu 807, Japan

The personal exposure to $NO_2$ generated from various heaters and cooking stoves were studied, using 85 university students. The students attached $NO_2$ filter badges to their chests or collars and wrote down the period of time for heating and cooking for 1 week. Types of heaters and smoking habits were described through a questionnaire. The urinary hydroxyproline/creatinine ratio (HOP/C) was examined as a biomarker for health effect. The outdoor $NO_2$ concentration during the study period was 13.5-13.7 $\mu g/m^3$. Smoking and the usage of electric heaters did not affect the personal exposure to $NO_2$. Personal exposure increased according to the length of time kerosene heaters or oil fan heaters were used. The $NO_2$ concentration during the heating by a kerosene heater and an oil fan heater was calculated to be 219 $\mu g/m^3$ and 474 $\mu g/m^3$, respectively. The correlation between the period of cooking and personal exposure was also observed. Neither smoking nor personal exposure to $NO_2$ were associated with the increase of urinary HOP/C. Abstract Text goes here.

A great majority of people perform sedentary activities indoors, whether at work, at leisure, or at home. In winter, non-ventilated type oil and gas fan heaters and kerosene heaters are used in most Japanese houses. These heaters may produce significant emissions of NO, $NO_2$, CO, $CO_2$, $SO_2$, aldehyde, volatile organics and respirable particles. It is reported that $NO_2$ concentration exceeded the Japanese Ambient Air Quality Standard (112.8 $\mu g/m^3$), when five different heating appliances, that is, a convection oil heater, reflection oil heater, oil fan heater, gas fan heater and propane gas heater, were used in a model room. The indoor heater is one of the main sources of $NO_2$ exposure. Although several reports (1 - 6) on the health effects of indoor pollution have been published, the health effect of indoor $NO_2$ emitted from heaters is still not clear. A common fault with these reports was that the quantification of the exposure to $NO_2$ was not sufficient. In this study, we selected university students as subjects because they lived in a small area around the university and spent the same amount of time in the same rooms in the university building during the day. After school they were in their apartments, which are usually one room, where they used only one heater, or at the most, two. Because the experimental period was during the pre-examination season, they hardly ever went out of their rooms at night. Thus, parameters other than heaters had little affect on the personal exposure levelsof the students. Using such a study group, we studied the personal exposure to $NO_2$ generated from various heaters and cooking stoves. Urinary hydroxyproline/creatinine ratio (HOP/C) was examined as a biomarker of health effects.

## Subjects and Methods

The subjects studied were 85 university students (76 males and 9 females) who lived within a 1.5 Km circle of the university. There were no sources of heavy air pollution, i.e., factories and roads with heavy traffic in this area. The students attached $NO_2$ filter badges (Filter badges $NO_2$, Advantec) to their chests or collars for 1 week and were asked to write down the length of time they used for heating and cooking. They also reported the kind of heater used. Determination of personal exposure to $NO_2$ was performed by the method of Yanagisawa and Nishimura (7). The concentration which was determined using the personal filter was expressed as the time-weighted average concentration. In order to study the $NO_2$ concentrations in outdoor air, $NO_2$ filter badges were also placed at three points outside of the building on the campus. At the end of the study, urine was collected before lunch in order to avoid dietary fluctuation. The urinary creatine concentration was determined by the method of Ogata and Taguchi (8, 9). Urinary hydroxyproline levels were also determined by the method of Prockop and Udenfriend (10).

## Personal Exposure Levels to Nitrogen Dioxide

The $NO_2$ concentration on the university campus was found to be 13.5 - 13.7 $\mu g/m^3$ during the experimental period. Of the 85 subjects studied, 20 students used oil fan heaters, 17 used kerosene heaters and only one student used a gas fan heater. Forty-seven students used electric air conditioners, electric heaters or electric hot carpets. These electric appliances were lumped together as clean heaters. There were no significant differences of exposure levels between the smokers and the non-smokers (Table 1). Table 2 compares the measured $NO_2$ levels and the Japanese Ambient Air Quality Standard which is formulated as: "the daily average of hourly values shall be within the range of 75.2 to 112.8 $\mu g/m^3$ (40 to 60 ppb) or below".

### Table 1  The effect of smoking on personal exposure

| Subject | Smoking | $NO_2$ concentration (ug/m$^3$) | | |
|---------|---------|---|------|------|
| | | N | Mean | S.D. |
| All subjects | Smoker | 21 | 55.7 | 63.3 |
| | Non-smoker | 64 | 51.7 | 59.0 |
| Clean heaters* | Smoker | 11 | 21.8 | 8.1 |
| | Non-smoker | 31 | 20.5 | 7.0 |

*Subjects who used clean heaters

**Table 2  Comparison of the personal exposure with the Japanese Ambient Air Quality Standard (75.2 - 112.8 $\mu g/m^3$)**

| Heater | Personal exposure to nitrogen dioxide | | | | | |
|---|---|---|---|---|---|---|
| | <75.2 $\mu g/m^3$ | | 75.2 - 112.8 $\mu g/m^3$ | | >112.8 $\mu g/m^3$ | |
| | N | (%) | N | (%) | N | (%) |
| Clean heater | 47 | (100.0) | 0 | (0.0) | 0 | (0.0) |
| Oil fan heater | 7 | (35.0) | 3 | (15.0) | 10 | (50.0) |
| Kerosene heater | 13 | (76.4) | 2 | (11.8) | 2 | (11.8) |
| Gas fan heater | 0 | (0.0) | 1 | (100.0) | 0 | (0.0) |

## Heating Time and Personal Exposure

Figure 1 shows the relationship between the heating periods and the personal exposure to $NO_2$ is shown in . The use of clean heaters did not increase the personal exposure to $NO_2$. In the case of students who used kerosene heaters or oil fan heaters, the personal exposure to $NO_2$ increased according to the length of time the heaters was used. As the total study period was 168 hours (24 hrs x 7 days), the average concentration of $NO_2$ can be calculated by substituting 168 hrs for x (see Fig. 1). The average concentration in the room was foud to be 219 $\mu g/m^3$ and 474 $\mu g/m^3$ when kerosine heaters and oil fan heaters were burning, respectively. Dassen et al. (11) studied the $NO_2$ exposure of 60 children and their mothers. Pesonal exposures to $NO_2$ of mothers and children and the $NO_2$ concentrations in the living room, kitchen, bedroom and outside the house were measured by means of filter badges. The mothers were also asked to write down the time they had used the heaters. In addition they kept diaries of their activities and those of their children. The exposure was calculated from the summation of the concentrations and the time which was spent by the mother and child in each room (the time weighted $NO_2$ concentration). They concluded that the exposures calculated from the time weighted $NO_2$ concentrations correlated very well with the measured personal exposures. However, they also reported that the time weighted concentrations differed significantly from the measured concentrations. In Dassen's study, the average concentration in each room during the study was measured regardless of whether the heater was on or off.

## Cooking Stove and Personal Exposure

The effect of cooking stoves on the personal exposure was also studied (Figure 2). All of the students used LNG (Liquid natural gas) fueled ranges. In the case of the electric stove, we did not count the time of usage because the electric stove does not emit $NO_2$. The correlation between the period of cooking and personal exposure levels was observed ($p<0.01$). From the equation of the regression line, the average exposure concentration during cooking was found to be 290 $\mu g/m^3$. Usually people turn on a heater or a stove when they are going to stay in the room or to cook in the kitchen. Wade et al. (12) found that in 5 minutes, the $NO_2$ concentration was up to 1,430 ug/m$^3$ in the kitchen.

## Urinary Hydroxyproline Level

The effect of smoking on the urinary HOP/C was examined. The urinary HOP/C

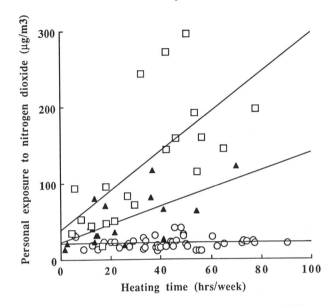

□ Oil fan heater     y = 2.59x + 37.5   r = 0.67

▲ Kerosene heater    y = 1.17x + 22.4   r = 0.66

○ Clean heater       y = 0.005x + 21.1   r = 0.00

Figure 1. The relation between exposure time and personal $NO_2$ exposure by heaters.

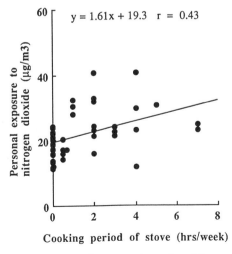

y = 1.61x + 19.3   r = 0.43

Figure 2. The relation between the usage time of cooking stove and personal $NO_2$ exposure

levels of smokers and non-smokers were not significantly different (Table 3). Table 4 shows the relationship between the personal exposure and urinary HOP/C. No significant difference was observed. In this study, personal exposure levels ranged from 11.3 to 295 $\mu g/m^3$. The concentration of $NO_2$ in a room where an oil fan heater was burning was estimated to be 474 $\mu g/m^3$. The occurrence of some respiratory disorders by $NO_2$ was suspected from such a high $NO_2$ exposure, but, no subjects of our study who were exposed to such high levels of $NO_2$ complained of respiratory symptoms.

Table 3   The effect of smoking on urinary hydroxyproline levels

| Smoking | N | HOP/C ($\mu mol/g$ creatinine) Mean | S.D. |
|---|---|---|---|
| Smoker | 20 | 117.9 | 33.8 |
| Non-smoker | 61 | 103.5 | 32.6 |

Table 4 Urinary hydroxyproline levels by personal exposure to $NO_2$

| $NO_2$ exposure | N | HOP/C ($\mu mol/g$ creatinine) Mean | S.D. |
|---|---|---|---|
| Less than 20 mg/m$^3$ | 21 | 119.9 | 37.5 |
| 20 to 50 mg/m$^3$ | 37 | 104.0 | 26.8 |
| Above 50 mg/m$^3$ | 22 | 105.0 | 30.5 |

References

1.  Azizi, B. H. O.; Henry, R. L. *Pediatr. Pulmonol.* **1990,** 9, 24-29
2.  Berkey, C. S.; Ware, J. H.; Dockery, D.W.; Ferris, B. G. Jr.; Speizer, F. E. *Am. J. Epidemiol.* **1986,** 123, 250-260
3.  Dijkstra, L.; Houthuijs, D.; Brunekreef, B.; Akkerman, I.; Boleij, S. M. *Am. Rev. Respir. Dis.* **1990,** 142, 1172-1178.
4.  Hoek, G.; Brunekreef, B.; Meijer, R.; Scholten, A.; Boleij, J. *Int. Arch. Occup. Environ. Health* **1984,** 55, 79-86
5.  Jones, J. R.; Higgins, I. T. T.; Higgins, M. W.; Keller, J. B. *Arch. Environ. Health* **1983,** 38, 219-222
6.  Koo, L.C.; Ho, J. H.; Ho, C.; Matsuki, H.; Shimizu, H.; Mori, T.; Tominaga, S. *Am. Rev. Respir. Dis.* **1990,** 141, 1119-1126.
7.  Yanagisawa, Y.; Nishimura, H. *J. Japan Soc. Air. Pollut.* **1980,** 15:316-323
8.  Ogata, M.; Taguchi, T. *Ind. Health* **1987,** 25, 103-112
9.  Ogata, M.; Taguchi, T. *Int. Arch. Occup. Environ. Health* **1988,** 61, 131-140
10. Prockop, D. J.; Udenfriend, S. *Anal. Biochem.* **1960,** 1, 228-239.
11. Dassen, W. G.; Matsuki, H.; Kasuga, H.; Misawa, K.; Yokoyama, H.; Shimizu, Y. *Tokai J. Exp. Clin. Med.* **1987,** 12, 83-95
12. Wade, III. W.A.; Cote, W. A.; Yocom, J. E. *J. Air Pollut. Control Assoc.* **1975,** 5, 933-939

# BIOMARKERS

# Chapter 16

# Biomarkers of Inherited and Acquired Susceptibility to Toxic Substances

## E. Nieboer

Department of Biochemistry and Occupational Health Program, McMaster University, 1200 Main Street West, Hamilton, Ontario L8N 3Z5, Canada

A brief overview is presented of genetic testing within the context of molecular epidemiology. Results are outlined for three pertinent studies completed in the author's laboratory: (*1*) the relationship of gastrointestinal absorption of aluminum to Alzheimer's disease and the allelic risk factor apolipoprotein E-∈4 for the latter; (*2*) inherent metallothionein induction capacities in peripheral-blood mononuclear leucocytes of non-smokers; and (*3*) characterization of nickel-induced mutations to illustrate compound-specific, acquired genetic damage. Potential applications in genetic screening (i.e., the detection of inherited susceptibility/protection factors) and genetic monitoring (i.e., determination of acquired changes in an individual's genetic material) are discussed.

Genetic testing may be subdivided into genetic screening and genetic monitoring (*1,2*). The emphasis in genetic screening is to detect inherited traits that predispose an individual to a pathologic effect if exposed to toxic agents (*1-3*). By contrast, genetic monitoring involves periodic examination of individuals to evaluate modifications within the genetic material that evolve over time, perhaps due to occupational or environmental exposures (*1,2*). Chromosomal damage or increases in DNA mutations might be the measured outcome (*4*). Within the context of the human genome project (*5*), with the goal of mapping and sequencing the entire genome, the application of genetic testing technologies in assessing the impact of exposures to both genotoxic and non-genotoxic agents may be expected to become routine. The objective of the current presentation is to share the result of three studies in the author's laboratory that illustrate this expectation.

EXPERIMENTAL CONSIDERATIONS

**GI Absorption of Aluminum in Healthy Volunteers and Alzheimer's Patients**

A critical review of the available literature suggests that age, Alzheimer's Disease (AD) and renal disease appear to enhance the gastrointestinal absorption of aluminum (6). Increased absorption with age and in patients with AD who were 76 years or younger was reported by Taylor et al. (1992). They assessed the increase in serum aluminum 60 minutes after administering an oral drink of aluminum citrate. On the basis of these findings, we hypothesized that the presence of the apoE4 allele (apolipoprotein, genetic variant ε4) might correlate with an increased capacity to absorb aluminum from the GI tract. The apoE4 allele appears to constitute an important risk factor for AD (7,8).

Serum samples for aluminum analysis (by electrothermal atomic absorption) were obtained from 20 healthy volunteers (age 15-59), 10 probable AD patients (age 64-84) and 7 healthy age-matched controls. For the AD patients, a blood sample was also taken and sent to the IMAGE Projet for apoE4 genotyping (see Acknowledgements). Immediately after these collections, 300 mg of aluminum (13.5 mL of aluminum hydroxide as Amphojel suspension) and 4.8 g of citrate in 100 mL of diluted lemonaid were ingested. One additional tube of blood for serum aluminum analysis was drawn 90 minutes after the ingestion of the aluminum/citrate solutions. Although a full statistical analysis of the data has not been completed at the time of writing, a number of tentative observations are rendered. (i) All aluminum levels before the Amphojel/citrate ingestion are in the accepted reference interval of 1-6 µg/L. (ii) The age-dependence of GI absorption of aluminum is not readily apparent. (iii) About 20% of all donors could clearly be considered "high absorbers", with final serum aluminum levels exceeding 150 µg/L, and this is independent of age. (iv) Six of the ten probable AD patients has at least one copy of the apoE4 allele, with a prevalence of 35%. (v) Even though the number of AD patients is small, there is no apparent link between aluminum absorption and the presence of apoE4 (only 1/10 had a final serum aluminum level > 150 µg/L). Our preliminary findings favour the interpretation that the weak epidemiologic link observed between AD and aluminum intake from drinking water (6) may well reflect a positive bias in the clinical diagnosis of AD. The latter may be due to independent neurologic symptoms arising from enhanced aluminum uptake in a significant proportion of individuals of AD age. One might also speculate whether the "high absorber" phenotype has a genetic component.

**Metallothionein Induction in Peripheral-Blood Mononuclear Leucocytes of Non-smokers and Smokers**

Besides roles of metallothionein (MT) in essential metal metabolism (10), it is generally accepted that elevated levels of MT can protect cells from acute exposure to metals (especially zinc and cadmium) or even organic species (including reactive electrophilic agents) (*11*). Since MT is an inducible protein, a pertinent question is whether inter-individual differences exist in MT expression. In 1989, we observed that MT induction capacity varied considerably in cells from 39 healthy, non-smoking volunteers aged 20-40 years (*9*). This was achieved by measuring both MT protein and MT messenger RNA (MT mRNA) in peripheral-blood mononuclear leucocytes (about 85% lymphocytes and 15% monocytes) before and after exposure for 12 or 6 hours to 10 $\mu$M $CdCl_2$ in culture. While the induction capacity of MT mRNA, defined as the ratio of induced levels observed in the *in vitro* incubation to constitutive (basal) levels, varied about 10-fold, the corresponding factor for MT protein was 2.6.

Recently, we have extended these studies to 20 smokers, aged 19-25. As before, basal levels were compared with *in vitro* concentrations induced during 16 hours with either 10 $\mu$M $CdCl_2$ or 125 $\mu$M $ZnSO_4$. The MT-protein induction capacity using $CdCl_2$ varied 2.7-fold in the smokers, compared to 1.7-fold for the pooled variation in 2 non-smokers whose mononuclear cells were sampled 13 times over a 4-month period. Differences in variances between the two induction-capacity categories were statistically significant according to the two-tailed F-test (p< 0.025). For MT induction in mononuclear cells with zinc, the corresponding differences in induction capacities were not significant (a 2.0-fold range for the smokers and 1.5-fold for the intra-individual comparison). The interpersonal variation in basal, Cd-induced and Zn-induced MT concentrations ranged one order of magnitude. It is concluded that smoking does not appear to affect constitutive MT levels nor the MT-induction capacity in human peripheral-blood mononuclear cells.

**Characterization of Nickel-Induced Mutations in AS52 Cells**

The evidence that nickel compounds are human carcinogens may be considered to be unequivocal for industrial exposure to water-soluble salts (e.g., nickel chloride or sulphate) and pyrometallurgical intermediates (mostly oxides and sulphides) (*12,13*). The relative carcinogenic potential of different nickel compounds or species has been extensively debated. In a recent publication (14), we have shown in culture using AS52 Chinese hamster ovary cells in a colony-forming assay that to give 50% survival, a 75-fold range in administered dose was required for three water-soluble salts (chloride, sulphate, acetate) and eight relatively insoluble particulate compounds (carbonate, hydroxide, oxides and sulphides). This compared to a 10-fold range in measured cytoplasmic and nuclear "dissolved" nickel concentrations in the same cells. The particulate compounds, taken up primarily by phagocytosis, raised the nickel levels in both intracellular compartments, while the water-soluble compounds preferentially increased the cytoplasmic levels. Most of the nickel compounds tested were shown to be weakly mutagenic when tested for mutations at the guanine-hypoxanthine phosphoribosyl transferase (gpt) gene. Some of the nickel-induced mutations have recently been characterized after clonal expansion of mutant cells and DNA amplification by the polymerase chain reaction (*15*). When compared to spontaneous or ethylmethane sulphonate (EMS) induced mutations, those generated by exposure to nickel compounds exhibited an increase in gene deletions relative to point mutations; the extent of which was compound specific: $Ni(SO_4) > Ni(OH)_2 > Ni_3S_2$. These results clearly suggest that more than a single genotoxicological mechanism is involved in the mutagenic activity of nickel compounds. Differences in intracellular compartmentalization are believed to be important.

## CONCLUDING REMARKS

A genetic-environmental model is the best etiological paradigm for AD (*6*). Multi-gene loci and a number of environmental factors appear to be involved. Linkages to loci on chromosome 1, 14 and 21 (early onset AD), and chromosome 19 (late onset) have been established; the apoE gene is located on the latter. Although more AD patients need to be studied, the apoE4 genotyping results reported here do not suggest an inherited susceptibility involving this locus with respect to the GI absorption of aluminum. Nevertheless, the identification of a significant proportion of high absorbers may well have toxicological implications since aluminum is known to be neurotoxic.

The 10-fold variation observed in basal and induced MT levels, as well as the 2.7-fold difference in induction capacity, likely reflect both environmental and experimental factors, and possibly genetic (inter-individual) differences. Since the variations in induction capacities were comparable in smokers and non-smokers, and a relationship between constitutive MT levels and number of cigarettes smoked was not evident, it appears that smoking was not a significant environmental factor in the young people examined. [On average, they smoked $12 \pm 7$ cigarettes per day; 30 was the highest]. Presumably, the exposure to Cd via tobacco smoke did not alter the effective cadmium concentration enough to affect the basal MT levels in the mononuclear cells. This is an important observation and facilitates the application of the biological monitoring techniques described in assessing environmental exposures to cadmium and mercury or perhaps in an evaluation of inherent susceptibilities/protection of exposed individuals to/from toxic effects. Reduced fertility reported in female dental assistants exposed to mercury vapour (*16*) might be a pertinent example.

The mutant profile analysis reviewed suggests that induction of mutations at levels only slightly above background may indeed be indicative of the mutagenic potential of nickel compounds. Molecular biology techniques can thus serve an important complementary role to more traditional mutagenicity testing. The results reported also illustrate that mutagenic profiles may well be compound specific. Future studies are needed to assess whether the experimental approach and findings described can be extended to assessments of compound-specific acquired genetic damage in humans.

## ACKNOWLEDGMENTS

The following colleagues participated in the study describing GI absorption of aluminum in healthy volunteers and Alzheimer's patients: J.S. Reiach (Honours Biology and Pharmacology Program, Department of Biology), and B.L. Gibson, M.D. (Department of Clinical Epidemiology and Biostatistics), McMaster University; D.W. Molloy, M.D. and D. Strang, M.D., Geriatric Research Group, Hamilton Civic Hospitals, Henderson General Division, Hamilton, Ont; and D. Gauvreau, Ph.D. and C. Bétard, Ph.D., Project IMAGE/Recherche Alzheimer, Centre Hospitalier Côte-des-Neiges, Montreal, Que. My co-investigators for the metallothionein induction work were: C.B. Harley, Ph.D. (Department of Biochemistry), P.S. Bhoi (Honours Biochemistry Program) and G.P. Doherty (Arts and Science Program), McMaster University. Co-workers in the characterization of nickel-induced mutations in AS52 cells were: F.E. Rossetto, Ph.D. and G.G. Fletcher, M.Sc. (Department of Biochemistry) and J.D. Turnbull, Ph.D., M.D. (Department of Medicine), McMaster University.

Financial assistance for the work described is gratefully acknowledged from a number of government agencies (NSERC, MRC, and the Ontario Workplace Health and Safety Agency), McMaster University (EcoResearch Chair's Program in Environmental Health) and industry (The Nickel Producers Environmental Research Association, Durham, N.C., and the International Lead Zinc Research Organization, Research Triangle Park, N.C.).

## LITERATURE CITED

1. Nieboer, E.; Rossetto, F.E.; Turnbull, J.D. *Toxicol. Lett.* 1992; *64/65*, 25-32.
2. Office of Technology Assessment. *Summary, Genetic Monitoring and Screening in the Workplace*, Report OTA-BA-456, Washington: Congress of the United States, 1990.
3. Calabrese, E.J. *J. Occup. Med.* 1986; *28*, 1096-1102.
4. Perera, F.P. *JNCI* 1986; *78*, 887-898.
5. Federation of American Societies of Experimental Biology. *FASEB J.* 1991; *5(1)*, 1-78.
6. Nieboer, E.; Gibson, B.L.; Oxman, A.D.; Kramer, J.R. *Environ. Rev.* 1995; *3*, 29-81.
7. Corder, E.H.; Saunders, A.M.; Strittmatter, W.J.; Schmechel, D.E.; Gaskell, P.C.; Small, G.W.; Roses, A.D.; Haines, J.L.; Pericak-Vance, M.A. *Science* 1993; *261*, 921-923.

8. Corder, E.H.; Saunders, A.M.; Risch, N.J.; Strittmatter, W.J.; Schemechel, D.E.; Gaskell, Jr. P.C.; Rimmler, J.B.; Locke, P.A.; Conneally, P.M.; Schmader, K.E.; Small, G.W.; Roses, A.D.; Haines, J.L.; Pericak-Vance, M.A. *Nature Genet.* 1994; *7*, 180-184.

9. Harley, C.B.; Menon, C.R.; Rachubinski, R.A.; Nieboer, E. *Biochem. J.* 1989; *262*, 873-879.

10. DaSilva, J.J.R.F.; Williams, R.J.P. *The Biological Chemistry of the Elements*, Clarendon Press, Oxford, UK, 1991.

11. Pattanaik, A.; Shaw, III C.F.; Petering, D.H.; Garvey, J.; Kraker, A.J. *J. Inorg. Biochem.* 1994; *54*, 91-105.

12. Doll, R. *Scand. J.Work Environ. Health* 1990; *16*, 1-82.

13. IARC Monographs on the Evaluation of Carcinogenic Risks to Humans. Vol. 49, *Chromium, Nickel and Welding.* International Agency for Research on Cancer, Lyon, France, 1990.

14. Fletcher, G.G.; Rossetto, F.E.; Turnbull, J.D.; Nieboer, E. *Environ. Health Perspect.* 1994; *102* (Suppl. 3), 69-79.

15. Rossetto, F.E.; Turnbull, J.D.; Nieboer, E. *Sci. Total Environ.* 1994; *148*, 201-206.

16. Rowland, A.S.; Baird, D.D.; Weinberg, C.R.; Shore, P.L.; Shy, C.M.; Wilcox, A.J. *Occup. Environ. Med.* 1994; *51*, 28-34.

Chapter 17

# Biological Monitoring and Genetic Polymorphism

Toshihiro Kawamoto and Yasushi Kodama

Department of Environmental Health, University of Occupational
and Environmental Health, P.O. Orio, Yahata nishi ku,
Kitakyushu 807, Japan

This study focused on clarifying the effects of the genetic
polymorphism of ALDH2 (low $K_m$ aldehyde dehydrogenase) on
toluene metabolism. The study subjects were 92 male toluene workers
and 234 non-exposed men. The genotype of ALDH2 was classified
into the homozygous genotype of normal ALDH2 (NN), the
homozygous genotype of an inactive one (DD) and the heterozygous
genotype of normal and inactive ones (ND) by means of PCR-RFLP.
The personal exposure levels to toluene were monitored using diffusion
type samplers and the measurement of urinary hippuric acid (HA) and
creatinine concentrations. In the toluene workers, positive correlations
between the personal exposure to toluene and the urinary HA levels
were observed in the NN, ND and DD groups in comparison to the
non-exposed groups. The urinary HA levels of the DD and ND groups
were significantly lower than that of the NN group.

Toluene is one of the most widely used organic solvents in the world. The exposure
levels of workers to toluene are estimated by their urinary hippuric acid (HA) level,
the main metabolite of toluene (1). In developed countries the measurement of
urinary HA level is one of the most important items during the medical check up of
toluene workers. Especially in Japan, employers must have urinary HA of toluene
workers measured every 6 month by law. Approximately 80 % of the absorbed
toluene is first converted to benzyl alcohol by the microsomal mixed function oxidase
system, then oxidized to benzoic acid by the alcohol dehydrogenase/aldehyde
dehydrogenase system, and finally conjugated with glycine to form HA (2). There
are three main isozymes of aldehyde dehydrogenase (ALDH, EC: 1.2.1.3), namely,
ALDH1 (high $K_m$ ALDH, of cytosolic origin), ALDH2 (low $K_m$ ALDH, of
mitochondrial origin) and ALDH3. It is believed that ALDH2 oxidizes most of the
acetaldehyde derived from the metabolism of alcohol. Genetic polymorphism in
ALDH2 activity is seen in Japanese and other Mongoloids, as well as in American

Indians. About half of the Japanese lack ALDH2 activity and experience facial flushing, tachycardia, nausea and hypotension owing to their low capability to metabolize acetaldehyde when they drink alcohol (*3*). The sequences of cDNA and genomic DNA of ALDH2 have been clarified (*4*). The catalytic deficiency is caused by a structural point mutation at the amino acid position 487 of the polypeptide subunit of the ALDH2 gene. At this position, a substitution of Glu to Lys resulting from the transition of G(C) to A(T) at DNA levels occurs (*5*). It is suspected that the metabolism of toluene in ALDH2-deficient workers is different from that in normal ALDH2 workers.

## Subjects and Methods

The study subjects were 92 male workers (ages 18-66 years) who handled toluene in a printing factory, an electric parts factory, and a painting workplace and 234 male university students and workers who were not exposed to toluene. Peripheral venous blood samples (7 ml) were obtained from all these individuals. Genomic DNA was isolated by the standard phenol/chloroform extraction method using a DNA extractor (Applied Biosystems Inc, model-340A) after complete digestion with 100 μg/ml of proteinase K (Wako Pure Chemical). The genotypes of ALDH2 were determined by artificial restriction fragment length polymorphism using *MboII* (*6*). The details were already reported by Kawamoto *et al.* (*7*).

The breathing-zone personal exposure levels to toluene were monitored a diffusion type sampler (Dupont Protek, G-AA). The workers attached the personal samplers to their chests or collars from the beginning to the end of their work shifts. The solvent collected on the sampler was desorbed in carbon disulfide (1.5 ml) overnight and quantified by gas chromatography with an frame ionization detector. Urine was collected in the afternoon from 3:00-5:00 PM. All samples were stored at -20°C until analysis. The urinary hippuric acid and creatinine concentration were determined by a high performance liquid chromatography (*8*). The measured HA concentration was corrected for creatinine.

The differences in the slopes and intercepts of the regression lines were examined by the analysis of Armitage and Berry (*9, 10*).

## Results

Fig. 1 shows the relationship between the personal exposure to toluene and the urinary HA by the genotypes of ALDH2. Positive correlations were observed in the NN, ND and DD groups. The 95% ranges of the slopes and intercepts for the regression lines, as well as the correlation coefficients, are shown in Table 1. The slopes of the regression lines decreased from NN, to ND to DD in this order. The intercepts for the regression lines were not significantly different for ALDH2 genotypes.

## Discussion

Benzoic acid, which is a precursor of hippuric acid, is contained in cinnamon, strawberry, etc. (*11*) and some beverages (*12*). However, the dietary variation of urinary hippuric acid is too small in comparison with the increase of urinary hippuric acid by toluene exposure (above 1 g/g creatinine) (*10, 13*). Kawamoto *et al.* (*14*) reported that the distribution of urinary hippuric acid measurements in the DD group were significantly different from those in both the NN and ND groups from a study of 890 urine samples from 253 toluene workers. Therefore, it can be said that the genetic polymorphism of ALDH2 should be taken into consideration when the biological monitoring of toluene is done during the medical check-up.

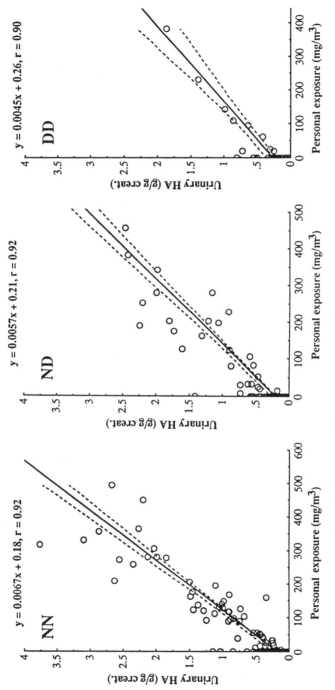

Fig. 1   Relationship between personal exposure to toluene and urinary hippuric acid by ALDH2 genotypes (The dotted curves mean the 95% confidence range for the mean hippuric acid.)

Table 1  Regression line between personal exposure level to toluene and urinary hippuric acid concentration by the genotypes of ALDH2, and smoking and drinking habits

| ALDH2 | Number of subjects | | Personal exposure | | Correction for Creatinine | | |
|---|---|---|---|---|---|---|---|
| | Exposed | Nonexposed | G.M.[1] | G.S.D.[2] | $\alpha$ (x 0.001) | $\beta$ (x 0.001) | r |
| NN | 55 | 116 | 85.8 | 3.30 | 6.7 ** (6.3 - 7.2) | 177 (131 - 222) | 0.92 |
| ND | 28 | 98 | 94.4 | 3.77 | 5.7 ** (5.2 - 6.1) | 209 (164 - 255) | 0.92 |
| DD | 9 | 17 | 74.2 | 3.00 | 4.5 (3.5 - 5.4) | 257 (168 - 346) | 0.90 |

$\alpha$ and $\beta$ are parameters of calculated regression lines of $y = \alpha x + \beta$, where x is the personal exposure concentration to toluene (in mg/m$^3$) and y is hippuric acid concentration in urine (g/g creat. for values corrected for creatinine concentration).

**P<0.01; All correlation coefficients are statistically significant (P<0.01)

[1]Geometric mean, [2]Geometric standard deviation

The biotransformation of toluene to hippuric acid is catalyzed by several enzymes, that is, cytochrome P-450, alcohol dehydrogenase (ADH), ALDH, and so on. Though we studied only the polymorphism of ALDH2, it is well known that some isozymes of cytochrome P-450 and ADH, that is, CYP1A1 (*15*), CYP2D6 (*16*), ADH2 (*17*) and ADH3 (*18*), have polymorphisms. We would like to examine the genotypes of those genes besides ALDH2.

## References

1.  Ogata, M.; Taguchi, T. *Int Arch Occup Environ Health* **1986,** *58,* 121-129.
2.  International Programme on Chemical Safety. *Environmental health riteria 52, Toluene;* World Health Organization: Geneva, 1985; 44-55
3.  Harada, S. *Electrophoresis* **1989,** *10,* 652-655.
4.  Hsu, L.C.; Tani, K.; Fujiyoshi, T.; Kurachi, K.; Yoshida, A. *Proc. Natl. Acad. Sci.* **1985,** *82,* 3771-3775.
5.  Crabb, D.W.; Edenberg, H.J.; Bosron, W.F.; Li, T.K. *J. Clin. Invest.* **1989,** *83,* 314-316.
6.  Harada S. *Alcohol and Alcoholism* **1993,** *Suppl 1A,* 11-13.
7.  Kawamoto, T.; Matsuno, K.; Kodama, Y.; Murata, K.; Matsuda, S. *Arch. Environ. Health* **1994,** *49,* 332-336.
8.  Ogata, M.; Taguchi, T. *Int. Arch. Occup. Environ. Health* **1988,** *61,* 131-140.
9.  Armitage, P.; Berry, G. *Statistical Methods in Medical Research;* 2nd ed; Blackwell Sci.: Oxford, 1987.
10. Kawamoto, T.; Koga, M.; Murata, K.; Matsuda, S.; Kodama, Y. *Toxicol. Appl. Pharmacol.* **1995,** *133,* 295-304.
11. Nagayama, T.; Nishijima, M.; Yasuda, K.; Saito, K.; Kamimura, H.; Ibe, A.; Ushiyama, H.; Naio, Y.; Nishima, T. *Food Hyg. Soc. J.* **1986,** *27,* 316-325.
12. Michitsuji, H.; Ohara, A.; Yamaguchi, K.; Fujiki, Y. *Matsushita Med. J.* **1987,** *26,* 106-116. (ISSN 0388-3734)
13. Kawamoto, T.; Koga, M.; Oyama, T.; Kodama, Y. *Arch. Environ. Contam. Toxicol.* **1996,** *30,* 114-120.
14. Kawamoto, T.; Murata, K.; Koga, M.; Hattori, Y.; Kodama, Y. *Occup. Environ. Med.* **1994,** *51,* 817-821.
15. Hayashi, S.; Watanabe, J.; Kawajiri, K. *Jpn. J. Cancer Res.* **1992,** *83,* 866-870.
16. Gough, A.C.; Miles, J.S.; Spurr, N.K. *Nature* **1990,** *347,* 773-776.
17. Yin, S.J.; Bosron, W.F.; Magnes, L.J.; Li, T.K. *Biochem.* **1984,** *23,* 5847-5853.
18. Bosron, W.F.; Magnes, L.J.; Li, T.K. *Biochem.* **1983,** *22,* 1852-1857.

Chapter 18

# Biomarkers Used for Assessment of Cancer Risk in Populations Living in Areas of High Potential Carcinogenic Hazard (Near Communal and Chemical Waste Dumping Sites)

J. A. Indulski and W. Lutz

Nofer Institute of Occupational Medicine, 8 Southwest Teresy Street, 90–950 Lodz, Poland

Biological monitoring of people exposed to environmental carcinogens using biomarkers provides measurable information on the intensity of absorption of specific carcinogens and the possibility to determine whether the agent is absorbed by the organism in a critical dose to produce specified health effects. Biomarkers enable the recording of very early health effects produced by environmental carcinogens and to determine cancer risk. Biomarkers also permit enhanced analysis of health risk in humans exposed to carcinogens and, because determinations are performed directly in human organism, uncertainties inherent in epidemiologic studies are avoided. A range of biomarkers are available enabling early detection of the disease and of the phenomena which precede it. It is now possible to assess cancer risk by laboratory test results.

The assessment of environmental changes caused by the emission of carcinogenic chemicals from industrial plants and waste dumping sites is usually effected by the determination of the concentrations of the individual chemicals in air, water and soil. Unfortunately, it is not easy to correlate the information obtained in this way with cancer risk in the exposed inhabitants of the chemically polluted areas. Such information fails to account for such essential factors as pollutant absorption rate of the human organism or individual human susceptibility to the harmful agents.

The measures of a hazard to an individual from exposure to environmental carcinogens are biomarkers, which are indices of abnormalities within different biological systems (at the level of the whole body or of particular organs, tissues or cells) that may be induced by carcinogens from various chemical or biological sources. Depending on the type of processes which accompany the effect of a specified carcinogenic agent on the human organism it may be determined with the aid of the biomarkers, which are then classified as exposure, effect and susceptibility biomarkers (1, 2, 3).

Biomarkers of exposure indicate the dose absorbed by quantitatively determining the carcinogenic agent (or its metabolites) absorbed, or the products of its interaction with endogenic substances (such as DNA and proteins). These are measurable indicators which appear in the organism after exposure to environmental carcinogens.

Biomarkers of effect indicate a measurable biochemical, physiological, behavioral or other alteration within an organism that, depending upon the magnitude, can be recognized as associated with an established or possible health impairment or disease.

Biomarkers of susceptibility are indicators of an inherent or acquired ability of an organism to respond to the challenge of exposure to environmental carcinogens. They may be used as measurable indicators of those biological factors of the organism which are present before exposure and are genetically related or display the features of acquired traits. The latter may result from diseases suffered in the past or from previous exposure to harmful (chemical, physical or biological) agents.

Biomarkers supply measurable information on the intensity of the carcinogenic environmental pollutant absorption and provide data on whether a carcinogen penetrates to a critical organ at a critical dose to produce changes at the molecular or subcellular level. The character of those changes is determined both by the type of the carcinogen, its quantity which has penetrated to the affected site, and the duration of the exposure. Biomarkers offer the possibility for early detection of the pathogenic chain phenomena, usually impossible to detect by other contemporary diagnostic methods *(4)*.

The application of biomarkers to monitor the carcinogenic effects of the environmental chemical pollutants on human organism, especially those which enable detection of the earliest carcinogenesis stages, demonstrates that their use is associated with the detection of the changes which precede overt clinical cancer stage rather than with the detection of the neoplastic disease itself. That is why the implementation of the specific biomarkers is so enormously important for the monitoring of the neoplastic process. The carcinogenic process comprises many stages, and lasts for several years. It is extremely important that early detection of the changes which start the process of the transformation of the cells from normal to the neoplastic ones offers the possibility for effective intervention when the changes are reversible so that it is possible to stop or retard cancer development *(5, 6)*.

Biomarkers of Exposure

The majority of the biomarkers which can be used to assess exposures to environmental carcinogens belong to the selective biomarker category. As a rule, a carcinogen is determined in the biological material in a non-modified form at the time when it has not yet been markedly biotransformed and when the exposure is still at a very low level or when a high degree of specificity is required *(7)*. Determination of native environmental carcinogens in the biological material is seriously hampered by the fact that the majority of the carcinogens, after they have been absorbed by the organism, are metabolized to water-soluble substances which can be easily removed to urine or bile. Consequently, exposure to those carcinogens is estimated by assessing the specific metabolites in urine. It is very important that the metabolites can be determined by a non-invasive method, which is relatively easy to accept by the patients (Table I).

Table I. Examples of carcinogenic chemicals (for man and/or animal) for which there is some basis for exposure tests involving measurements in biological media

| Chemical | Substance measured |
|---|---|
| Acrylonitrile | Acrylonitrile in urine, isothiocyanates in urine |
| Aflatoxin | Aflatoxin in urine |
| 4-Aminobiphenyl | 4-Aminobiphenyl in urine |
| Arsenic (inorganic) | Sum of inorganic arsenic, mono-methylarsenic acid, cacodylic acid in urine |
| Asbestos | Asbestos bodies in sputum |
| Benzene | Phenol in urine |
| Benzidine | Benzidine and N,N'-diacetylbenzidine in urine |
| Chromium | Chromium in urine, chromium (VI) bound to erythrocyte proteins |
| 4,4'-diaminodiphenyl-methane | Methylene dianiline in urine |
| 3,3'-dichlorobenzidine | 3,3'-dichlorobenzidine and N-acetyl-3,3'-dichlorobenzidine in urine |
| 4,4'-methylene-bis-(2-chloroaniline) | 4,4'-methylene-bis (2-chloroaniline) in urine |
| 2-naphtylamine | 2-naphtylamine in urine |
| Polycyclic aromatic hydrocarbons | 2-hydroxypyrene in urine |
| Vinyl chloride | Thiodiglycolic acid in urine |
| Bischloroethyl nitro-urea | Bischloroethyl nitrourea in plasma |
| 2,6-dinitrotoluene | 2,6-dinitrobenzoic acid in urine |
| Nickel | Nickel in urine and plasma |

In spite of its several inherent disadvantages, reference should be made to mutagenicity tests, especially bacterial mutagenicity tests, which may be considered to be valuable screening tests in a set of methods used to determine potentially mutagenic and/or carcinogenic effects of chemicals on populations living in the vicinity of chemical plants or waste dumping sites. Mutagenicity assessment using bacteria makes it possible to determine in a relatively simple way whether or not the chemicals present in the analyzed biological liquid (usually urine) are mutagenic to the bacteria. This test cannot be used to determine whether the chemicals present in urine are mutagenic or carcinogenic also to humans. Yet the positive result of the test makes it possible to assume that the chemicals involved may be considered to be potentially carcinogenic to humans, unless proof to  the contrary has been obtained *(8)*.

A method is available involving the determination of the adduct formed by the carcinogens (or their metabolites) with cellular macromolecules, such as DNA and proteins, which is very useful in assessing exposures to environmental carcinogens *(9)*. Adducts formed with blood proteins and DNA may represent only a small fraction of the total internal dose but, because they have a long half-life in the body (relative to exposure frequency), may accumulate to detectable levels with continued exposure.

Considering the results of recent research, feasibility of using DNA adducts as biomarkers to detect and assess the extent of the exposure to environmental carcinogens seems undeniable. There are, however, some uncertainties in the interpretation of the results for the assessment of cancer risk. In that case, the adducts are not considered as exposure biomarkers but rather as the biomarkers of early health effects.

Subject to some reservations, assessment of DNA adducts and of the adducts of some proteins constitutes one of the best, if not the best, biological dosimeter for evaluating exposure to environmental carcinogens. The use of the adducts as the biomarkers of exposure to the environmental carcinogens offers the possibility to determine high cancer risk populations. Assessment of the adducts makes it possible not only to confirm that a patient has been exposed to a specific environmental carcinogen (or environmental carcinogen mixture), but also to indicate that potentially carcinogenic damage to the genetic material in the cells of the people exposed to the carcinogens has taken place *(10)*.

The adducts implemented to assess the exposure to the carcinogenic action of environmental chemical toxins make it possible to evaluate the dose at which the substances have been absorbed by human organism, thereby enabling mathematical determination of cancer risk *(11)*. Furthermore, the implementation of adduct determination makes it possible to detect and identify heretofore unknown environmental mutagens capable of triggering the neoplastic process and to include these in the process of mathematical risk analysis *(12)*. Different adduct levels detected in the exposed and non-exposed populations confirm that adduct determination may constitute a useful tool in epidemiological studies (Table II). Enormous progress in the implementation of new ultrasensitive quantitative techniques of adduct determination has made chemical (molecular) dosimetry determinations almost as sensitive as those of radiation dosimetry. Adduct analysis implemented in the studies on human populations exposed to environmental genotoxic substances constitutes a very promising research tool. At present, we can say that adduct determinations, irrespective of whether they are performed on DNA or proteins, not only enable confirmation of exposure to a specified genotoxic substance, but also point to the occurrence of potentially carcinogenic damages in the genetic material.

Genetic factors seem to play an essential role in the individual assessment of the exposure dose and consequent assessment of cancer risk *(18)*. They may contribute to the increase or decrease in the number of the adducts formed. The confirmation of the relationship between the dose of the genotoxic substance and the number of the determined adducts is thereby made more difficult. The same problem, although much less evident, is present in population studies. Monitoring of the temporary changes of adduct profiles seems to be more useful in the assessment of individual exposures to genotoxic substances.

Table II. DNA adduct levels in workers exposed to PAHs determined by ELISA or by the $^{32}P$ postlabeling

| Source | Method | Occupation/process | No. of subjects Asseyed | Positive | Adducts/10⁸ nucleotides Mean SD | Range | Reference |
|---|---|---|---|---|---|---|---|
| PBL | ELISA | Roofers | 28 | 7 | | 1.2-72 | Shamsuddin et al. 1985 (13) |
| PBL | ELISA | Foundry | 20 | 7 | | 1.8-12 | Shamsuddin et al, 1985 (13) |
| WBC | ELISA | Foundry/high | 4 | 4 | | 24-48 | Perera et al, 1988 (14) |
| WBC | ELISA | Foundry/medium | 13 | 13 | | 3-60 | Perera et al, 1988 (14) |
| WBC | ELISA | Foundry/low | 18 | 13 | | 0.9-25.8 | Perera et al, 1988 (14) |
| WBC | ELISA | Controls | 10 | 2 | | 0.9-9.0 | Perera et al, 1988 (14) |
| WBC | ELISA | Firefighters | 43 | 15 | 56.1+48.6 | | Liou et al, 1989 (15) |
| WCB | ELISA | Controls | 38 | 13 | 34.5+13.8 | | Liou et al, 1989 (15) |
| WBC | $^{32}P$ | Coke oven (Silesia, Poland) | 63 | 63 | 11.6/24.5 | Adduct levels were determined in two different laboratories | Hemminki et al, 1990 (16) |
| WBC | $^{32}P$ | Controls/urban (Silesia, Poland) | 19 | 19 | 10.2/21.1 | | Hemminki et al, 1990 (16) |
| WBC | $^{32}P$ | Controls/rural (population in Eastern Poland) | 15 | 15 | 8.4/4.4 | 0.4-7.1 | Hemminki et al, 1990 (16) |
| WBC | $^{32}P$ | Aluminium | 21 | 21 | 3.08+1.69 | 0.2-2.4 | Schoket et al, 1987 (17) |
| WBC | $^{32}P$ | Controls | 29 | 29 | 1.30+0.53 | | Schoket et al, 1987 (17) |

Microscopically visible damage in the genetic material of peripheral blood cells, such as structural chromosomal aberrations, sister chromatid exchanges (SCE) or micronucleated cells, has been used to monitor exposure to genotoxic compounds for a long time. Workers in various occupations have an increased number of chromosomal changes after extended exposure to suspected carcinogenic compounds. The disadvantages of the cytogenetic approach are the lack of specificity and the uncertain biological relevance of these parameters in relation to cancer development. It has never been proven that there is an increased risk of developing cancer resulting from exposure, although the data from a large prospective study suggest that people with a large number of SCE developed cancer more easily than people with a lower number of SCE *(19)*. However, cytogenetic aberrations observed in peripheral lymphocytes provide presumptive evidence that a genotoxic effect may have occurred in the target organ. Cigarette smoking is one of the exposures that can easily be detected by the cytogenetic tests *(19)*.

Analysis of mutations of the hypoxanthine phosphoribosyltransferase (HPRT) gene in lymphocytes, glycophorin in erythrocytes, and haemoglobin mutants in erythrocytes offers a new approach to detect human exposure to genotoxic compounds in the work place or in the general environment *(9)*. However, these tests do not give any information on the source of exposure, and the present sensitivity of the technique is quite low. No significant difference in the level of HPRT mutants in lymphocytes could be detected between smokers and non-smokers. One of the difficulties in the mutation assays is that a great number of wild-type mutations exists. Previous difficulties in quantification of the mutants may in part be overcome by the application of the new molecular-biological techniques, such as the polymerase chain reaction in combination with efficient electrophoretic systems, e. g. denaturing gradient gel electrophoresis or efficient DNA sequencing techniques *(20)*.

Biomarkers of Effect

The review of the exposure biomarkers presented earlier shows that bio-chemical and cytological methods are currently available which can be successfully used to monitor populations exposed to environmental mutagens and carcinogens. Nonetheless, not all of these enable quantitative assessment of the carcinogen dose which has reached the cells of the target organ to produce specific genome changes (in DNA or chromosomes). The exposure biomarkers presented above, irrespective of their true usefulness in the assessment of exposure to environmental carcinogens, point to the occurrence of some changes capable of triggering the process of carcinogenesis. Their suitability for use in specified projects intended to ensure chemical safety for the people inhabiting chemically polluted areas results mainly from the fact that their determinations enable taking steps intended to eliminate mutagenic or carcinogenic agents from the environment or reduce their concentrations. From the viewpoint of a single healthy inhabitant of a chemically polluted area, the scope of such steps is very limited, as the changes detected at the level of his/her genome are in principle irreversible. The promotion stage, which is the next stage of the carcinogenic process, offers wider scope for intervention. The average time which had elapsed from blood sample collection and storage in the blood bank and cancer detection was 14 months. At least one oncoprotein was detected in the serum samples collected earlier from as many as 7 out of the 9 patients with pulmonary carcinoma. Furthermore, the two negative cases that subsequently developed lung cancer had their last blood samples drawn more than two years prior to the time of diagnosis and had no samples available at the time of diagnosis.

From the above data and from the data presented earlier it may be concluded that the detection of increased oncogene expression (or oncogene mutation) by oncoprotein determination in blood serum may serve as a valuable biomarker of cancer risk in the populations exposed to environmental carcinogens.

There have been a number of reports confirming that occupational exposure to asbestos dust, vinyl chloride, and/or aromatic amines is associated with increased incidence of elevated blood serum tumor marker concentrations without accompanying clinical symptoms of cancer *(24)*.

The analysis of the data on the possibilities of using tumor markers for the monitoring of cancer risk in people living in the areas of particularly high exposure to harmful chemical agents (waste dumping sites, chemicals depots), indicates that the determinations of two tumor markers, namely tissue polypeptide antigen (TPA) and carcinoembrional antigen (CEA) are the most effective for this purpose *(25)*. Both selected biomarkers, TPA and CEA, are considered to be universal for various cancer types. Therefore, simultaneous TPA and CEA determination is recommended. In this way, the chances for detecting cellular changes associated either with the intensified cell division processes (TPA) or with intercellular communication disturbances (CEA) can be increased.

To confirm feasibility of determining both the biomarkers, a series of tests was performed at the Nofer Institute of Occupational Medicine, Lodz, Poland, on workers exposed to occupational carcinogens *(26, 27)*. In an investigation on workers with previous history of occupational exposure to benzidine, almost four times higher incidence of blood serum TPA concentrations above the cut-off level was detected as compared with the incidence in the control group which had not been occupationally exposed to carcinogenic chemicals (Table III). Extremely meaningful were the data indicating that in the so-called slow acetylator patients that were more predisposed to develop bladder cancer when exposed to aromatic amines, elevated blood serum TPA concentration levels were observed more frequently than in the fast-acetylator subjects exposed to those amines. We have also confirmed that tobacco smoking and long period of employment under conditions of exposure to aromatic amines significantly increase the incidence of pathologic blood serum TPA concentrations.

When investigating blood serum TPA and CEA concentrations in workers occupationally exposed to asbestos/cement dusts, we have determined that increased incidence of both marker levels above the cut-off value was observed in workers with periods of work exceeding 20 fiber/year/cm$^3$ (Table IV). Simultaneous increase in the concentration of both biomarkers was not detected in any of the workers examined.

**Table III. Percentage of people with a TPA concentration exceeding the cut-off concentration (90 U/L) according to smoking and acetylation status**

| Test group | Percentage of people with TPA concentration exceeding 90 U/l |
|---|---|
| Nonsmokers not occupationally exposed to any toxic chemical | 2.4% |
| Smokers not occupationally exposed to any toxic chemical | 4.6% |
| Smokers exposed to aromatic amines | 27.3% |
| Nonsmokers exposed to aromatic amines | 22.2% |
| Slow acetylators exposed to aromatic amines | 29.3% |
| Fast acetylators exposed to aromatic amines | 14.3% |
| Slow acetylators smokers exposed to aromatic amines | 33.4% |

**Table IV. Number of workers exposed to asbestos-cement dust with the TPA and CEA serum concentrations exceeding the cut-off level**

| Asbestos exposure and smoking status | Tumor marker investigated | | |
|---|---|---|---|
| | TPA | CEA | TPA+CEA |
| Workers with periods of work < 20 fiber-year/cm$^3$: | | | |
| Non-smokers (n=19) | 1 (5.3%) | 1 (5.3%) | 2 (10.5%) |
| Smokers (n=15) | 1 (6.7%) | 2 (13.4%) | 3 (13.3%) |
| Workers with period of work > 20 fiber-year/cm$^3$ | | | |
| Non-smokers (n=31) | 4 (12.9%) | 1 (3.2 %) | 5 (16.1%) |
| Smokers (n=24) | 3 (12.5%) | 5 (20.8%) | 8 (33.3%) |
| Workers exposed to asbestos-cement dust (n=89) | 9 (10.1%) | 9 (10.1%) | 18 (20.2%) |

As the biochemical mechanisms causing an increase of concentrations of both the aforementioned biomarkers (TPA and CEA) in blood serum are different, some authors *(25)* are of the opinion that, in studying the effects of environmental agents on the human organism, the increase in their concentration can be considered independently as a possible indicator of neoplastic process triggering. Taking this fact into account, it should be noted that in our population of 89 workers, as many as 18 had concentrations higher than the cut-off level in both the markers studied. This constitutes nearly 20% of the population studied. Accordingly, every fifth worker employed under conditions of exposure to asbestos/cement dust showed tumor marker concentration changes which could point to increased risk of cancer.

These results indicate that simultaneous determinations of blood serum TPA and CEA concentrations may constitute one of the essential diagnostic procedures for determining the risk of cancer under conditions of exposure to environmental carcinogens present at workplace.

When concluding on the results of cancer marker determinations it should be borne in mind that they are feasible mainly in population studies - their reliability in assessing cancer risk in a specified single patient is very limited. The reliability of those determinations can be significantly increased by combining them with the determinations of the concentrations of other biomarkers belonging both to the exposure and early health effects categories *(28, 29)*.

Conclusion

At present it can be stated that a more thorough analysis of the problem of health risk in a human being exposed to environmental agents has become possible. The assessments performed directly in the human organism make it possible to eliminate the uncertainties often associated with epidemiologic studies.

Currently, there is an urgent need for cancer risk assessment under conditions of exposure to environmental carcinogens to be based on good knowledge of molecular mechanisms involved in the cancer process. Only then it will become possible to base cancer risk assessments on such biological circumstances which occur in human organisms exposed to environmental carcinogens.

Cancer is a disease which results in adverse health effects, often in death. Therefore, early detection not only of the disease itself, but also of those phenomena which may increase cancer risk, is extremely important. At present, a number of good laboratory tests (biomarkers) is available which make this task feasible, including those which assess the exposure to the environmental carcinogens and those which make it possible to detect the earliest stages of cancer process before clinical symptoms manifest themselves. It seems that cancer risk assessments based on particular laboratory determinations have become feasible.

**Literature cited**

1.   Fowle, J.R.; Sexton, K. *Environ. Health Perspect.* 1992, 98, 235-241.
2.   WHO. *Biomarkers and Risk Assessment: Concepts and Principles.* Environmental Health Criteria 155, Geneva, 1993.

3.    Lowry, L.K. In: *International Symposium on Human Health and Environment: Mechanism of Toxicity and Biomarkers to Assess Adverse Effects of Chemicals*; Mutti, A.; Chambers. P.L.; Chambers, C.M,. Eds.: Elsevier: Amsterdam, 1995, pp. 31-38.
4.    Silbergeld, E.K.; Davis, D.L.; *Clin. Chem.* 1994, 40, 1363-1367.
5.    Rhodes, N.; Pasules, R.S.; Roberts, J.D. *Environ. Health Perspect.* 1995, 103, 504-506.
6.    McMahon, G. *Environ. Health Perspect.* 1994, 102, 75-80.
7.    Bernard, A.; Lauwerys, R. In: *Indicators for Assessing Exposure and Biological Effects of Genotoxic Chemicals*; Aitio, A.; Becking, G.; Berlin, A.; Bernard, A.; Foa, V.; Kello, D.; Krug, E.; Leonard, A.; Nordberg, G., Eds.; Office for Official Publications of the European Communities: Luxembourg, 1988, Chapter I.
8.    WHO. 2. *Summary Report on the Evaluation of Short-Term Tests for Carcinogens (Collaborative Study on In Vitro Tests)*. Environmental Health Criteria 47, Geneva, 1985.
9.    Wogan, P.B. *Environ. Health Perspect.* 1992, 98, 167-178,
10.   Farmer, P.B. In: *Use of Biomarkers in Assessing Health and Environmental Impacts of Chemical Pollutants*. Travis, C.C., Eds., Plenum Press, New York, 1993, pp. 53-62.
11.   Ehrenberg, L. In: *Use of Biomarkers in Assessing Health and Environmental Impacts of Chemical Pollutants*. Travis C.C., Eds., Plenum Press, New York, 1993, pp. 1-7.
12.   Tornqvist, M. In: *Use of Biomarkers in Assessing Health and Environmental Impacts of Chemical Pollutants*. Travis C.C., Eds., Plenum Press, New York, 1993, pp. 17-30.
13.   Shamsuddin, A.K.M.; Sinopoli, N.T.; Hemminki, K.; Boesch, R.R.; Harris, C.C. *Cancer Res.* 1985, 45, 66-68.
14.   Perera, F.P.; Hemminki, K.; Young, T.L.; Brenner, D.; Kelly, G.; Santella, R.M. *Cancer Res.* 1988, 48, 2288-2291.
15.   Liou, S.H.; Jacobson-Kram, D.; Poirier, M.C.; Nguyen, D.; Strickland, P.T.; Tockman, M.S. *Cancer Res.* 1989, 49, 4929-4935.
16.   Hemminki, K.; Grzybowska, E.; Chorąży, M.; Twardowska-Saucha, K.; Sroczyński, J.W.; Putman, K.L.; Randerath, K.; Phillips, D.M.; Hewer, A.; Santella, R.M.; Young, T.-L.; Perera, F.P. *Carcinogenesis*, 1990, 11, 1229-1231.
17.   Schoket, B.; Phillips, D.H.; Hawer, A.; Grover, P.L. *Mutat. Res.* 1988, 204, 531-541.
18.   Shields, P.G. *Environ. Health Perspect.* 1994, 102, 81-87.
19.   WHO. *Guide to Short-Term Tests for Detecting Mutagenic and Carcinogenic Chemicals*. Environmental Health Criteria 51. Geneva, 1985.
20.   Albertini, R.J.; Nicklas, J.A.; Fuscoe, J.C.; Skopek, T.R.; Branda, R.F.; O'Neill, J.P. *Environ. Health Perspect.* 1993, 99, 135-141.
21.   Brandt-Rauf, P.W. *Scand. J. Work Environ. Health*, 1992, 18, 27-30.
22.   Brandt-Rauf, P.W. *Scand. J. Work Environ. Health*, 1992, 18, 46-49.
23.   Brandt-Rauf, P.W.; Smith, S.J.; Niman, H.L.; Goldstein, M.D.; Favata, E. *J. Soc. Occup. Med.* 1989, 39, 141-143.

24. Bernard, A.; Lauwerys, R. In: *Indicators for Assessing Exposure and Biological Effects of Genotoxic Chemicals*; Aitio, A.; Becking. G.; Berlin, A.; Bernard, A.; Foa, V.; Kello, D.; Krug, E.; Leonard, A.; Nordberg, G., Eds.; Office for Official Publications of the European Communities: Luxembourg, 1988, Chapter VI.
25. Pluygers, E.P.; Beauduin, M.,; Baldewyns, P.E. *Cancer Detec. Prev.* 1991, 1, 57-68.
26. Lutz, W.; Indulski, J.; Krajewska, B. In: *Selected Papers from the XX Medichem Congress Occupational Health in the Chemical Industry*, 6-9 October 1992, London, United Kingdom. WHO Regional Office for Europe. Copenhagen, 1992, pp. 127-131.
27. Indulski, J.A.; Lutz, W.; Krajewska, B.; Więcek, E. In: *International Symposium on Human Health and Environment: Mechanisms of Toxicity and Biomarkers to Assess Adverse Effects of Chemicals*, September 25-30, 1994, Salsomaggiore Terme (Parma), Italy. Associazione Universitaria Italiana di Medicina del Lavoro "Bernardino Ramazzini", Parma, 1994, p. 76.
28. Indulski, J.A.; Lutz, W.; Krajewska, B. *Pol. J. Occup. Med. Environ. Health*, 1993, 6, 149-156.
29. Indulski, J.A.; Lutz, W. *Int. J. Occup. Med. Environ. Health*, 1995, 8, 11-16.

# Chapter 19

# Molecular Epidemiology in Cancer Risk Assessment

## J. A. Indulski and W. Lutz

Nofer Institute of Occupational Medicine, 8 Southwest Teresy Street,
90–950 Lodz, Poland

The rapid development of molecular biology techniques has significantly
aided the assessment of the detrimental consequences of chemically polluted
environment on humans. It has also resulted in the development of molecular
epidemiology, a new branch of epidemiology. Molecular epidemiology has
become an important science which finds increasing applications in
environmental monitoring. Nevertheless, a number of problems have
emerged which seriously limit its practical application in epidemiologic
research. Correct interpretation of test results seems to be one of the most
difficult stages of assessing, by means of molecular epidemiology techniques,
exposures to environmental pollutants and the resultant health effects. This
refers in particular to those instances of the assessment of exposures (to
environmental carcinogens for instance) for which formal exposure limits are
not available. In many instances, molecular epidemiology tests are performed
only because specified methods are available, but in many cases the results
fail to provide any useful information. Also, the implementation of these new
research methods has given rise to completely new ethical, social and legal
problems. This is particularly true about research on the individual
susceptibility to the adverse effects of environmental chemicals. And finally,
it is extremely important to ensure that the enthusiasm about the feasibility
of using the most recent molecular biology technologies does not
overshadow the epidemiological problems associated with the necessity of
providing suitable health care to the inhabitants of chemically polluted areas.

Cancer is a long-lasting multi-stage process, and the epidemiologic data indicate that
environmental chemical agents (chemical carcinogens) play an important role in this
process. A chain of certain events must take place prior to carcinogenic cell
transformation. The events include contact of the organism with the carcinogen, its
absorption into the organism, transportation into the cells, transformation to a reactive
metabolite (not all carcinogens become necessarily activated), and interaction with the
critical (DNA) molecule in the target cells. The interaction results in the formation
of carcinogen-DNA adducts, which in turn may lead to point mutations or breaks
and/or rearrangements in the chromosomes. The changes in cellular genome DNA are
not random. Usually specific genes or gene groups, which participate in the process

of transformation of normal cells to carcinous ones, are affected. Among these genes, there are some whose protein products stimulate cellular growth and division, and some others whose products inhibit those processes. The first group includes protooncogenes, the other comprises suppressor genes. The tremendous progress in elucidating the genetic mechanisms of cancer initiation and development and the advances in the methods for the qualitative and quantitative assessment of the tumour markers have paved the way for biological monitoring of the populations exposed to environmental carcinogens. Methods employed by the molecular biology for investigating the changes on the level of human genome have become the primary tool in assessing the risk of cancer by molecular epidemiology - a rapidly developing branch of science. Molecular epidemiology attempts not only to indicate the populations with high risk of cancer but also to determine high-risk individuals. Risk assessment has been based on the determinations of the internal exposure component, and also (as in the case in the conventional epidemiology) on the measurements of the external exposure component *(1, 2, 3)*.

In assessing exposure to environmental carcinogens, distinction is made between the external dose defined as the quantity of carcinogen in contact with the organism determined through personal or external monitoring, and the internal dose, which is defined as the total quantity of chemical absorbed by the organism during a specified time. Out of the total quantity of absorbed carcinogen molecules, only a portion reaches the target tissue. A portion will reach the critical site on the macromolecules, with only a fraction of the latter amount acting as the biologically effective dose. Determination of each of these doses may be helpful to the epidemiologist in assessing cancer risk. The most effective assessment of cancer risk is obtainable from the determination of the biologically effective dose.

The methods of exposure assessment employed in molecular epidemiology fall into two categories:

1)    determination of the level of chemical carcinogens, their metabolites and/or adducts formed by them with the macromolecules in the easily available biological liquids or in the cellular material;

2)    determination of biological effects, such as changes in gene structure and expression, and also cytogenetic changes.

## Adducts formed by environmental carcinogens with DNA and proteins

Among the methods included in the first category, determination of the adducts formed by the carcinogens with such macromolecules as those of DNA and proteins is particularly valuable for the assessment of cancer risk under conditions of exposure to environmental carcinogens. It is worth noting here that adduct determinations are used by the epidemiologist both for assessing the extent of exposure to carcinogens (the first category) and for assessing the genotoxic potential of the carcinogens to assay the effects of exposure to the carcinogens (the second category). In epidemiologic studies, the assessment of DNA or protein adducts may be used either as an independent variable (exposure biomarker) or as a dependent variable (effect biomarker) and the choice depends on the problem posed by the epidemiologist.

The principles of applying carcinogen-DNA adduct determinations in epidemiology are based on the observation that the majority of environmental carcinogens are electrophilic compounds or are transformed into electrophilic metabolites. The electrophilic compounds may be covalently bound to the nucleophilic sites in the DNA bases to form respective adducts (Table I). Thus, new-formed adducts, if they are not removed from DNA during repair processes, or if errors occur in the course of the repair processes (before DNA replication has taken place), may lead to point mutations in genes. It is believed that gene mutations may initiate the process of transforming a normal cell into a carcinous one, which may be associated with protooncogene activation or suppressor gene inactivation. Formation of genotoxic carcinogen adducts with DNA seems to play an essential role in subsequent (promotion and progression) stages of carcinogenesis. It should be stressed that, although point mutations constitute an essential factor in the carcinogenesis initiation, formation of DNA adducts may also cause other changes in the genetic material of the cell. Formation of the adducts may lead to DNA chain break, gene amplification and rearrangement, and may also affect the occurrence of chromosomal aberrations *(4, 5, 6)*.

Feasibility of using DNA adduct determinations as biomarkers of exposure to environmental carcinogens seems to be indisputable when we consider the results of recent research. However, some doubts have been expressed in the use of carcinogen-DNA adduct determinations for the assessment of cancer risk *(7)*. In the latter instance, the adducts are used as early effect biomarkers. Not all adducts are of equal significance for inducing gene mutations and, consequently, development of cancer.

### Table I. Example of DNA adducts

| | *Carcinogen* | *Adduct* |
|---|---|---|
| 1. | 4-methylnitrosamino-1-(3-pyridyl)-1-butanone | N-7-methyldeoxyguanosine |
| 2. | ethyl methanesulphonate | N-7-ethyldeoxyguanosine |
| 3. | ethylene oxide | N-7-(2-hydroxyethyl) deoxyguanosine |
| 4. | urethan | N-7-(2-oxoethyl)deoxyguanosine |
| 5. | benzo(a)pyrene | $N^6$-[7,8,9-triolbenzo(a)pyrene]-deoxyadenosine |
| 6. | N-nitrosodiethylamine | $O^4$-ethyldeoxytymidyne |
| 7. | 4-aminobiphenyl | N-(deoxyguanosin-8-yl)-4-amino-biphenyl |
| 8. | 4-methylnitrosoamino-1-(3-pirydyl)-1-butanone | N-7-methyldeoxyguanosine |
| 9. | Carcinogens which are generating free oxygen radicals | 8-hydroxydeoxyguanosine |

An environmental genotoxic substance may form several different adducts with DNA, differing in the speed of their formation and elimination in the course of the repair processes, their location in the genome and in their biological effects. A high number of adducts is not necessarily a factor in increasing the biological consequences. "Hot spots" particularly susceptible to adduct formation have been demonstrated to exist in the animal and human DNA. Their location and number may vary, depending on type of cells and tissue to which the cells belong *(8)*.

Metabolic equilibrium between genotoxic substance activation (formation of electrophilic derivatives) and deactivation processes constitutes an important factor which may affect the number of formed adducts. The equilibrium is directly affected by polymorphism of P-450 cytochromes and other enzymes participating in the metabolism of the genotoxic substances, such as glutathione transferase, glucuronyl transferase, or epoxide hydrolase. This is why it is necessary, when interpreting the results obtained from DNA adduct assessments, to consider interindividual differences in the activity of the enzymes which participate in the cellular metabolism of the carcinogens. Interindividual differences in the cellular activity of repair enzymes which remove modified bases from DNA must be also taken into account *(9)*.

The majority of tissues contain numerous cell types, each with a different ability to transform chemical substances into DNA-reactive derivatives. For example, lungs comprise a great number of cell types in which the relative concentrations of various P-450 cytochrome isoenzymes vary, depending on cell type *(10)*. Thus, a compound may have high concentrations of promutagenic adducts in one cell type, without causing similar effect in some other type of the cells. The concentration of a DNA adduct determined in the whole tissue homogenate may be substantially higher or lower than that in a specified cell type *(10)*.

Some DNA adducts are repaired quickly, while others are repaired slowly, and the reduction of adduct number correlates with the rate of cell transformation. Besides, concentration and gene location of DNA adducts change with time which has elapsed from exposure to genotoxic chemicals. Different repair mechanisms in the genome also make it difficult to use total DNA repair ability as an indicator of cell susceptibility to carcinogens *(11)*.

Difficulties in obtaining target tissue which has not undergone cancer changes is another factor making the use of DNA adduct determinations for the assessment of cancer risk in people exposed to occupational and environmental carcinogens less popular. Consequently, DNA adduct determinations in humans are performed usually in peripheral blood lymphocytes. However, these determinations enable only indirect estimate of the number of adducts which may be formed in the cells which constitute the object of carcinogen target activity. For obvious reasons, the assessments of dose-response relationship for chemical carcinogens carried out on animal models using radioisotope-labelled carcinogens cannot be performed in humans. The determination of DNA adducts in people is limited not only because of the restricted availability of test material but also because the level of carcinogen binding is much lower than that obtained in model systems. Only a limited number of data is currently available on the feasibility of using the assessment of adduct concentrations in human lymphocytes for the determination of their concentrations in the cells of the target tissues and organs *(12)*.

Decent Advances in DNA Methods

Very sensitive detection methods are currently used for DNA adduct analysis intended to assay human exposure to environmental carcinogens. These include fluorescence, immunoassay, and $^{32}$P-postlabelling test methods *(13, 14, 15)*.

Fluorescence methods are used chiefly for adducts formed by polycyclic aromatic hydrocarbons (PAH). However, due to low level of PAH-DNA *in vivo* binding and as a result of the similarity of excitation and emission spectra of various PAH, use of conventional fluorescence spectrophotometry is difficult. The difficulties have been largely overcome by using synchronous fluorescence spectroscopy (SFS), involving fluorescence spectrum recording at variable excitation and emission wavelengths. The sensitivity of this method is equivalent to the detection of 3 PAH adducts per $10^8$ nucleotides. The relatively high DNA sample size (about 100$\mu$g) required in this method is a serious limitation *(16)*.

Immunoassay techniques applied to DNA adduct analysis include Enzyme-Linked Immunosorbent Assay (ELISA) and ultrasensitive enzyme radioimmunoassay (USERIA). The starting point for the development of these methods was the observation that the weak immunogenic properties of nucleic acids are changed as a result of the modifications caused by the carcinogen. Many monoclonal and polyclonal anti-carcinogen:DNA adduct antibodies have been recently obtained. These immunoassay techniques enable detecting 1 PAH adduct per $10^7$ - $10^8$ nucleotides, which means that the adduct detection limit is at the level of femtomoles. The quantity of DNA required for the analysis is about 50$\mu$g. The limitation of the immunoassay method is that specific antibodies directed against specified adducts must be available *(17)*.

Recently, carcinogen:DNA adduct analyses are being increasingly carried out using the $^{32}$P-postlabelling test *(13)*. The characteristics which have contributed to the popularity of this method include the versatility and sensitivity of the quantitative determinations. The method is extremely sensitive and, after suitable modifications, can detect 1 adduct per $10^{10}$ normal nucleotides. It has proved to be particularly useful for detecting non-polar PAH adducts, for example, 7,8-diol-9,10-oxybenzo(a)pyrene deoxyguanosine *(15)*. As a result of limitations inherent in the chromatography systems, alkyl DNA adducts are difficult to detect by the $^{32}$P-postlabelling technique. In the latter instance, a combination of radioisotope method and immunochemical precipitation technique is employed *(18)*.

A number of substances known to exhibit genotoxic and carcinogenic properties display properties resulting from the generation of free oxygen radicals which react with DNA bases to form corresponding oxygen adducts *(19)*. Combination of gas chromatography, mass spectrometry, and immunoassay techniques is used for detection and determination of these oxygen adducts. Unfortunately, these methods are feasible only when high concentrations of modified bases are present in the biological test material, for example, as a result of exposure to ionizing radiation. They cannot be used to analyze biological liquids, such as urine, of patients exposed to occupational or environmental carcinogens. Besides, they have proved to be insufficiently sensitive and too time-consuming and labour-intensive *(20)*.

The problem of analysing low concentrations of DNA base oxygen derivatives has been in part solved by determination of 8-hydroxydeoxyguanosine, an oxygen adduct. The adduct can be determined at high precision and selectivity by electrochemical detection following separation of enzymatic DNA hydrolysate, using reverse-phase HPLC *(21)*.


Use of Proteins Instead of DNA

As obtaining DNA from cells and procuring of the cells themselves during examinations on large human populations has proved difficult, work has been started to test whether determinations of the adducts formed with some easily available proteins could be used instead of DNA adduct determinations (Table II). The information obtained during observations of human subjects and from experiments on test animals shows that good correlation exists between the quantity of the adducts formed with DNA in the genetic material of peripheral blood leucocytes and the number of the adducts formed with erythrocyte haemoglobin. The relationship has been observed to occur within a wide range of carcinogen doses. Blood serum albumin has been also found useful for the indirect assessment of the formation of the adducts with DNA. Both proteins contain in their structures several nucleophilic groups capable of forming adducts with the electrophilic, native carcinogens or their metabolites. These include nitrogens at the ends of haemoglobin and albumin polypeptide aminoacid chains. With haemoglobin, valine is the N-terminal aminoacid. The majority of the nucleophilic groups are, however, contained in the aminoacids found inside the polypeptide chain. The following can be included in the groups: nitrogen in the amine group of lysine and in the imidasol ring of histidine, sulphur in the thiol group of cysteine and methionine, and oxygen in the carboxylic groups of the aspartic and glutamic acids and in the hydroxyl groups of serine and threonine *(22, 23)*.


**Table II.  Example of haemoglobin adducts**

| | *Carcinogen* | *Adduct* |
|---|---|---|
| 1. | methylmetane sulphonate | S-methylcysteine |
| 2. | N-nitrosodimethylamine | S-methylcysteine |
| 3. | urethan | N-(2-oxoethyl)valine |
| 4. | ethylene oxide | N-(2-hydroxyethyl)valine |
| 5. | styrene oxide | N-(phenylhydroxyethyl)valine |
| 6. | benzene | S-phenylcysteine |
| 7. | 4,4'-methylenebis(2-chloroaniline) | Sulphinamide with cysteine |
| 8. | Acrylamide | S-(2-carboxyamidoethyl)cysteine |
| 9. | Acrylonitryle | S-(2-cyanomethyl)cysteine |
| 10. | Benzo(a)pyrene | Benzo(a)pyrene tetrol esters |

The analytical procedure to be employed for the identification and quantitative determination of protein adducts depends on the type of adduct formed and on its location in the protein (whether it is a terminal aminoacid or an aminoacid located within the polypeptide chain of protein). Adducts located inside the polypeptide chain can be determined only after complete hydrolysis in acid medium or by proteolytic enzymes. Such procedures are, unfortunately, very time-consuming and may often result in artifacts *(22)*.

Adducts formed with the carboxylic groups of the aminoacids located inside the polypeptide chain of the glutamic and aspartic acids are relatively easily released as a result of mild acid hydrolysis. Similar procedure is employed for releasing haemoglobin aromatic amine adducts bound as sulfinamides to cysteine residues.

In analysing the adducts formed with valine located at the amine end of haemoglobin polypeptide chain, a modified Edman degradation method is used, which has been designed to determine protein aminoacid sequence. During this procedure, haemoglobin separated from erythrocytes and containing an adduct with N-terminal valine, as a result of treatment with the Edman reagent (pentaphenyloisocyanate), releases the valine adduct from the globin polypeptide chain as a corresponding derivative of phenylthiohydantoin. The derivative, after extraction, can be subjected to further analysis by means of combined gas chromatography/mass spectrometry (GC/MS) technique. The latter procedure is one which is most frequently used for protein adduct analysis. Fluorometry and immunology methods are used less extensively. The method employing Edman degradation enables detection of 1 pg adduct per gram haemoglobin and is characterized by high repeatability *(22)*.

The procedure for the determination of adducts formed with aminoacids located inside the polypeptide chain employing enzymatic protein hydrolysis and quantitative analysis of the adduct released during the hydrolysis by the GC/MS method is considerably less sensitive. The sensitivity of this method is 10 to 100 times lower than that obtained by the application of Edman degradation. The procedure, during which lipophilic and aromatic hydrocarbon adducts are formed with aminoacids in the globin polypeptide chain, is characterized by relatively high sensitivity of the order of femtomoles/g haemoglobin. The high sensitivity is associated with the facility with which those adducts can be extracted from protein hydrolysate *(23)*.

Generally speaking, hopes associated with the introduction of adduct determination, to be used as a nuclear dosimeter, have been essentially fulfilled. Conditions have been provided which make it possible to determine the dose of the genotoxic substances (environmental carcinogens) and to calculate the risk of cancer associated with those substances. The implementation of adduct determinations made it also possible to detect and identify heretofore unknown mutagens responsible for initiation of carcinogenesis. The discovery of the differences between exposed and non-exposed populations has confirmed that adduct determinations may serve as a useful tool in epidemiologic studies. Individual genetic factors seem to play an essential role in the assessment of individual dose and, subsequently, in the assessment of cancer risk. These differences may favour either stimulation or inhibition of adduct formation. As a result, the confirmation of the relationship between genotoxic substance dose and the number of determined adducts becomes difficult. This problem is considerably less evident in population studies. It seems that, in order to assess individual exposures to genotoxic substances, monitoring of the temporary changes of adduct profiles is more useful. Actually we can say that adduct determinations, independent of whether they are performed on DNA or on proteins, not only confirm that exposure has taken place but also point to the development of potentially carcinogenic damages in the genetic material. Notwithstanding the enormous progress in the methodology, further methodological improvements are necessary.

The changes in gene structure and expression

It should be stressed that the methods described above for the detection of changes in the human genome as an effect of environmental carcinogens on human cells, and the generation of DNA adducts of the carcinogens or their active metabolites should be used for early detection of precarcinogenesis. These changes may lead to gene mutations or other defects of the genetic material that are responsible for initiating carcinogenesis. This is the reason why molecular epidemiology applies laboratory methods based on molecular biology techniques which may help identify abnormalities in the gene sequence or expression that are known to affect the growth, division and diversification of the cells. The impairment of these processes in turn is thought to initiate the transformation of the normal cell into carcinogenic cells. Biochemical systems regulating cellular growth and division in different organs and tissues are in many ways connected with protooncogenes and suppressor genes *(24, 25)*.

Protooncogenes have the potential for coding oncoproteins which participate in different cellular functions. They include proteins which act as growth factors, membrane receptors for these factors, proteins transmitting the transduction signal from membrane receptor to the inside of the cell, kinase-like proteins regulating vital processes within the cell, and nuclear proteins which influence expressions of the genes activating DNA synthesis and accelerating cellular division. Environmental carcinogens may transform protooncogenes into oncogenes via different routes. The most prevalent mechanisms include the induction of point mutations, protooncogene translocation and/or amplification within the chromosome *(26)*.

As evidenced by the experience gathered so far, carcinogenesis is a multi-step process. Thus it can be assumed that the abnormalities of single oncogenes are not sufficient to induce a complete carcinogenic transformation. It seems plausible that each of the stages, from the normal state to the complete carcinogenic transformation of the affected cellular clone, may be conditioned by the acquisition of a specific protooncogene-related genetic deviation *(27)*.

The activity of another category of genes, termed suppressor genes or anti-oncogenes, consists in counteracting, or even stopping, the carcinogenic processes by the inhibition of cell proliferation. So far the data on the suppressor genes are scarce. The type best described in the literature is the p53 gene whose mutations or deletions have been found in a variety of human cancer types. Under physiological conditions, p53 is an indispensable element of the control and regulation of the cellular division, DNA replication and cell diversification and it determines the built-in cell decay facility (apoptosis). The p53 gene as a growth inhibitor serves the role of a "molecular gendarme" defending the integrity of the cellular genome and inhibiting proliferation of the cells with DNA damage. Consequently the repair mechanisms are activated before a new cell division cycle starts. When the repair processes fail, p53 induces apoptosis thus preventing the growth of abnormal cells. The loss of suppressive properties is brought about by the inactivation of p53 in one allele and deletion of the other allele, which would mean that the suppressor genes exhibit a recessive activity. The blockage of the normal function of p53, apart from an impaired regulatory mechanism of the transcription, may inhibit the built-in apoptosis. However, it is not certain to what extent other molecular changes may play a part in this process. Nevertheless, it can be stated at present that a defect within the p53 gene may be crucial for the development of different types of human cancer *(28, 29)*.

For molecular epidemiology, the finding of special significance was that the activation of ras protooncogenes or erbB-2 protooncogenes is associated with point mutations of these genes induced by common environmental carcinogens such as PAH, nitrosamines or aromatic amines. It was equally important to discover that the activation of protooncogenes and the related increased number and changes of the cellular oncoproteins can be monitored by determining their content in easily accessible biological fluids, blood and urine *(26)*. More serious problems are encountered when one tries to determine proteins whose synthesis is controlled by suppressor genes. One of the reasons is the short half-life of the proteins, amounting to 6-20 min, which limits their accumulation in the cells. Thus they cannot be detected either in blood or urine. The mutated form of p53, with the half-life ranging from 6-12 hours is detectable in blood and can be used as a direct marker of mutations at the p53 gene level *(29)*.

A large number of the mutations within protooncogenes and suppressor genes was observed in the tissues affected by cancer. It remains to be determined whether the mutated cells within the group of the normal ones can be detected prior to the clinical diagnosis of cancer.

An interesting example of the applicability of determining serum oncoproteins for the assessment of cancer risk are the studies conducted by Brandt-Rauf et al *(26)*. The measurements were carried out using blood serum samples collected from 46 workers exposed to asbestos or silica and stored in a blood bank. Fourteen of these persons have subsequently developed cancer, including the cancer of respiratory tract. Of the nine respiratory tract cancer cases, seven were found to be positive for serum oncoproteins, and on the average their serum specimens were positive 14 months prior to the time of the clinical diagnosis. The two negative cases that subsequently developed lung cancer had their last blood samples drawn more that two years prior to the time of diagnosis. These results, as well as the results of some other studies confirm that detection of increased expression of protooncogenes or their mutation through blood serum oncoprotein determination may serve as an important diagnostic indicator of cancer risk in populations exposed to occupational and environmental carcinogens.

Cytogenetic changes

Several cytogenetic tests, such as incidence of structural chromosomal aberrations, incidence of micronuclei, or incidence of sister chromatid exchanges are used in the monitoring of the mutagenic and carcinogenic effects of environmental factors on humans *(30)*. Single-chromatid aberrations are more frequent in people exposed to chemical mutagens. Ionizing radiation induces mainly formation of chromosomal aberrations, and both chromatids of a specified chromosome are affected *(31)*. The micronucleus test detects the effects of genotoxic agents causing chromosome breaks or mitotic spindle damage *(32)*. The sister chromatid exchange test serves to determine mutagen-DNA interactions which lead to the occurrence of DNA homologous fragment displacements between two chromatids in one chromosome *(33)*. Peripheral blood lymphocytes are usually used in those tests. However, the interpretation of the results is difficult because of varying starting parameters, the variation being due to endogenous (sex, age, diseases suffered in the past) and exogenous (lifestyle, tobacco smoking, alcohol drinking, dietary habits) factors *(33)*. Unlike ionizing radiation tests, for which corresponding chromosomal markers (dicentric chromosomes) have been identified, the tests performed with chemical carcinogens have failed to identify any such specific markers. Therefore, results of chemical exposure tests can be used to assess group exposure only; they are not suitable for assessing exposure of individuals. If, for example, increased incidence of chromosomal aberrations in peripheral blood cells is observed in a group of subjects potentially exposed to environmental carcinogens as compared with chromosomal aberration incidence in the controls, it can be concluded that the test group has been exposed to carcinogenic agents, but the determination of exposure levels for individual subjects is not feasible. Similarly, the absence of differences in chromosomal aberration incidence between the test subjects and the controls does not exclude a history of the exposure in the test group. The same is true about the assessment of the incidences of micronuclei and sister chromatid exchange in peripheral blood cells.

Identification of genetically-related susceptibility to cancer induced
by environmental chemical carcinogens

Identification of individuals with genetically related increased susceptibility to carcinogenic activity of environmental carcinogens has been one of the crucial problems of molecular epidemiology.

Although individuals may experience similar exposure to environmental carcinogens, the differences in cellular metabolism of these carcinogens may result in evidently different quantities of carcinogen or its active metabolites reaching the target sites, causing response levels to be different. Such differences in response levels may also occur when target carcinogen doses are similar. Behaviours like those are genetically-related or result from the interaction of numerous external factors, such as age, dietary habits, diseases or exposures to harmful environmental agents (other than the current one) suffered in the past. The various factors responsible for

**Table III. The cytochromes P-450 (Cyp) involved in carcinogen metabolism**

| Cyp gensubfamily | Tissue | Suspect carcinogen |
|---|---|---|
| 1A | liver, lung | benzo(a)pyrene, arylamines, nitrosamines |
| 2A | liver | aflatoxin $B_1$, nitrosamines |
| 2B | liver, lung, gastrointestinal tract | aflatoxin $B_1$, cyclophosphamides |
| 2C | liver, gastrointestinal tract | benzo(a)pyrenes |
| 2E | liver, brain, leucocytes | nitrosamines, ethanol, benzene |
| 3A | lung, gastrointestinal tract | benzo(a)pyrenes, cyclosporin |
| 4A | lung | aflatoxin $B_1$, aromatic amines |

differences in the response of individuals to carcinogenic agents are present even prior to exposure assessment and remain independent of it.

The processes of carcinogen metabolic activation and detoxication are gene-controlled. The differences in the expression of those genes, or occurrence of mutations, lead to the polymorphism of the enzymes which metabolize environmental carcinogens (Table III). The polymorphism is responsible for the differences in the speed of cellular activation or detoxication processes. It leads to the differences in individual sensitivity, increasing or reducing the biologically active dose of the environmental carcinogen. N-acetyltransferase may serve as an example of enzyme polymorphism *(34)*. The enzyme catalyses introduction of acetyl substituent to the amine group of aromatic amines, including also those known to be carcinogenic, such as benzidine or 2-naphthylamine. The differences in the cellular activity of this enzyme is associated with the fact that human population comprises so-called slow and fast acetylators. Epidemiologic studies have confirmed that the risk of bladder cancer development in the slow acetylators is higher, while that of colon carcinoma is lower under conditions of exposure to aromatic amines. It has been also demonstrated that CYP1A2 N-oxidation polymorphism is associated with increased incidence of colon carcinoma *(34)*, while S-glutathione transferase polymorphism is positively correlated with the pulmonary carcinoma, adenocarcinoma in particular *(35)*.

Tobacco smoking provides another example illustrating the effect of enzymatic polymorphism on environmental carcinogen-induced response of the organism. Tobacco smoke is known to promote cancer development, but all smokers do not necessarily develop pulmonary cancer. A situation like this is in part due to the genetic variability of aromatic hydrocarbon hydroxylase (CYP1A1 enzyme) activity, causing remarkable differences in binding of benzo(a)pyrene hydroxy-epoxy derivatives to DNA (adduct formation) in tobacco smokers *(10)*.

During long years of research on the metabolic phenotype of human individuals, resulting from enzyme polymorphism, it had been sometimes necessary to administer a suitable marker drug to human test subject and observe its clearance. Recently, techniques based on polymerase chain reaction have been implemented which use DNA isolated from peripheral blood lymphocytes or from other cells (for example bladder epithelium cells found in urine). These techniques make it possible to detect genotypes of already known polymorphisms which involve various xenobiotic-metabolizing enzymes, including S-glutathione transferase and also the isoenzymes of P-450 cytochrome (CYP1A1 and CYP2D6) *(36, 37)*.

## Literature cited

1.   Correa, P. *Cancer Epidemiol Biomarkers Prev.* 1993, 2, 85-88.
2.   Perera, F. P., Santella, R. In: *Molecular Epidemiology: Principles and Practices.* Schulte, P.; Perera, F. P., Eds., Academic Press, New York, 1993, pp. 277-300.
3.   Schulte, P. A. In: *Molecular Epidemiology: Principles and Practices.* Schulte, P.; Perera, F. P. Eds. Academic Press, New York, 1993, pp. 3-44.
4.   Farmer, P. B. *Clin Chem.* 1994, 40, 1438-1443.
5.   Rhodes, N.; Paules, R. S.; Roberts J. D. *Environ. Health Perspect,* 1995, 103, 504-506.

6. Sampson, E. *Clin. Chem.* 1994, 40, 1376-1384.
7. Pearce, N.; de Sanjose, S.; Boffetta, P.; Kogevinas, M.; Saracci, R.; Savitz, D. *Epidemiology*, 1995, 103, 504-506.
8. Anderson, M. W.; Reynolds, S. H., You, M., Maronpot, R. M. *Environ. Health Perespect*. 1992, 98. 13-24.
9, Shields, P. *J. Occup. Med.* 1993, 35, 34-41.
10. WHO. *Biomarkers and Risk Assessment: Concepts and Principles.* Environmental Health Criteria 155. Geneva, 1993.
11. Dresler, S. L. In: *The Pathobiology of Neoplasia*. Sirica, A. E., Eds., Plenum Press, New York, 1989, po. 173-197.
12. Hemminki, K. *Toxicol. Letters*, 1995, 77, 227-229.
13. IARC. *Sci. Publ.* Vol. 124, Lyon, 1993.
14. Talaska, G.; Roh, J. H.; Getek, T. *J. Chromatogr.* 1992, 580, 293-323.
15. Beach, A. C.; Gupta, R. C. *Environ. Health Perspect.*, 1992, 98, 139-141.
16. Wielgosz, S. M.; Brauze, D.; Pawlak, A. L. *Mutat. Res.* 1991, 205, 271-282.
17. Santella, R. M. *Mutat. Res.* 1998, 205, 271-282.
18. Kang, H-J.; Konishi, C.; Kuroko, T.; Huh, N-K. *Environ. Health Perspect.* 1993, 99, 269-271.
19. Pryor, W. A. *Br. J. Cancer*, 1987, 55, 19-23.
20. WHO. *Adducts: Identification and Biological Significance.* Hemminki, K.; Dipple, A.; Shuker, D. E. G.; Kadlubar. F. F.; Segerback, D.; Bartsch, H., Eds.; IARC Scientific Publications No 125, Lyon, 1994.
21. Tagesson, Ch.; Kallberg, M.; Leanderson, P. *Toxicol. Methods*, 1991, 1, 1-10.
22. Tornquist, M. In: *Use of Biomarkers in Assessing Health and Environmental Impacts of Chemical Pollutants.* Travis, C. C., Eds.; Plenum Press, New York, 1993, pp. 17-30.
23. Perera, F. P. *Mutat. Res.* 1988, 205, 255-269.
24. Bos, J. L.; Krejl, C. F. *IARC Sci. Publ.*, 1992, 116, 57-65.
25. Cooper, G. M. *J. Cell. Biochem.* 1992, 166, 131-136.
26. Brand-Rauf, P.W. *Scand. J. Work Environ. Health*, 1992, 18, 46-49.
27. Hennings, H.; Glick, A. B.; Greenhalgh, D. A.; Morgan, D. L.; Strickland, J. E.; Tennenbaum, T.; Yuspa, S. U. *Proc. Soc. Exp. Biol. Med.* 1992, 202, 1-8.
28. Sugimura, T.; Terada, M.; Yokoda, J.; Hirohashi, S.; Wakabayashi, K. *Environ. Health Perspect.* 1992, 98, 5-12.
29. Prosser, J.; Evans, J. *Environ. Health Perspect.* 1992, 98, 25-37.
30. Sorsa M.; Wilbourn, J; Vanio, H. In: *Mechanisms of carcinogenesis in risk identification*. IARC Scientific Publications No 116, Lyon, 1992, pp. 543-554.
31. Morris, S. M.; Domon, O. E.; Mc Garrity, L. J.; Kodell R. L.; Casciano, D. A. *Cell Biol. Toxicol.* 1992, 8, 75-87.
32. French, M.; Morley. A. A. *Mutat. Res.* 1985, 147, 29-36.
33. Sorsa, M.; Ojajarvi, A.; Salomaa, S. *Teratog. Carcinog. Mutahen*, 1990, 10, 215-221.
34. Kadlubar, F.F.; Butter, H.A.; Kaderlik, K.L.; Chou, H-S; Lang, N.P. *Environ. Health Perspect.* 1992, 98, 69-74.
35. Seidegard. J.; Pero. R.W.; Markowitz. M.M.; Rousch. G.; Millerd, G.; Beattlee, J. *Carcinogenesis*, 1992, 11, 33-36.
36. Hirvonen, A., Husgafvel-Pursiainen, K.; Karjalainen, A., Anttila, S.; Vainio H. *Cancer Epidemiol. Biomarkers Prev.* 1992, 1. 385-489.
37. Hollstein, M. C.; Wild, C. P.; Bleicher, F.; Chutjmataewin, R. *Int. J. Cancer.* 1992, 53, 1-5.

# SPECIMEN BANKING

Chapter 20

# Environmental Specimen Banking and Analytical Chemistry
## An Overview

G. V. Iyengar[1] and K. S. Subramanian[2]

[1]Biomineral Sciences International Inc., 6202 Maiden Lane, Bethesda, MD 20817
[2]Environmental Health Directorate, Postal Locator 0800B3, Health Canada, Tunney's Pasture, Ottawa, Ontario K1A 0L2, Canada

This chapter provides a concise account of the specimen banking concept in relation to analytical chemistry and precautions to be observed during chemical analysis of such stored specimens. An attempt will also be made to present practical guidelines for setting up a human tissue bank, and to consolidate the experience gained in this field in a few countries presently operating such tissue banks. Where relevant, some aspects of non-human environmental specimen collection will be included.

### Specimen Bank: An Emerging Discipline

Historically, monitoring the environment dates back several centuries (1). The main environmental specimens involved in the earlier efforts were sediments, ice, peat and tree rings which were recognized as sources capable of reflecting the changing environmental patterns. Later, materials of biological origin obtained from museum collections were also used. These included herbarium specimens of mosses and lichens, zoological specimens of fish, bird feathers and eggs, animal hair, and human remains such as bones, teeth, hair and preserved organs. Among these specimens, mosses and bird feathers were found to be reliable indicators, a fact that supports the current use of mosses as biological indicators of atmospheric metal pollution.

Thanks to the increased awareness of environmental problems in the 70s, the topic of clean environment assumed a special meaning. Many initiatives surfaced to combat problems of environmental safety and to assess the changes that are occuring in the environment. As a natural tool to aid this endeavour, the value of preserving appropriate and unaltered specimens for biological monitoring was rediscovered. This necessity gradually led to the development of systematic approaches and the concept of Environmental Specimen Banking (ESB) began to assume importance. Thus, the concept of specimen preservation (and analysis) with its objective to safeguard the originality of the sampled specimen and the technical aspects of perfection to achieve these goals have been the bane of several studies dealing with specimen banking *(2-14)*. In the past two decades, there have been several efforts to explore various aspects of biomonitoring to improve environmental health. The concept of specimen banking has been in the forefront in many of these efforts.

An important aspect of specimen banking is that both organic and inorganic constituent information can be obtained on the same specimen. This has particular relevance to human ecological problems when multiple environmental parameters such as the food chain on the one hand, and industrial exposures on the other, are simultaneously involved.

## Objectives

ESB is primarily concerned with the long-term preservation of representative environmental specimens (e.g., soil, sediments, air, water and biotic material) for deferred analysis (i.e. retrospective chemical analysis) and evaluation *(11)*. The specimens are carefully collected and systematically archived for future use. ESB differs from real-time monitoring (RTM) in terms of time interval between sample collection and analysis. Samples collected for RTM are stored for a relatively short period of time and are analyzed as soon as practical; samples collected for specimen banking are intended for long-term monitoring (LTM) and therefore, they are selected and collected with the intent of storage for decades. In all cases, the quantity of sampled material required for RTM and the quantity desirable for specimen banking (i.e. LTM) are different.

Specimen banking enables tracing newly recognized pollutants and permits reevaluation of an environmental finding when new and improved analytical techniques become available. This is particularly relevant in the context of the estimated 60,000 or so chemicals of industrial consequence, of which only a handful are currently believed to be examined for their environmental impact *(5)*. Further, the role of ESB is enhanced by the possibility of providing additional samples from a parent pool to answer questions raised by RTM, and this is a crucial component of a specimen banking program. Carefully designed and wisely applied ESB can help to identify and to characterize the nature and extent of dispersion of potentially harmful chemicals in selected environments. As elucidated by Lewis *(11)*, if the capacities of ESB are linked to appropriately designed monitoring and predictive tests, an assessment of potential harm is possible, and reasonable regulatory or management practices can be recommended. Through routine fingerprint analyses of chemicals in the specimens to be stored, it is likely that some contaminants will be identified early, before serious impacts are observed.

## Establishment of Specimen Banks

**Infrastructure.** The parameters that influence the establishment of a human tissue specimen banking facility are funding considerations, environmental health activities, and proximity to a large research centre (preferably a geographically convenient location in the country) with bioanalytical and related technical facilities. To develop an analytical facility as a part of the specimen banking activity would be too expensive, and therefore access to a research centre with the above mentioned infrastructure is critical. An efficient transport system with the possibility of a mobile sample collection facility (a small work station with filtered air, a cold storage and related equipment to carry out a few basic laboratory operations if necessary) for field work will be a definite advantage. Uninterrupted electrical power, liquid nitrogen supply and institution of a constant observation system are some of the other requirements. The proximity to a major biomedical centre provides the much needed multidisciplinary input and helps to develop a systematic user group for the environmental surveillance projects. This kind of infrastructure has been a crucial point responsible for the success of existing major specimen banking facilities.

Assembling together an effective team of analysts to provide for the analytical needs of a specimen banking project is equally challenging as one of the infrastructure requirements. In the inorganic analysis area access to different analytical methods is indispensible. Similarly in the organic area, analysis of environmental samples such as water, soil, sediment and biomaterials requires dealing with complex mixtures (e.g. organohalogenes). Therefore, establishment of an analytical scheme to screen for a large number and types of organic compounds is difficult because of the requirements of selective extraction, digestion and isolation procedures required to detect various organic species. This is a major factor in planning organic analysis in any specimen banking program.

Experience has revealed that it will probably take the entire first year to prepare the program to develop protocols and to identify suitable sampling locations and project paticipants. During the second year, establishment of the bank and collection of a select number of samples to test the sampling and analytical systems and an intermediate evaluation of the project can be accomplished. The third year should enable normal specimen banking process and related analytical work.

**Scope and Size.** The scope of an intended biomonitoring activity bears direct influence on the size of the facility. Basically two systems come into the picture. i) Efforts that are predominantly dedicated to real time monitoring (RTM) where the focal point would be periodic investigations to assess the impact of suspected harmful situations. Routine monitoring activities and occupational health problems are typical examples. ii) A situation that calls for establishing a link between the original status and probable changes in the environment, a unique provision accomplished by retaining a part of the original sample for subsequent evaluation, technically termed specimen banking. This is a typical situation when exploring unexpected events such as unsuspected exposures to a multitude of chemicals. Although these two monitoring systems differ in their basic features, some requirements are the same for both cases. For example, there is no difference in choosing specimens and the analytical precautions required to handle them. RTM is less complicated since extended storage (and preservation) of specimens is not anticipated. The need to preserve the specimen for an unknown length of time introduces technical as well as biological sample (structural) integrity and related difficulties, and changes the scope of the problem. The costs involved are formidable and therefore, the question of affordability may require a reevaluation.

Also to be considered is the potential for a possible information gap if too many samples are collected and are not analyzed at some time or the other. As an on going expenditure, the analytical part is the most expensive component of the specimen banking budget, more so if the purpose is a comprehensive coverage of constituents. Therefore, it may be prudent to collect a modest number of samples and devote more attention to improving the quality of the stored specimen as well as the analytical output.

**Multidisciplinary Needs.** A multidisciplinary approach is an essential component of many biomedical investigations *(15)* and specimen banking is no exception. In dealing with specimen banking problems, the multidisciplinary angle includes both the scientific and the administrative components. The ultimate success of a banking facility depends upon the coherence achieved among all the participants. This becomes obvious when one realizes that specimen banking facilities can and need to address transboundary problems and therefore, extend far beyond national perspectives. The addressing of scientific issues of a total environment naturally involves inter- and intra-agency interactions *(16)*. A workable forum has to be provided for this activity as a part of the overall specimen banking concept. Thus it becomes obvious that several agencies have to participate in sharing the burden of conceptual, financial and organizational matters. After this phase is put in place, a panel of experts drawn from ecologists, toxicologists, analysts, statisticians and biomedical researchers (including cryobiology expertise) would be required to outline the issues involved. Environmental contamination being such a broad subject, the final compositon of the expert group depends upon the objectives of a particular banking facility. For example, a human tissue bank would typically need more biomedical, bioanalytical, biostatistical and toxicologcal expertise. This scientific committee has to interface with the administrative panel drawn up from chief supporters and users of the banking facility. These considerations should begin as the bank is planned and continue throughout its operation.

On the negative side, it would be prudent to anticipate problems in coordinating a shared effort of this kind. Obviously, the individual institutions pursue differing priorities and differences in opinion may not always be avoidable. This may also reflect in inconsistencies in funding obligations. Therefore, a strong, resourceful and forceful coordinating system will be needed to steer the activities of the specimen bank towards its initial goals.

**Biostatistical Considerations.** The general biostatistical and epidemiological issues relating to specimen banking can be broadly catergorized as design and measurement problems and analysis and interpretation problems. The design of the specimen banking system should be motivated by the ultimate application of the results or products derived from the system. From a statistical perspective there should be considerations of validity, reliability, efficiency and cost. The first design should be to define the target population. Then the sample should be ideally a random sample of the defined target population. Logistical considerations may preclude a strictly random sample being selected. It is important to avoid selection bias and obtain a "representative sample" for valid generalization using probabilistic models. The size of the sample is directly related to the ability to detect important differences among subgroups in the target population using statistical tests of significance. The ability of statistical tests to detect the difference of a given size is called the "statistical power" of the test. The sample design of the specimen bank should be versatile to detect clinically meaningful differences.

The process of measurement depends on the nature of the variable and the level of measurement. These may be nominal, ordinal, interval or ratio. Instrumental sophistication alone is not adequate to dictate the level of measurement since on may occasions cost and efficiency considerations dictate the choice of the level of measurement. It is important to check the statistical model appropriate for this choice which again depends on the final policy decisions. When the aim is to increase the overall precision of any variable, the reduction of total variance should take into account the components of variation that can typically be explained by analysis of variance models. The partioning of variance also permit the estimation of the reliablility of the component systems. The correlation structure of the variables is useful in formulating prediction equations based on multiple regression equations. When a large number of variables are observed it may be possible to efficiently shrink the number of analytic determinations by using strategic regression equations. Problems of multicolinearity arising from using highly correlated variables should be kept in mind. The correlation structure also permits analyzing the measurements using factor analytic models which permit meaningful interpretation as well as efficient data compression techniques. While a single measurement however precise may not permit discrimination among significant risk groups, the method of discriminant analysis strategically exploits the correlation structure among the variable to obtain optimum discrimination with efficiency and cost reduction.

From epidemiologic considerations the analytic data derived from the specimen bank should be linked to important epidemiologic variables such as demographic, socioeconomic, lifestyle and occupational variables. It may be useful to design the system with future considerations of computerized record linkage in mind. Or eventually, there should be a Specimen Data Bank Registry. The following are some of the typical epidemiologic studies that can be potentially conducted using the specimen data bank: descriptive or prevalance studies, case control studies, and/or cohort studies . A detailed account of the biostatistical aspects are summarized in a special report *(17)*.

**Medico-legal and Ethical Issues.** With respect to regulations, customs and restrictions that should be considered while planning collection of biological samples, there are differences among various countries, in some cases even between different states within a country. These regulations can be severely restrictive with respect to human samples. For example, in the U.S. program, the three states involved (Maryland, Washington and Minnesota) in the human liver archiving project had different regulations. After about 10 years of liver collections, sampling has stopped in Seattle, Washington, U.S.A., due to changes in state laws. These things have to be considered in finalizing a sampling outline.

"Informed consent" should be sought from donors to obtain samples such as blood, and for collecting biopsies during surgery. The following information should be provided to the donors before obtaining their consent: procedure to be followed for taking the specimen; description of any attendant discomforts and risks which might reasonably be expected for the individual and the community; an assurance that the data and results will be kept confidential but accessible to the donor; and the donor is at liberty to withdraw from the investigation at any stage if the process involves repeated sampling. Besides obtaining informed consent, there are few other restrictions to be overcome when collecting specimens from living subjects. For collecting a specimen such as blood, both sterile and non-contaminating conditions have to be met. This also involves training medical personnel in trace analysis problems especially those related to contamination during collection, and educating the specimen bank personnel to observe safety precautions in handling infected specimens.

Human Tissue Banking is a relatively new phenomenon from medico-legal and ethical perspectives. Although established technical, legal and social guidance is available in the context of reproductive technology for banking ova, sperm, fertilized ova, embryos and zygotes, this approach is not directly applicable to specimen banking problems. Concerning the technical aspect, the collections for reproductive technology are handled differently than those intended for specimen banking. As for the legal aspect of human tissue banking, in the absence of any specific legal guidance presently available, the considerations applicable to the banking of specimens such as ova and sperm for reproductive purposes, provide an appropriate basis to highlight the problems. The assumption here is that these tissues are but a subset of the class of human tissues in general. Concerning the ownership aspect, Canada, for example, does not currently recognize property rights in human tissue, but the ethical consensus appears to be that individuals from whom tissue is collected for reasons other than those expressly and specifically mandated by law should have the right to decide, within applicable ethical and legal limits, the use to which this tissue will be put in. In context of human tissue banking, medico-legal restrictions are basically dependent on the country in question, and the existing specifications have to be considered in preparing the protocols. Transboundary situations, if prevalent, would probably complicate the situation. However, acceptable norms for shipping biological reference materials from one country to the other provides a basis where radiation sterilized dry biologicals and dietary materials qualify for transport.

Whether or not this is applicable to a specimen banking situation has to be evaluated on a case by case basis.

In the United States, for autopsy samples obtained for the NIST Specimen Bank, the pathologist is responsible for the medio-legal aspects, and for obtaining permission from next-of-kin, if and when necessary. The identity of the autopsied subject remains confidential. The laboratory receiving the autopsied material is expected to abide by the norms of treating the samples responsibly and make the findings available to the pathologist on request. Similar conditions would apply to biopsy and other specimens procured from a hospital and the physician concerned would have access to the results of tests carried out. The situation is similar in Germany where permission to acquire post mortem specimens is more difficult to obtain than ante mortem specimens due to legal complications.

For further information on this subject, the reader may consult chapter 23 in this book.

**Access to Banked Specimens.** Two factors govern this issue: the needs of the supporting agencies and the requirements of the scientific community. Obviously, to a great extent the interest of the participating agencies will set the overall course for the use of banked specimens. However, a few general guidelines may be offered in the light of the existing experiences in the field of specimen banking. These remarks are restricted to a central system.

Optimally, the specimen banking facility management should have independent authority in controlling the specimens in the bank. This means an agency that provides funds and participates in sampling and related activities of the banking program, accepts the bank management as the primary custodian of the banked specimens. This will not only facilitate smooth day-to-day working of the facility, but also enable the coordinator to alter priorities in times of unexpected budget cuts. Further, it ensures the much required continuity of approach in establishing strategies for looking far into the future, by not making the banking program a pawn to changing administrative (at the agency level) whims. In context of budget cuts, a facility owned by a single agency is likely to face severe setbacks in the event of fiscal stringencies compared to multiple agency supported system. In allocating the activities of the banking facilities, obviously, the needs of the supporting agencies get preference. An advisory committee of the supporting agencies operating through the program coordinator can be a solution.

Specimen banking as a discipline is still evolving and therefore, more research work needs to be carried out. A prudent use of the facility would be to arrive at a solution that calls for roughly a third of the collections to be made available to each of the three representations: the supporting agencies; the scientific community; and the activities of the banking facility for archival purposes to address unexpected events and to continue quality assurance measures.

**Database Management.** Often, the time and effort required for data processing is not accounted for in planning multianalyte studies. These investigations generate large amounts of numerical information. Data reduction is a major task and the budget should provide for this. Long-term biomonitoring studies should look upon this as a long-term commitment.

Databases (natural and exposure related) are essential for monitoring status and trends in environmental health. Non-uniformity in developing databases has been a limiting factor affecting the value of exposure information for risk assessment, risk management, surveillance of status and trends, and epidemiologic studies. According to Sexton *(18)*, the current and future exposure related databases should incude the following: (i) standardized procedure for the collection, storage, analysis, and reporting of data; (ii) an enhanced ability to compare data over time, i.e. conduct comparison studies of "old" and "new" methods; (iii) mechanisms for coordination and cooperation among public and private sector organizations with respect to the design, maintenance, exchange, and review of information systems; (iv) measurement of actual exposures and change, and review of information systems; and (v) data collection, storage, and retrieval methods that permit easy manipulation of information for both model building and testing. References (13, 46) provide further insight into the role of databases in environmental exposure studies.

As such, no internationally accepted structure for data management in specimen banks has yet emerged. As for the prevailing practices, in the German programme a computer system 4633 (DATAPOINT Deutschland GmbH) with a 20 MB hard disc permits the use of a custom computer programme (namely DATABUS) as the workhorse. Each sample is coded with an unmistakable identity and entered into the data system. This is a 14-digit code containing information on matrix, location of storage, storage temperature, storage container, weight of the sample, what it is analyzed for, laboratory where analyzed and running number for that particular sample. A separate working group has been established at the University of Muenster in the same complex where the Muenster Bank is situated.

In the U.S. bank, the day to day data management is essentially managed by the bank supervisor with the help of a standard Personal Computer with Word Perfect 5.1 and Paradox 3.0 (Data Manager) As shown in the protocol section, information on the human liver samples is collected, appropriately coded and entered into the computer. This is the basic source of information that contains details collected in the field, and therefore, an important piece of documentation. The rest of the data handling is as in any established laboratory.

## Existing Specimen Banks

Several countries such as Canada, Germany, Japan, Sweden, United Kingdom and the United States are presently engaged in biomonitoring activities through specimen banking. However the level of activity is not the same in all of these countries due to the scope of the monitoring activities. Thus, the specimen banking facilities located in Germany and the United States are the most well established. Therefore the technical and operating features of these three major specimen banking facilities are presented in some detail in the following sub-sections.

**Canada.** As part of the Great Lakes Monitoring Program, three agencies are involved in operating specimen banks. These are: Canadian Wildlife Services (CWS); Department of Fisheries and Oceans (DFO); and National Water Research Institute (NWRI). These banks are not formally coordinated. Coordination is by direct contact between the scientists and through various ad hoc committee meetings. These three facilities complement each other with the task of monitoring contaminants in the Great Lakes, and participate in external activities as and when the need arises. The major project objectives are *(10, 11)*: to fill in gaps in temporal or spatial trend analyses; to augment previous sample sizes; to verify previous results through re-analysis of samples using modern methodology; and to measure for the presence of chemicals of current interest in samples collected in the past.

CWS operates on a fairly large scale and banks specimens of birds, eggs, mammals, fish, reptiles and amphibians in support of surveillance and monitoring of wildlife for exposure to toxic chemicals. This facility has been actively carrying out toxic chemical monitoring since 1980 following a restructuring that recognized the potential of the existing specimens numbering more than ten thousand. Measures were undertaken to archive a select number of well preserved specimens and transforming the facility into a full scale specimen bank. The bulk of the material is stored at -40°C while a small portion is preserved at -80°C. For example, homogenized aliquots of whole fish, invertebrates and plankton are archived at -80°C. Recently, a limited number of specimens have also been preserved at -140°C. The DFO preserves samples of whole fish, invertebrates, and plankton at -80°C whereas, NWRI stores sediment samples at room temperature.

The specimen bank of the CWS, in collaboration with other Canadian Agencies, has initiated several long-term monitoring (LTM ) projects covering the Atlantic Coast (Seabirds), Eastern Forest (Woodcock Wings), Prairie Provinces (Prairie Falcons), Pacific Coast- Fraser Estuary (Great Blue Heron) and Great Lakes (Herring Gulls). Since 1970, several studies have been carried out to study organic pollutant trends in the Great Lakes area on specimens of birds, eggs, mammals, fish, reptiles, amphibians and Herring Gulls. Herring Gull eggs collected between the years 1971 and 1982 have been examined for chlorinated organics such as PCBs and DDE to establish contaminant levels. A substantial decline in both PCB and DDE levels has been demonstrated over time. The remedial measures introduced in Canada and the United States since the 1970s, limiting the use and disposal of PCBs are believed to have contributed to the observed decline in their levels.

Thus the CWS has over several years, acquired considerable data on the stability of organic chemical residues in Herring Gull egg contents stored at -30°C and -40°C since 1979. The reanalysis exercise carried out in 1985 has demonstrated that some organochlorine residues remain stable in egg samples; these include heptachlor epoxide, dieldrin, oxy-chlordane, hexachlorobenzene, p,p'-DDE and PCB. Studies have also been carried out to test specimen stability over extended periods of time. There are plans to extend these biomonitoring activities by including human tissue collection *(19)*.

**Germany.** The German program is a joint task of several units. Thus the facility at the Nuclear Research Center in Juelich acts as a central bank with satellite banks in Hamburg, Ahrensburg, Berlin, Kiel and Muenster. The governing principles for establishing the German environmental monitoring program are as follows: 1) real time information as to the distribution of man made chemicals and perhaps some of their decomposition products in the environment; 2) trends with respect to increasing threats posed by certain environmental chemicals believed to be deleterious to the environment including man; and 3) the long term preservation of aliquots of such samples which were originally analyzed for mapping out the present day distribution of known harmful chemicals and interpreted to determine any trends which may exist. These facilities archive a variety of environmental specimens. The purpose is to establish a comprehensive chain of materials (human, aquatic and marine organisms, terrestrial ecosystems and food chain) that are accumulators of environmental pollutants.

The Juleich bank was designed and constructed during 1980-81. The entrance to the storage and laboratory parts is controlled through an anteroom. Only prefiltered air is allowed to enter the facility. In addition, each storage section and the laboratory rooms are equipped with class 100 clean room installations. This facility contains liquid nitrogen vapor (LN2) freezers (-120°C at the top to -196°C at the bottom) as well as compressor cooling systems (-80°C). In all, there are 18 LN2 freezers (6 with 1.4 m$^3$ and 12 with 1 m$^3$ each). The average price of each LN2 freezer was 17,000 DM in 1981. The facility also has 4 compressor cooling systems (1.7 m$^3$ each) and a big refrigerator (1 m$^3$) . The total capacity storage amounts to 20.4 m$^3$ at LN2 temperature and 6.8 m$^3$ at - 80°C. These two cooling facilities are used for regular storage and for comparative purposes while the refrigerator is utilized for storing dried reference and control materials. Other features include a 24-hour alarm system, access to back-up power and a supply of LN2 for emergency purposes, etc. For technical details consult ref. *(20)*.

Juelich is responsible for the collection of non-human (carp, marine algae, soil, etc.) samples, specialized transportation under cryogenic temperature conditions and is responsible for inorganic analyses. However, the Juelich facility includes some parallel storage of human liver specimens for storage stability and related investigations. This facility operates on LN2 cooling.

The Muenster facility is entirely devoted to human specimen banking, and crucial access to medical expertise is assured through the University of Muenster. The 34 m$^3$ volume cold cell of the Muenster facility is housed adjacent to the Institute of Pharmacology and Toxicology. The operating temperature of the central storage part of the cell is maintained at -85 to -90°C using large capacity electrical cooling compressors. In all there are 3 such compressors, and each compressor has the required capacity to sustain all the cold storage operational needs of the facility. At any given time two compressors are in service while the third one functions as a stand-by. As a back up measure, the facility is also supplied with electrical power by a dedicated generator. In addition to these precautions, a supply of liquid nitrogen is also maintained for direct cooling of the cell, if necessary. The system has been operating without any major problems since 1980.

The cell is made of special steel (V-4A quality) and all the soldered areas and related critical spots are coated with teflon. The main part of the cell is equipped with fabricated storage compartments. One part of the cell contains two clean benches and a facility for LN2 storage (1.9 $m^3$ for comparative study purposes). The entrance to the specimen bank is maintained at -20°C and is separated from the main cell by insulated doors. As a part of established practice, the samples to be stored are precooled (by storing them in a spare cooling system for a few hours) to the required temperature in order to keep the cell under stable temperature conditions. Special safety outfits capable of protecting the workers at - 80°C and below are worn by operating personnel, and nobody is allowed to enter or work alone inside the cold cell. Additional features include a 24h-watch security system and related emergency measures. Precise cost estimates of this facility are difficult to obtain since this facility is supported by multiple means. The Muenster facility is staffed with multidisciplinary expertise designated to fulfill several functions in addition to maintainig the specimen bank. For further details consult ref. *(7)*.

The pilot phase at Muenster started in the 1970s with collections of liver, fat, and whole blood as target specimens. Subsequently, the bank was expanded to include over 20 major body components such as kidney, lung, muscle, thyroid, ovary and adrenal, among other specimens, from the same subjects as a part of an extended monitoring program. Besides the collection of over 13,000 autopsy samples (whole blood and various organs) for the banking program, over 266,000 samples (13) have been obtained for RTM. The monitoring program for organic compounds is extensive and several classes of compounds such as halogenated hydrocarbons and polychlorinated biphenyls (PCB), polycyclic aromatic hydrocarbons (PAH), aromatic amines and phenolic compounds have been determined. Pesticide monitoring of whole blood, blood serum, and urine is carried out routinely as part of the RTM. Over a period of 5 years between 1982 and 1987, a decrease in the level of pentachlorophenol (PCP) in urine and whole blood has been demonstrated. This trend has been linked to the 1979 German legislation restricting the use of PCP as a wood protectant medium. The monitoring for inorganics (mainly metals) is primarily aimed at Al, As, Be, Cd, Hg (including methyl-Hg), Pb, Sb, Se, Sn, and Tl. A few other elements are determined periodically to establish reference levels. Both Cd and Pb are determined in whole blood on a routine basis.

Human liver was selected for inclusion since it is an organ in which both inorganic and organic species accumulate; the macroscopic pathology of the liver is relatively homogeneous as compared to other human tissues; and sufficient sample can be obtained from one individual specimen. The fatty tissue is of interest for the lipophilic compounds. Milk and whole blood are ideally suited as real-time monitoring specimens. The choice of mussels was influenced by their worldwide availability and the extensive experiences regarding bivalves as indicators of water quality. The carp and the butterfish are good indicators for the changes in the composition of sediments and water. Extensive analytical work has been carried out in environmental materials collected by various German facilities participating in the specimen banking project. In all, as of 1988, the databank in Muenster contained approximately 20,000 individual results of organic and inorganic analysis for the entire spectrum of specimens in the banking project *(7)*.

**Japan.** The Japanese effort initiated in the early 80s to explore preservation of a variety of matrices such as atmospheric particulate matter, water, sediment, human blood serum, human hair and mussel tissue has remained on a modest scale. Storage facility consists of clean rooms at 5 and 25°C and freezers maintained at -20 to -80°C. LN2-cooled facilities providing -196°C storage conditions are also available. Since there are no immediate plans to update the facility to operate as a full scale bank, older samples are routinely replaced by newer and more valuable samples as and when they are procured. General interest recognizing the importance and usefulness of specimen banking for the long-term environmental monitoring of toxic chemicals is the basis for initiating the modest activities that are currently in progress. The pilot specimen bank program located at the National Institute of Environmental Sciences in Tsukuba is focussing on understanding procedural problems in documentation and retrival of stored specimens, maintenance problems encountered with storage facilities and evaluating safety measures relevant to specimen banking to counter unexpected situations  such as earthquakes *(21)*.

**Sweden.** The Swedish effort is based on freezing the specimens at -30°C. Due to cost considerations freezing below -30°C has not been considered. Some alteration in the composition due to this condition of storage is anticipated and accepted as a compromise. Samples for storage are collected from well defined, ecologically homogeneous areas representing three different ecosystems: terrestrial, fresh water and marine. Each ecosystem is represented by about 10 areas from which biological materials is collected yearly and kept in the specimen bank. About 50 individuals of each species are collected in each habitat. The following species have been chosen where different areas are represented by different species. Terrestrial- rabbit, reindeer, moose, fox, starling hearlings; fresh water- perch, roach, char and pike; marine- blue mussels, flounder, cod, young Baltic herring and guillemot. A total of 3500 samples are collected yearly. By choosing appropriate intervals for collection and by the process of elimination and substitution at any given time 10-25 years of perspective is feasible. However, storage instability at -30 degree centigrade is the primary limiting factor for the duration *(22)*. Although the general aim is to set up a biomonitoring system, the emphasis is on DDT and PCB research. The trends of many environmental contaminants in the Baltic Sea through evaluation of materials stored in the bank are being followed. Also, development of a human tissue bank at the Karolinska Institute is envisaged.

**United Kingdom.** The concept of preserving specimens for extended future use appears to have been practised in the U.K. over one hundred years ago by experimenting with soils at the Rothamsted Experimental Station, Harpenden *(23)*. The effect of various chemical treatments on soils was followed by investigating nutritive composition of agricultural produce (plants and grains). Specimens of soils and plants have been stored in glass bottles in dry form at room temperature. There have been other efforts to bank biological materials for environmental monitoring purposes. These include the Tolworth Laboratory near London (50 samples of owl species, selected specimens stored at -18°C and investigated for chlorinated pesticides, PCBs, mercury and cadmium), the Institute of Terrestrial Ecology in Huntington (internal organs of birds and eggs of wild birds stored in domestic freezers and investigated for phosphorous, organic chlorides, herbicides and heavy metals, and concluded that the storage system was unsuitable for retention of mercury and preservation of the chemical integrity of DDT), and the Fisheries Laboratory at Burnham-on-Crouch (mainly shellfish stored at -18°C , the entire fish collection amounted to about 220 cu ft by 1982, monitoring strategies are being evolved). The main purpose of the U.K. banks has been to study effects of agricultural technology and to investigate certain specific industrial pollutants. The facilities existing at different sites are associated with and ancillary to, a specific project to study contaminant levels. There are no plans in this country to set up large tissue banks and even human tissue collections for environmental monitoring purposes are not considered *(23)*.

**United States of America.** In the United States systematic activity in specimen banking area is attributable to the National Biomonitoring Specimen Bank (NBSB), which began as a joint project of the U.S. Environmental Protection Agency (EPA) and the Department of Commerce (DOC). The idea of a specimen bank emerged in 1972 *(24)* to ensure the availability of analytically and biologically valid specimens for environmental monitoring purposes. Subsequently, in 1976, a pilot project was conceived as a joint effort between the EPA and the National Institute of Standards and Technology (NIST) (formerly National Bureau of Standards) and a comprehensive program was developed to acquire working experience in all aspects of specimen banking. Four types of samples representing a broad range of environmental accumulators were recommended for the initial phase of the program. These were: soft human tissues from organs with accumulator function (liver); accumulators of aquatic origin (bivalve shellfish such as mussels or oysters); a typical food chain (human diet); and an indicator of airborne pollutants (lichen or moss to indicate long-term trends in atmospheric pollutants).

The NIST specimen banking facility was completed in 1979 and contained an area measuring 27 ft x 32 ft of class 100 laboratory space, of which 15 ft x 32 ft was designated as sample handling and preparation area. The storage section has been recently expanded by the addition of another unit measuring 17 ft x 27 ft modular class 1000 type. Therefore, the total area occupied by the bank is approximately 1300 square feet, divided into three parts: (i) class 100 sample preparation laboratory; (ii) class 100 homogenization area; and (iii) a class 1000 area containing all of the LN2 freezers. Currently, 10 LN2 freezers and 4 electric freezers (for -25 and -80°C storage) hold all the samples in the program. The LN2 freezers are automatically replenished from external LN2 reservoirs. The biological storage freezers (compressor type) are provided with LN2 back-up measures and the whole facility is under the surveillance of a 24 hour monitoring system that indicates freeaer fsilres and power failure or fluctuations, if any, in the temperature conditions.

The laboratory part is equipped to handle both organic and inorganic analytical sample prepartion requirements. The organic sample preparation can be carried out on stainless steel bench top, while the area designated for inorganic analysis has non-metallic (e.g. plastic) bench tops or plastic covered bench tops. For further technical details consult ref. *(25)*.

Since the inception of this project over 600 human liver samples have been collected from various regions in the U.S. (Baltimore MD, Minneapolis MN and Seattle WA). In 1985 the National Oceanic and Atmospheric Administration (NOAA) initiated activities connecting its National Status and Trends (NS and T) to NBSB. Thus two monitoring projects, namely the Mussel Watch Project and the Benthic Surveillance Project evolved to include banking *(26)*. Under these new initiatives samples of sediment, mussels, oysters and fish tissue are collected for banking *(27)* and analyzed for chlorinated pesticides, PCBs, selected polycyclic aromatic hydrocarbons and a number of trace elements *(28-32)*. In 1987, another joint effort (NOAA, NIST and the Mineral Management Service) began collecting and banking Alaskan marine mammal tissue (blubber, kidney and liver). In the human nutrition and health front two initiatives are in progress at the NIST. These include human serum and total diet composites *(33)* archived and evaluated for storage stability. The human serum project is sponsored by the National Cancer Institute and the diet program is a joint effort by the U.S. Food and Drug Administration, the U.S. Department of Agriculture and the NIST in collaboration with the International Atomic Energy Agency in Vienna. The serum samples are for monitoring Beta-carotene and vitamins A, C and E. As for diets, biomonitoring of several trace elements of biological significance and storage stability of selected organic nutrients are foreseen.

A major role assigned to the pilot EPA-NIST specimen banking facility was to conduct research on all aspects of preservation of bioenvironmental specimens for long-term use and develop systematic protocols. The research included the following key aspects: i) develop and evaluate protocols for contamination-free sampling, processing and storage of environmental specimens; ii) evaluate and improve analytical methods for determination of trace elements and organic pollutants; iii) develop and evaluate conditions that permit long-term storage of samples without change in pollutant concentrations; and iv) evaluate a specimen bank as a means of storing samples for monitoring pollutant trends over time.

In the United States, during the first 18 months, 300 liver samples were collected from the three locations mentioned earlier. For the remaining 2-3 years of the pilot phase the samples were received at a rate of approximately 100 per year. This phase of sample collection was also used to evaluate the sample collection protocol with respect to sample procurement costs, efficiency of transport, initial set-up costs for each site, time required to recieve hepatitis and AIDS tests results, histological slides and suitability of donor selection criteria.

The United States program is operated from a single location based at the National Institute of Standards and Technology in Gaithersburg. Policy matters concerning protocols, expansion of the banking facility, use of the storage space, and the data generated are jointly managed by the participating institutions such as the NIST, National Oceanic Atmospheric Administration and the Environmental Protection Agency. The NIST facility is set up for human liver, marine specimens and food composites. The responsibility for running the bank on a day to day basis is entirely left to the NIST. Presently, major efforts are directed towards marine mammal specimens and the human liver program. The liver specimens have been preserved for over 12 years now. Since the inception of the program over 600 human liver samples have been collected and stored. The protocol for collecting the liver samples and the step by step working procedures have been documented. The storage conditions adopted are liquid nitrogen temperature for preservation and -80C and -25C cooling facilities (electrical freezers) for stability studies. The organic contaminants that have been determined in human livers include: hexachlorobenzene, the beta isomer of hexachlorocyclohexane, heptachlor epoxide, trans-nonachlor, dieldrin, p,p'-DDE, and p,p' DDT. The results showed that the most abundant pesticide residue was p,p'-DDE, the metabolic derivative of p,p'-DDT. In the inorganic analysis program, up to 30 elements are being determined in livers to establish baseline concentrations. Significant among the findings for the concentration ranges are, more closely grouped values of many essential trace elements and lower than previously reported concentrations for Al, As, Sn and Pb.

## Funding and Management

Until now the funding for all specimen banks has been almost exclusively from federal government sources. Information available on the management issues are mainly from the German and the U.S. banks. In Germany, the facility in Juelich acts as a center for coordinating the activities of the satellite banks. It is also responsible for quality assurance (mainly inorganic measurements), development of sampling protocols and sample transport. In the United States, the responsibility for running the bank on a day to day basis is entirely left to the NIST. Policy matters concerning protocols, expansion of the banking facility, use of the storage space and the data generated are jointly managed by the participating institutions such as the NIST, NOAA, and the EPA.

## Analytical Aspects

**Sampling and Sample Handling Strategies.** For a comprehensive bioenvironmental monitoring program, the basic planning should take into consideration the requirements of both inorganic and organic pollutants. Although preferences for any particular group of pollutants can be accomplished by modifications within the system, the working procedures should be carefully evaluated. If the focal point is organic pollutants, then retention of the biochemical integrity of the specimen (e.g., by cryogenic preservation) is of highest consideration during pre- and post- sampling stages. On the other hand, sampling tools (non-contaminating or specific tools made of Ti etc) and ambient conditions play a crucial role if inorganic pollutants are of primary concern. Therefore, a combination of both types of working procedures is necessary for a successful and broadly based biomonitoring program.

Sample selection and collection are critical components of a biomonitoring system and a compromise at this stage would seriously eclipse the biomonitoring potentials of an investigation. A critical consideration in evaluating the adequacy of sampling and processing conditions should take into account that specimen bank is for long-term preservation of samples that are representative of the location and biological features prior to the time of collection. Further, processing and storage conditions should minimize extraneous contamination of the samples and safeguard against any changes in chemical composition. In this context, the detailed sampling and processing protocols developed by the NIST for liver *(34)*, sediments, fish and selected marine organisms *(35)* and marine mammal tissues *(36)* are good examples of the efforts required for such operations. These documents focus on the need for collection of duplicate samples, selection of materials for preparing working tools to minimize the hazard of contamination, and the need for freezing the samples as soon as possible.

The development of protocols with specific details for collection, transportation, processing, banking and analysis is indispensable for a successful tissue archival project. Protocol development is one of the most essential components of a specimen bank. The protocol begins before the actual tissue collection takes place, and should include the following features: the physical set-up of the collection site (cleanliness conditions, danger of contamination from tools and other sampling aids); sampling kit and cleaning procedures; instructions for anatomical incisions; and instructions for handling the tissue when it is removed from the source; details for further processing of the sample; safe container for transport; labeling instructions; and sample information sheet to be filled by the person collecting the sample . Arrangements should be in place for transport of field samples to the laboratory under frozen conditions (transport containers with dry ice) and for permanent storage conditions *(3)*. This involves coordination of the activities of the individuals responsible for various segments in the chain of operations. Without a well coordinated system, the sample validity can be totally lost.

**Storage and Preservation Strategies.**   For biochemical and technological reasons, there is a difference between specimen storage and specimen preservation. Preservation essentially relates to the process of retaining morphological and cellular integrity of tissues or organs. Therefore, under ideal preservation conditions, specimens so protected may be expected to retain the natural chemical forms of various constituents with the ability to restore all the "functional" properties when required. This is the kind of preservation (with "total" functional properties) that is of great interest to the developmental biology discipline where cells are preserved and recovered for various applications *(37)*. In the case of specimen banking, preservation primarily relates to the structural integrity of biochemical moities (e.g. enzymes, antigens), and therefore, rupture of the cells, if any, during banking may not necessarily disqualify the specimen. Storage, on the other hand, especially from the trace element studies point of view, basically implies an overall safe, clean, uncontaminated and non-degraded containment.

The standard technique used for preservation is at cryogenic temperature which is considered to be below -80°C. The science and technology of cryopreservation (creating an inactive state of metabolism by subjecting the cells of an organ to ultra low temperatures, namely at liquid nitrogen cooling) is an established tool in the field of developmental biology where cells, tissues and embryos are routinely preserved and regenerated for various applications *(37)*. The success of preservation by this technique depends upon the prevention of recrystallization of the ice, which is lethal.

To overcome these shortcomings, the biologists use what are known as cryoprotectants (e.g. dimethyl sulfoxide), and immerse the tissue in a few mL of the protectant solution of appropriate strength and gradually cool the container by retaining it under liquid nitrogen. Depending upon the circumstances and requirements, temperatures between 0 and -80°C have also been used *(37)*. Because of the simplicity, convenience and the low temperature attainable associated with liquid nitrogen, use of this medium for cryopreservation is quite common. Obviously, in a specimen banking program aiming at biomonitoring the environment, cryopreservation without the use of cryoprotectants alone is valid (and adequate), and therefore, the general recommendation is to use ultra low temperatures, i.e., at liquid nitrogen conditions. Preservation of a sampled specimen is largely dependent upon the matrix properties, the time span available between collection and transport to the laboratory and subsequent handling for analysis. For example, preserving bone, hair or nail is relatively simple when compared to the conditions required to preserve soft tissues such as brain, liver, lung or kidney.

During the pilot phase of the NIST specimen banking project, the storage scheme was designed to evaluate the question of appropriate temperature for storing biological samples and to bank well-characterized reference samples. The liver samples of the left lobe are received at the NIST as sections A and B. All of the A sections are placed at LN2 vapour temperature for long-term storage and B sections are used for analytical work. For example, during one phase approximately 30 of the B sections were homogenized using the cryogenic homogenization technique to provide about 20 aliquots of 6-8 g each per aliquot. The sample is retained in frozen form during grinding and sample aliquots are transferred to teflon storage jars inside a cold nitrogen atmosphere glove box to minimize water condensation on the frozen samples. Currently, the A sections represent part of a valuable bank of well characterized and documented samples. In addition, a large quantum of analytical data are available from the analysis of

B sections. This kind of parallel storage is a very important concept for a specimen bank. At the U.S. facility, the A and B sections of a sample are stored in separate freezers. Therefore, in the event of failure of a freezer, only one part of the sample would be lost and the other section of the sample would still be available. To investigate aspects of long-term storage, the sample aliquots were stored under four different conditions: room temperature after freeze drying, frozen at -25°C, frozen at -80°C, and frozen at liquid nitrogen vapour temperature (-125 to -190°C). These aliquots have been reanalyzed during the pilot study phase and compared to data from real-time analysis (i.e., analysis performed soon after homogenization) to determine if changes in the concentration of trace elements or trace organics (e.g., organochlorine pestcide residues) have occured. Samples stored at -25°C and -150°C (LN2 vapour temperature) have been analyzed after 7 years of storage. Selenium and zinc among the inorganics and 4,4'-DDE and PCB 153 among the organic pollutants have been tested. Comparison of the results for 24 samples for selenium and zinc from each of the two conditions with baseline data showed no indication of any change of zinc concentration *(38)*. However, the analytical uncertainty in the determination of selenium was greater compared with that of zinc. In the case of the organics, 9 aliquots from each of the two storage conditions were compared and as with trace elements, no signigicant changes were observed. Unfortunately, no organic baseline data from the 1980 liver samples were available to determine whether analyte concentrations had changed at both of the storage conditions since the initial storage *(27)*. Extensive work has been carried out at the Juelich facility to examine the storage stability of all the sample systems of the banking project through coordinated efforts. For example, among the human specimens both blood and liver have been investigated for the stability of cadmium, mercury and lead and found to be satisfactory *(39, 40)*. The Canadian group has accumulated specimen stability information over five and a half years of storage at -30 to -40°C. For example, several chemical residues have been shown to be stable in Herring Gull egg homogenate by systematic reanalysis *(41)*.

In the US study, even though the chemical analysis of the samples stored at -25 and -150°C indicated no significant changes in composition, there was some physical evidence of changes in the sample aliquots. At -25°C, the aliquots of frozen liver homogenate had formed ice crystals under the container lids and on the sample surface (i.e., moisture in the sample had separated), and the aliquots were no longer powdery but clumped. On the other hand, the samples stored at -150°C had remained powdery, as they had been at the time of homogenization *(27)*.

**Determination of Organic Constituents.** Analytical procedure used in the U.S. pilot project for the determination of selected individual polychlorinated biphenyl (PCB) congeners and chlorinated organic pesticides in human liver is based on HPLC. Two liquid chromatographic fractionations are utilized to remove analytical interferences and to separate the analytes of interest into two fractions prior to quantification by high resolution gas chromatography with electron capture detection. The contaminants identified include: hexachlorobenzene, the beta isomer of hexachlorocyclohexane (B-HCH), heptachlorepoxide, trans-nonachlor, p,p'-DDE, dieldrin and p,p'-DDT. The results showed that the most abundant pesticide residue was p,p'-DDE, the dehydrochlorinated derivative of p,p'-DDT. A composite of about 10 livers from the German collection was also analyzed as part of an interlaboratory comparison of methods. Compared to the US sample, it is of interest to note the large auantity of hexachlorobenzene (equivalent to 2 $\mu$g/g extractable fat) in the German specimen *(42-44)*. This example illustrates the potential differences in baseline values which may be discernable on a large international scale. Recently, frozen whale blubber and frozen mussel specimens (reference materials) have been analyzed for a variety of trace organic constituents using gas chromatography with electron capture detection and gas-chromatography-mass spectrometry.

Concerning the German program Ballschmitter has presented a list of over 30 individual or classes of organohalogen compounds (C1 to C26) in his survey article *(42)* to provide an idea of the analytical efforts involved. An example of the complexities involved is described with reference to the ubiquity of organohalogens in air, chemicals, solvents and technical equipments, a challenging problem for specimen banking programs. Several German institutions are involved in various aspects of organic analysis *(45)*. In one approach, multi-step separation schemes based on liquid chromatography procedures to isolate PCBs and DDT, coupled with high resolution capillary gas chromatography linked to a electron capture detection system has been developed for application to tissue samples. These procedures have been verified by exchange of samples between the German bank and the NIST facility for the determinination of organochlorine pesticides. Results reported for hexachlorobenzene, Beta-HCH and 4,4'-DDE are in very good agreement suggesting good standaridization steps in both the laboratories *(40)*. In another analytical scheme, also using capillary gas chromatography and use of two internal standards, methods were developed for preconcentration and determination of chlorinated hydrocarbons, chlorobiphenyls and other trace organic substances *(46)*. Elimination of fat content was achieved by high pressure chromatographic methods (HPLC), supported by gel permeation to clean up coextracted macromolecules, if any. Recently, World Health Organization (WHO) carried out an intercalibration study on dioxins and furans in human milk and blood *(47)*. The study included 19 laboratories from 14 countries. The qulaity control (QC) exercise was designed for direct comparison of laboratory and method performance related to clean up methods, instrumental parameters, analyte concentration, laboratory quality assurance/quality control (QA/QC) programs and laboratory experience. The study revealed that laboratory is the single most important determinant of data precision and accuracy. The mode of analyte enrichment (sample clean up), analyte measurement (gas chromatography/mass spectrometry; GC/MS protocol), and analyte concentration showed weaker correlations with data quality.

In the German program besides the liver samples, pesticide monitoring of whole blood, blood serum and urine is carried out as part of RTM. The monitoring program for organics in human tissue samples is more extensive than the U.S. pilot program and currently, several activities are taking place. Over 3000 human fat tissue collections have been analyzed for a number of pollutants and the data are stored in the central databank of the Environmental Specimen Bank at Muenster *(48)*. For example, the following organochloro compounds were determined: pp-DDT, hexachlorobenzene, pp-DDT, op-DDT, alpha-hexachlorocyclohexane, pp-DDD, beta--hexachlorocyclohexane, op-DDE , gamma--hexachlorocyclohexane, pp-DDE, pentachlorophenol, PCB28, heptachloroepoxide, PCB52, Dieldrin, PCB101, PCB 138, PCB 153, PCB 180.

A shift in the distribution of PCP towards the lower end of the scale is clearly seen between 1982 and 1987. A similar tendency was also noticed for PCP in human whole blood *(43)*. These were effects following a legislative action in 1979 restricting the use of PCP as a wood protectant medium. Further observations on the measurements of other organics such as PCP, DDT and DDE in whole blood and blood plasma have been made *(43)*. These observations reveal a continuous decrease in the mean concentration of DDT (which has a very slow excretion rate from the body) and its biotransformation product DDE over a period of 1982 to 1986. Of importance is the fact that the ratio of DDT/DDE has also changed, indicating a definite reduction in recent exposure to DDT. Another observation relates to the serum PCP concentration which was shown to correlate well (especially at lower concentration levels) with hexachlorobenzol. Based on this, these investigators theorize that the PCP burden in normal human subjects does not correspond to the direct exposure to PCP (as commonly believed) from the external environment, but is related to the exposure of biotransformation product of hexachlorobenzol *(43)*.

**Determination of Inorganic Constituents.** In developing analytical schemes for inorganic constituents in a specimen banking project, it is prudent to plan for a broad range of elemental coverage since the aim is environmental surveillance and it is necessary to establish baseline values for as many elements as possible. This means access to at least one multielement analytical technique is very helpful.

In the U.S. specimen banking program, the pilot stage involved use of several analytical techniques: atomic absorption spectroscopy (AAS), isotope dilution mass spectrometry (IDMS), instrumental neutron activation analysis (INAA) and voltametry (VOLT) were used to determine a total of 31 elements. In addition to the four techniques which were used routinely, radiochemical neutron activation analysis (RNAA) procedures for Sn and Pt were developed. Details of these analytical procedures have been discussed elsewhere *(49)*. In the U.S. program both INAA and RNAA have been consistently used for on-going analytical work. In context of the multielement survey potential, NAA offers excellent possibilities as demonstrated for a variety of matrices such as bovine liver *(47)*, human milk *(50)* serum, blood and urine *(51)*, among others.

In the U.S. program, up to 30 elements have been determined to establish baseline concentrations in human livers in 96 specimens collected in 1980, 1982 and 1984. These livers cover all the three collection locations namely Baltimore, MD; Minneapolis, MN; and Seattle WA *(46)*. One observation from the inorganic baseline data for the human livers is that many of the pollutant trace element concentrations are on the low end of or below previously reported ranges. Specially, the levels of Al, As, Sn and Pb are lower than previously reported concentrations of these elements in human liver from 1940 to 1972 *(52)*. The data for lead concentrations in the human liver specimens illustrate the use of baseline data in monitoring environmental trends in pollutant levels. The results for lead in three pools of human livers from 1980, 1982 and 1984 show mean values of 0.55, 0.39 and 0.47 $\mu g/g$, respectivley, indicating reductions as compared to earlier findings *(52)*. These changes can be linked to improvements in analytical methodology, improvements in control of sample contamination and a decrease in the lead levels in the environment. This decrease in the lead levels is best illustrated by comparing the median values for the three locations (Baltimore, Minneapolis and Seattle), 0.46, 0.34 and 0.26 $\mu g/g$, respectively. These data more accurately reflect the decrease in lead levels in the environment from 1980 through 1984 which may be attributed to the decrease in the use of leaded gasoline and lead-containing paints *(10)*. Data for arsenic in the 1984 human liver collections (2-20 ng/g) illustrate the improvements in the analytical procedures and have set the basis for establishing reliable reference values *(53)*. Another significant finding from this study is the very narrow range of values obtained for many essential trace elements. Thus, there is only a factor of 1.8 difference between the highest and lowest values of selenium.

In the German program AAS has been used predominantly for elemental analysis while voltammetry has proved to be a valuable alternative when a second method was needed for verification purposes. Detailed procedures for handling blood plasma, whole blood and hair are described *(51)*. For other specimens such as liver and urine, the steps are modified appropriately.

In the German program, inorganic analysis is designed to cover the following three groups of elements: aluminium, antimony, arsenic, barium, beryllium, cadmium, lead, mercury, silver, strontium, thallium and tin (environmental exposure related heavy metals); chromium, copper, iron, manganese, nickel, selenium, vanadium and zinc (essential elements); and calcium, magnesium, phosphorous, potassium, sodium and sulphur (essential minor elements). Analysis of 192 blood samples and 234 livers have been completed so far and the data are accumulated in the central databank of the Environmental Specimen Bank at Muenster *(48)*. No review or summary of the information has been reported. Determination of cadmium and lead in blood as a part of the RTM is in progress through the Muenster facility. Specimens for monitoring organic pollutants.

**Chemical Speciation.** The identification of the chemical forms (speciation) of biochemically active components of trace elements in biological systems is gaining importance. Measurement of total element concentrations do not provide information on bioavailability and toxicity. Cryopreservation of biological specimens (without the use of cryoprotectants) meets the sample quality requirements for both organic and inorganic pollutant species. Several precautions are needed to handle the chemical speciation problems including a sound biochemical approach. Typically speciation problems are studied by combining a separation step (e.g. high pressure liquid chromatography) with an atomic absorption or emission spectrometer or a mass- spectrometer. In biological systems, elements often occur as element-ligand complexes with considerable affinity to exchange among various chemical forms. Therefore, the integrity of a certain element-species during sampling, preparation and processing is unpredictable and may lead to conflicting claims. The field of trace element speciation is still too young and more research efforts are needed. Biomonitoring programs assisted by Specimen Banks provide excellent opportunities to address speciation problems since the preservation conditions are compatible for protecting the biochemical integrity of the organic matrix.

**Analytical Quality Assurance.** Bioenvironmental investigations require provisions for a "total" Analytical Quality Assurance (AQA) to yield meaningful results. These are measures that encompass all stages of an investigation such as experimental design, collection of analytically and biologically valid specimens, analytical measurement processes and proper evaluation of analytical data, including data interpretation. Further, one of the critical decisions that the analyst needs to make is the degree of quality (quality standard), or tolerance limits, required for the purpose of the investigation for which the analyses are being made. If the tolerance limits are set narrower than the investigation really requires or not feasible under practical laboratory conditions, it can cause unnecessary expense and loss of time.

One of the most effective tools in initiating an AQA program is to use two or more independent analytical methods to verify the accuracy of an analytical finding. This, for example, is the situation in the inorganic (trace element) analysis area since alternative methods of determination are applicable to a majority of trace elements. On the other hand, in the area of organic analysis, procedures for AQA of many constituents is still evolving. Therefore, in a specimen banking program  the state of the art of analysis permits acquisition of reasonably well-founded inorganic baseline data, whereas additional efforts may be needed to reach the same stage for the organics.

Another approach involves the use of certified reference materials (CRM) which offer the most effective and direct approach for method validation in analytical chemistry. This has been recognized by many countries leading to several new initiatives. Literature survey reveals that impressive progress has been achieved in developing a series of CRM for inorganic constituents in a variety of matrices *(54)*. This has been possible since in most cases two or more independent analytical methods required for such certification programs is readily available. In the area of organic analysis, procedures for AQA for many constituents are still evolving *(55)*. There is ample awareness among investigators recognizing the need for generating reliable results, but this perception has not translated into action other than the recognition of inconsistencies stemming from intercomparison trials. There is no swift follow-up by identification of the sources of discrepancy, initiation of the remedy, and subsequent AQA exercises. One obvious reason for this slow progress is the paucity of funding for basic research in analytical methodology, thus shifting a major fraction of the developmental work to a few institutions such as those dealing with RM programs.

The factors mentioned above are only partially responsible for the lack of rapid progress in developing a wide variety of urgently needed organic RMs. There are of course genuine methodological and technical hurdles, by far the most challenging step being preparation of a suitable natural matrix while retaining the compositional integrity of the organic material, without excessive cost for extended preservation.

Established agencies such as the European Community Measurement and Testing Program (MTP, Belgium), NIST (USA), International Atomic Energy Agency (IAEA Austria), National Institute of Environmental Studies (NIES Japan), and the National Research Council of Canada (NRC, Canada) have developed several certified RMs ranging from simple solutions for calibration of analytical instruments to complex natural matrix materials. Some of these are fortified, suitable for validation of methods adopted for nutritional and environmental biomonitoring and related programs.

## CONCLUSIONS

Environmental specimen banking (ESB) enables a systematic collection and careful preservation and storage of an array of environmental samples for deferred chemical characterization and evaluation. ESB permits identification of environmental trends, retrospective evaluation, identification of new chemicals in the environment and development of baseline data for use in public health issues. Long-term financial commitment is mandatory to ensure uninterrupted operation of the banking and monitoring programs. The access for utilization of samples in the specimen bank and the ownership of the archived portion of the collections are issues that should be resolved at the planning stage of the project. Biostatistical and epidemiolological considerations have to be built into the protocols of future specimen banking programs. There is a need for multidisciplinary involvement in setting up a tissue bank. Although the ultimate success of a specimen banking facility rests on being able to offer well preserved specimens certified for their storage stability, not many investigations have exclusively focussed on this subject. Efforts are needed to harmonize environmental measurements. Specimen banking offers an excellent opportunity to initiate action in this context (56). No tangible mode of data management (documentation procedures, data collection and processing) appears to be in place to serve as a model for evaluating specimen bank analytical output, especially in connection with human tissue data collection. One reason for this deficiency can be seen in the lack of representation of biostatistical experts in the

working groups. Many chemometric measures *(49)* are now available as effective
tools to facilitate better data interpretation and this should be given consideration
in future plans. However, there can be little doubt that the research and
applicational potential of ESB is practically inexhaustible and is limited only by
the imagination of the investigators.

**Literature Cited**

1.   *Historical Monitoring. MARC Report Number 31,* Monitoring and
     Assessment Research Centre; University of London: London, 1985.
2.   *Human Tissue Monitoring and Specimen Banking;* Environ. Health Persp.
     Supplement, Vol. 103, Suppl. 3, 1995.
3.   *Progress in Environmental Specimen Banking;* Wise, S. A.;  Zeisler, R.;
     Goldstein, G.M., Eds.; National Bureau of Standards (presently National
     Institute of standards and Technology): Gaithersburg, MD, 1988, NBS
     Special Publication No. 740.
4.   *The Use of The Biological Specimens for the Assessment of Human
     Exposures to Environmental Pollutants;* Berlin,  A.; Wolff, A. H.;
     Hasegawa, Y., Eds.; Martinus Nijhoff: The Hague, 1979.
5.   *Monitoring Environmental Materials and Specimen Banking;* Luepke, N. -P.,
     Ed.; Martinus Nijhoff: The Hague, 1979.
6.   Boehringer, U. R.; Schmidt-Bleek, F. In *The Use of The Biological
     Specimens for the Assessment of Human Exposures to Environmental
     Pollutants;* Berlin, A.; Wolff, A. H.; Hasegawa, Y., Eds.; Martinus Nijhoff:
     The Hague, 1979, pp 13.
7.   Kemper, F. H. In *The Use of The Biological Specimens for the Assessment
     of Human Exposures to Environmental Pollutants;* Berlin, A.; Wolff, A. H.;
     Hasegawa, Y., Eds.; Martinus Nijhoff: The Hague, 1979, pp 342.
8.   Stoeppler, M.; Duerbeck, H. W.; Nuernberg, H.W. *Talanta.* 1982, *29,* 963.
9.   *Environmental Specimen Banking and Monitoring as Related to Banking;*
     Lewis, R. A.; Stein, N.; Lewis, C.W., Eds.; Martinus Nijhoff: The Hague,
     1984.
10.  Wise, S. A.; Zeisler, R. *Environ. Sci. Tech.* **1984,** *18,* 302A.
11.  Lewis, R. A. *Environ. Prof.* **1986,** *8,* 138.
12.  Commission on Life Sciences. *Ecological Knowledge and Environmental
     Problem-Solving: Concepts and Case Studies;* National Research Council,
     National Academy Press: Washington, DC., 1986.
13.  Lee, R. E. *Biol. Trace Elem. Res.* **1990,** *26/27,* 321.
14.  *Biological Environmental Specimen Banking;* Stoeppler, M.; Zeisler, R.,
     Eds.; Science of the Total Environment, Elsevier: Amsterdam, 1993, Vol.
     139/140.
15.  *Biological  Trace  Element  Research:  Multidisciplinary  Perspectives;*
     Subramanian, K. S.; Iyengar, G. V.; Okamoto, K., Eds.; ACS Symposium
     Series No. 445, American Chemical Society: Washington, DC., 1991.
16.  Iyengar, G. V. In *Biological Trace Element Research: Multidisciplinary
     Perspectives;* Subramanian, K. S.; Iyengar, G. V.; Okamoto, K., Eds.; ACS

Symposium Series No. 445, American Chemical Society: Washington, DC., 1991, pp 1.

17.  Sexton, K. *Arch. Environ. Health.* **1992,** *47,* 398.

18.  Schladot, J. D.; Backhaus, F. W.; Reuter, U. *Juelich Report, Juel-Spez-330,* 1985, Juelich, Germany.

19.  Subramanian, K. S. *Sci. Total Environ.* **1993,** *139/140,* 109.

20.  Zeisler, R.; Harrison, S. H.; Wise, S. A. *Biol. Trace Ele. Res.* **1984,** *6,* 31.

21.  Ambe, Y. *NBS Special Publication No. 740;* National Bureau of Standards: Gaithersburg, MD, 1988, pp 22.

22.  Olsson, M. *NBS Special Publication No. 740;* National Bureau of Standards: Gaithersburg, MD, 1988, pp 26.

23.  King, N. In *Monitoring Environmental Materials and Specimen Banking;* Luepke, N. -P., Ed.; Martinus Nijhoff: The Hague, 1979, pp 74.

24.  Wise, S. A.; Fitzpatrik, K. A.; Harrison, S. H.; Zeisler, R. In *Monitoring Environmental Materials and Specimen Banking;* Luepke, N. -P., Ed.; Martinus Nijhoff: The Hague, 1979, pp 108.

25.  Becker, P. R.; Wise, S. A.; Koster, B. J.; Zeisler, R. *Alaskan Marine Mammal Tissue Archival Project: Revised Collection Protocol.* NISTIR 4529, U.S. Dept. of Commerce, National Bureau of Standards, Gaithersburg, MD, 1991, p 39.

26.  Farrington, J. W.; Risebrough, R. W. *Hydrocarbons, Polychlorinated Biphenyls and DDE in mussels and oysters from the U.S. coast, 1976-78: The Mussel Watch.* Woods Hole Oceanographic Institute Technical Report WHOI-82-42, NTIS PB83-133371, October 1982.

27.  Wise, S. A.; Koster, B. J.; Parris, R. M.; Schantz, M. M.; Stone, S. F.; Zeisler, R. *Int. J. Environ. Anal. Chem.* **1989,** *37,* 91.

28.  Zeisler, R.; Stone, S. F.; Sanders, R. W. *Anal. Chem.* **1988,** *60,* 2760.

29.  Becker, P. R.; Wise, S. A.; Koster, B. J.; Zeisler, R. *Alaskan Marine mammal tissue archival project: A project description, including collection protocols.* NBSIR 88-3750, National Bureau of Standards: Gaithersburg, MD, 1988.

30.  Zeisler, R.; Harrison, S. H.; Wise, S. A. *NBS Special Publication No. 656,* National Bureau of Standards: Gaithersburg, MD, 1983.

31.  Stone, S. F.; Koster, B. J.; Zeisler, R. *Biol. Trace Ele. Res.* **1990,** *26/27,* 579.

32.  Schantz, M. M.; Benner, B. A. *Fres. J. Anal. Chem.* **1990,** *338,* 501.

33.  Iyengar, G. V.; Tanner, J. T.; Wolf, W. R.; Zeisler, R. *Sci. Total Environ.* **1987,** *61,* 235.

34.  Lillestolen T. I.; Foster, N.; Wise, S. A. *Sci. Total Environ.* **1993,** *139/140,* 97.

35.  Zeisler, R.; Greenberg, R. R.; Stone, S. F.; Sullivan, T.M. *Fres.Z. Anal. Chem.* **1988,** *332,* 612.

36.  Stoeppler, M. In *Biological Reference Materials;* Wolf, W.R., Ed.; Wiley: New York, NY, 1985, pp 281.

37.  Kartha, K. K. *Cryopreservation of Plant Cells and Organs;* CRC Press: Boca Raton, FL, 1985.

38.   Reuter, U.; Ballschmitter, K. *Umweltprobenbank: Bericht und Bewertung der Pilotphase;* Bundesministerium fuer Forschung und Technologie, Springer Verlag: Bonn, 1988, pp 86.

39.   Schantz, M. M.; Koster, B. J.; Wise, S.A.; Becker, P.R. *Sci. Total Environ.* **1993,** *139/140,* 323.

40.   Parris, R. M.; Chesler, S. N.; Wise, S. *NBS Special Publication No. 706,* National Bureau of Standards, Gaithersburg, MD, 1985. pp 171.

41.   Elliott, J. E. *NBS Special Publication No. 740;* National Bureau of Standards (presently National Institute of Standards and Technology): Gaithersburg, MD, 1988, pp 4.

42.   Ballschmitter, K. In *Environmental Specimen Banking and Monitoring as Related to Banking;* Lewis, R. A.; Stein, N.; Lewis, C.W., Eds.; Martinus Nijhoff: The Hague, 1984, pp 264.

43.   Kemper, F. H.; Eckard, R.; Bertram, H. P.; Mueller, C. *Umweltprobenbank fuer Human-Organproben;* Internal Report of the Muenster Specimen Banking Facility, Muenster University, 1990.

44.   Korte, F.; Gebefuegi, I.; Oxynos, K. *Umweltprobenbank: Bericht und Bewertung der Pilotphase;* Bundesministerium fuer Forschung und Technologie, Springer Verlag: Bonn, 1988, pp 92.

45.   Stephens, R. D. *Anal. Chem.* 1992, *64,* 3109.

46.   Zeisler, R.; Greenberg, R. R.; Stone, S.F. *NBS Special Publication No. 706;* National Bureau of Standards: Gaithersburg, MD, 1985, pp 82.

47.   Zeisler, R.; Wise, S.A. In *Biological Reference Materials;* Wolf, W.R., Ed.; Wiley: New York, NY, 1985, pp 257.

48.   Krieg, V.; Wisniewski, R. *Umweltprobenbank: Bericht und Bewertung der Pilotphase;* Bundesministerium fuer Forschung und Technologie, Springer Verlag: Bonn, 1988, pp 107.

49.   Currie, L.A. In *Biological Trace Element Research: Multidisciplinary Perspectives;* Subramanian, K. S.; Iyengar, G. V.; Okamoto, K., Eds.; ACS Symposium Series No. 445, American Chemical Society: Washington, DC., 1991, pp 74.

50.   Minoia, C.; Sabbioni, E.; Apostoli, P. *Sci. Total Environ.* **1990,** *95,* 89.

51.   *Guidelines for Environmental Specimen Banking in the Federal Republic of Germany (Working Document);* Federal Environmental Agency: I3.1-93061-1/1, Berlin, 1989.

52.   Iyengar, G. V.; Kollmer, W. E.; Bowen, H. J. M. *The Elemental Composition of Human Tissues and Body Fluids;* Verlag Chemie: Weinheim, Germany, 1978.

53.   Iyengar, G. V.; Woittiez, J. *Clin. Chem.* **1988,** *34,* 474.

54.   Ihnat, M. In *Quantitative Trace Analysis of Biological Materials;* McKenzie, H. A.; Smythe, L. E., Eds.; Elsevier: Amsterdam, 1988, pp 739.

55.   Iyengar, G. V.; Wolf, W.R. *Fres. J. Anal. Chem.* 1995.

56.   Keune, H. *Sci. Total Environ.* **1993,** *139/140,* 537.

# Chapter 21

# Environmental Specimen Banking

## Contributions to Quality Management of Environmental Measurements

**Rolf Zeisler[1]**

**International Atomic Energy Agency, Laboratories Seibersdorf, P.O. Box 100, A–1400 Vienna, Austria**

Specimen Banking involves the systematic collection and archiving of ecologically relevant biological and environmental samples. It is an effective tool for pollutant trend monitoring by real time and retrospective analysis. Some significant archives of samples have been established during the past decades by several industrialized nations, with the earliest samples dating from the 1960s. The value of these archives lies in the planned and evaluated processes for collection, preparation, storage and analysis of the samples. In particular the incorporation of pre-sampling, sampling and transport and storage in the analytical process has provided accurate and traceable data. The value of these quality measures is exemplified with applications in the US National Biomonitoring Specimen Bank, where program performance was studied and where samples of human livers and of marine organisms and sediments were analyzed. Selected results of these studies are included in the discussion.

Environmental data are collected with substantial efforts around the globe. Their purpose is multi-fold: They serve the elucidation of environmental phenomena and are the basis of scientific research on the changes in nature through anthropogenic activities. They are used to identify problems that may threaten ecological and/or human health. They are the basis of regulatory control in the environment and they can support detective actions for the identification of illicit and unknown or unintended input in the environment. The basic requirements for all these uses of environmental data are the same: the qualitative identification must be possible with an accepted degree of certainty and the quantitative value must be generated within accepted limits of uncertainty. In short, environmental data must be of such quality that makes them useful for a given purpose, that assure the absence of defects and that is worth the price.

The quality of the environmental data is impacted by many factors that are not necessarily controllable through individual efforts. Therefore considerations of these factors during the planning of such programs and projects are necessary. A prominent factor is obviously the cost of providing the data, but obtaining higher quality (for a

[1]Current address: National Institute of Standards and Technology, REACT–B125, Gaithersburg, MD 20899

price) is certainly a cost-effective measure. Unrealistic regulations, such as the requirement for detection of extremely low levels may increase the cost or alternately result in low quality of data. In general, the need to determine ever lower levels of substances impacts the cost and the quality of the relevant results. Furthermore, political factors may significantly impact the quality of environmental data through pressure for rapid answers or frequently changing priorities, and last but not least, the socio-economic environment may suppress the desire for quality data.

To further complicate the task, it has become necessary in environmental measurements, for example in ecosystem monitoring and biological monitoring, to involve many scientific disciplines in a measurement program. An approach to the measurement would involve the design of a model, the selection of an appropriate sample population, the sampling, transport and storage, the preparation for analysis and the actual determinations, the calculation of results, and dissemination of data and their interpretation. Figure 1 displays, on an arbitrary scale, an individual error potential in relation to neighboring steps, if each steps would be controlled separately. It must be also considered that nowadays quality assurance guidelines are readily available for the analytical steps ( *1* ), therefore considerably minimizing the error potential for the central

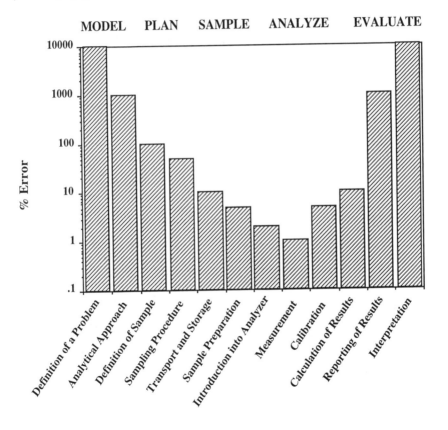

Figure 1 The environmental measurement process and likely extent of error, on an arbitrary scale, for the individual steps of the process compared to the analytical measurement error, most likely the smallest error.

"classical" analytical activities. Obviously, a sectoral approach bears the risk of deviating from the controllable small error potential of the central analytical steps to encounter large uncertainties in the resulting environmental data. Therefore interdisciplinary co-operation is the key to minimize the errors in environmental measurements, together with an expansion of the principles of quality assurance to all steps of the environmental measurement approach. These considerations on the quality of environmental measurements have been a paramount factor in the early steps of the development of environmental specimen banking approaches (2).

Since the early discussions on specimen banking more than twenty years ago (3), the archiving of environmental samples has become an accepted concept in environmental monitoring strategies. Many scientists, institutions, and several countries are using environmental specimen banking as an attractive and cost-effective contribution to the investigative tasks in a system of environmental monitoring. A specimen bank provides a systematic historical record for the evaluation of samples collected in environmental, ecological, or human studies. Specifically, a specimen bank fulfills the following objectives:

-   relevant and well characterized samples are preserved for extended time periods without changes;
-   samples are collected in anticipation of changes in the environment and/or changes in anthropogenic input for deferred assay;
-   current pollution studies can refer to historical samples and baseline data to determine trends over time;
-   a comprehensive approach is implemented for the selection, collection, analysis, and evaluation of samples;
-   sufficient material is available to allow "matrix search", "fingerprint analysis", or "key component" directed comprehensive evaluation; and
-   banked sample portions are available for re-analysis to assure previous results.

In these capacities, a specimen bank can serve as an integral part of an ongoing monitoring program or develop its own strategies for validated environmental monitoring that should be made part of the assessment of environmental status. The association of such a program with leading analytical institutions and environmental organizations will further assure the relevance of data and samples.

The above discussions have been the guiding principles during the scientific investigations on the feasibility of the specimen bank concept. In pilot programs for the establishment of specimen banks, all aspects of the analytical approach (Figure 1) have been considered in interdisciplinary research (4, 5). These investigations resulted in guidelines and protocols or standard operating procedures for each step from the development of the investigative model to the evaluation of the data. Examples of these achievements are presented below; many of the procedures have found their way as standard procedures in environmental analytical chemistry world wide for technical assurance of the quality of analytical data. It should be noted that today's principles of total quality management have been developed over a similar time frame as the technical achievements and quality measures of the specimen bank approach; their combination in the implementation of regular specimen banking as part of ecological monitoring is the basis of documented and validated environmental data that are provided to national monitoring efforts (6).

## Technical Concepts for Improved Quality

**The NIST Specimen Bank.** In 1979 a special clean room facility was completed at the National Institute of Standards and Technology (NIST), then the National Bureau of Standards (NBS) (7). This facility was among the first of its kind to be implemented for analytical chemistry. It consists of a class 100 clean room that is divided in two parts for consideration of inorganic and organic analytical work, both parts equipped with work benches and ancillary equipment made from materials considered non-contaminating for the respective analyses, i.e., acrylic and plastics in the inorganic section, glass and stainless steel in the organic section. The storage area is adjacent to the analytical clean room and complies to a class 10000 standard. It is equipped with compressor type freezers for storage at $-25°C$ and $-80°C$ and with liquid nitrogen freezers for storage at about $-150°C$. The selection of materials and implements for sampling and storage was based on extensive research work concerning the transfer to and loss from the investigated samples of components (pollutants) of interest. The concept of the protection of the composition of the sample through technical installations, advanced procedures, and careful choices of materials was made a guiding principle in the specimen bank. Some of these are described in more detail below. Similar facilities have been put into operation by other specimen banks (*6, 8*).

**Sample collection, transport and storage.** Because of the extremely low levels of trace elements and organic pollutants found in most environmental samples, extreme caution must be exercised during sample collection and processing to avoid contamination and losses. The guiding principle is to attain control of the sample as quickly as possible when removing it from the natural environment. This is done under "clean and sterile conditions" that are to be achieved even in remote sampling locations. Obviously, gradual differences exist between sampling indoors in an autopsy room or in a marine research lab on board a ship, or in the field on an ice flow; consequently, the risks for contamination are different. But, by minimizing the time the sample is exposed to the environment, subsampling under controlled condition, and sealing the sample in the sample container as well as flash freezing after this process, the risk of change in the composition of the original sample is significantly minimized. Once frozen, the sample is transported at liquid nitrogen temperature (or in some instances on dry ice) to the specimen bank facility and stored. The sample never leaves this low temperature until the actual preparation for analysis. Special attention has been given to the sampling implements and containers. Implements made from pure metals or compounds (e.g., titanium, fused silica, Teflon) minimize possible contamination to one element or compound and exclude most other common contamination risks. In particular, Teflon was found to be well suited for purposes of minimizing inorganic and organic contamination. It also exhibits excellent properties for cold storage, as it does not become brittle, and it is commercially available in a wide variety of containers, bags, and solid bulk material for the fabrication of implements.

**Sample preparation and analysis.** Sampling for trace analysis is a major concern when quality analytical results are required. The reduction of a bulk sample to a laboratory sample (test portion) suitable for the analytical technique employed often introduces errors caused by contamination, loss, or sample inhomogeneity. These errors may become the limiting factor in achieving precise and accurate analytical results. An efficient and contamination-free homogenization procedure for biological materials was found in the brittle fracture technique (*9*), the cryogenic homogenization of bulk samples with various types of mills. This technique permits the use of Teflon, for example in ball and disk mills, that will not introduce additional sources of contamination (*10*). Other types of mills for larger size samples and continuous operation, fitted with titanium or Teflon are also used in specimen banking (*11*). Employment of these contamination-free devices for size reduction and homogenization of biological tissues at cryogenic temperatures also reduces loss of volatile components and possible changes in composition during the size reduction step. The size reduction and homogeneity is essential for the effective use of analytical techniques and the preservation of specimen bank material.

The determinations of inorganic and organic constituents may involve significantly different sample preparations for the respective analytical measurements, but in taking the test portions from the same deep-frozen aliquots of a sample, the sample history being the same to this point, the end results are assured to be compatible. In the NIST specimen bank analytical work, comprehensive approaches are used to arrive at very complete data sets of high quality. In the inorganic analytical work, combinations of multi-element techniques, often applied to the same small test portion (typically 250 mg dry sample) of a specimen sample have covered all elements of environmental concern, as well as a large number of the biological micro and trace elements as possible indicators of health or nutrition (*12, 13*). Care has been taken to assure that element concentrations determined by more than one technique come to acceptable agreement. This involved the application of proven methods that can fulfill the requirements of quality assurance measurements. This agreement of different techniques on many individual measurement points allows inference as to the quality of results that may be obtainable only by one technique. The organic determinations focussed initially on organochlorine pesticide residues and PCBs; later, polycyclic aromatic hydrocarbons, dioxins, furans, etc., have been included. Because organic methods are not as varied as inorganic (e.g., in extraction or digestion methods), quality is mostly assured through the exchange of test portions of the same samples among qualified laboratories (*14*). In any case, the analyst is in a position to draw from the banked material for additional analytical work and quality assurance, since preparation of the test portions and storage conditions assures that the newly-analyzed material is equal to the previous.

## Guidelines and Protocols to Assure Quality

A specimen is a sample representative of an ecosystem, population, site, or individual organism selected for inclusion in the specimen bank based on the following criteria:
- specimen bears ecological and biological significance;
- specimen is a (long-term) accumulator;
- samples are available over long time periods and can be collected in many regions;
- undisturbed samples are available; and
- specimens may be individual or pooled samples.

Each type of specimen considered for monitoring and banking has its own requirements for the analytical process, consequently, the analytical protocol must be developed and evaluated in conjunction with the purpose, design, and implementation of the monitoring program. To exemplify this approach, this paper refers to the initial work on human liver specimens of the US National Environmental Specimen Bank (*15*). The procedures and protocols developed in this early stage of specimen banking have found widespread use, e.g. in the collection of tissue samples from marine mammals (*16*) or roe deer (*17*).

As recognized before (Figure 1), the largest error potential exists in the numerous pre-analysis steps. The selection of human liver (or marine mammal and deer livers in the later programs) as an environmental indicator was driven by the fact that a sizable sample of macroscopic homogeneity and good anatomical description could be taken in a manner that excludes or minimizes initial intrusion into the sample. The procedure to excise the whole liver without distortion was developed with medical examiners and implemented with appropriate training of autopsy personnel on handling and sub-sampling. The sampling protocol was designed to avoid contaminating the sample by either inorganic or organic constituents. The protocol specifies all items that were used, such as pre-cleaned, dust free Teflon sheets, bags and storage jars, titanium knives for dissection, dust-free vinyl gloves, high purity water; it also specifies the sequence of all steps from excision to sealing the sample in its container, flash freezing, and recording all relevant information. The frozen samples were then kept at -150°C

until used for analysis. The field sampling procedures are supported by strict procedures for the preparation of implements and auxiliary materials (*18*). Sample information as well as sampling, storage, and handling records accompany the sample in all
steps of banking and analytical work. Reporting of results is therefore assured to be
traceable to the original sample.

The active specimen bank programs have prepared publications and manuals that
make these procedures available to researchers and monitoring programs world wide.
Thus harmonized approaches for the particularly critical pre-analysis operations are
available that will provide  comparable analytical data which fulfill many quality
requirements. The available information extends from procedures for individual sample
types as mentioned above to manuals describing the collection of a large number of biota
and other environmentally-related materials, such as the Manual for Nordic Countries:
Nordic Environmental Specimen Banking (*19*).

## Quality Management in Environmental Analytical Chemistry

The specimen banks have played a leading role in implementing quality management
principles in environmental analytical chemistry by examining all steps of the analytical
process and applying the principles of documentation, repeatability, and traceability to
these steps. It has been essential that premier research institutions in analytical
chemistry participated in the development of the quality principles, long before the
implementation of guidelines for analytical laboratories have become common instruments in the daily work of environmental assessment. The specimen banks have created
inventories of samples that are meticulously characterized and preserved and have
provided environmental baseline data against which trends in environmental pollution
can be reliably measured.

The achievements in improved data quality are illustrated with the data set
obtained on 120 human livers from the NIST Specimen Bank. The main features of
these high quality data are illustrated in Figure 2. Data were obtained with similar
precision and accuracy for more than thirty elements ranging in concentration levels over
nine orders of magnitude, including several elements where limited or no data had been
available until this work. It was found that concentration ranges for some essential trace
elements were narrower than previously reported and also that certain pollutant elements
were found to be an order of magnitude lower than reported before. Even small trends
in environmental exposure, such as the declining levels of lead in the human lead body
burden can be traced based on such reliable data, though the number of investigated
specimens is relatively small (*20*).

Specimen banking has made a fundamental contribution to environmental
analytical chemistry by including the history of a sample in the analytical process to
provide controlled and traceable results. It preserves samples for retrospective analysis
for quality assurance and the provision of missing data. Thus specimen banking is a
self-sustaining management system for high quality environmental data. The overall
picture of changes in chemical composition in the environment no longer remains
murky, unclear and fuzzy because of a significant lack of data especially from the past.
Relevant questions concerning a potentially hazardous chemical such as: "When and
where did it first enter the environment?", and "Where did it come from?" will find
answers since data, samples or approaches for the analytical investigation exist. In view
of the hundreds of new chemicals that are introduced every year and the tens of
thousands that are in use in industry, agriculture, human consumption, etc., it is
necessary to complement the routine approaches for monitoring with a system that
provides for trend and anticipatory monitoring, as well as action plans and activities that
are adequate for any test in environmental assessment. The existing Specimen Banks
have illustrated their capacity in this field fulfilling all requirements for quality of
environmental data.

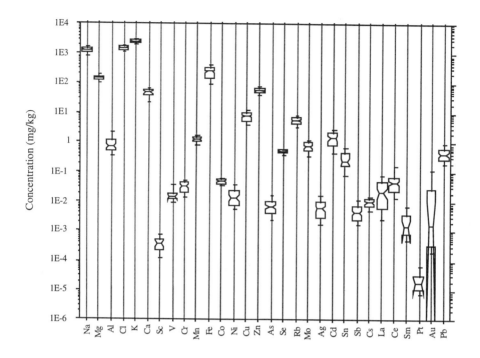

Figure 2   Notched box plots of the results of element determinations in 120 human livers from the NIST Specimen Bank.

## Literature Cited

1.   ISO/IEC Guide 25, *General Requirements for Testing Laboratories,* 3rd rev., Geneva, Switzerland, 1990.
2.   Wise, S. A.; Zeisler, R. *Environ. Sci. Technol.* **1984,** 18, 302A-307A.
3.   Luepke, N. P., Ed. "Monitoring Environmental Materials and Specimen Banking"; Martinus Nijhoff Publishers: The Hague, the Netherlands, 1979.
4.   Berlin, A.; Wolff, A. H.; Hasegawa, Y., Eds. "The Use of Biological Specimens for the Assessment of Human Exposure to Environmental Pollutants"; Martinus Nijhoff Publishers: The Hague, the Netherlands, 1979.
5.   Lewis, R. A.; Stein, N.; Lewis, C. W., Eds. "Environmental Specimen Banking and Monitoring as related to Banking; Martinus Nijhoff Publishers: The Hague, the Netherlands, 1984.
6.   Kemper, F. H.*Sci. Total Environ.* **1993,** 139/140, 13-26.
7.   Moody. J. R. In "Pilot Program for the National Environmental Specimen Bank - Phase I"; Harrison, S. H., Zeisler, R.; Wise, S. A., Eds.; EPA-600/1-81-025, 1979, pp. 8-18.
8.   Schladot, J. D.; Stoeppler, M.; Schwuger, M. J. *Sci. Total Environ.* **1993,** 139/140, 27-36.
9.   Iyengar, G. V.; Kasparek, K. *J. Radioanal. Chem.* **1977,** 9, 301-16.
10.   Zeisler, R.; Langland, J. K.; Harrison, S. H. *Anal. Chem.* **1983,** 55, 2431-34.
11.   Schladot, J. D.; Backhaus, F.W. In: Progress in Environmental Specimen Banking NBS/SP-740 **1988,** U.S. Government Printing Office, Washington, DC, pp 184-193.

12.  Zeisler, R., Stone, S.F., and Sanders, R.W. *Anal. Chem.* **1988,** 60, 2760-2765.
13.  Zeisler, R., Ostapczuk, P., Stone, S.F., Stoeppler, M. *Sci. Total Environ.* **1993,** 139/140, 403-410.
14.  Wise, S. A. *Proc. Coastal Zone '93,* **1993,** 2531-2541.
15.  Zeisler, R.; Harrison, S. H.; Wise, S. A. *Biol. Trace Elem. Res.* **1984,** 6, 31-49.
16.  Becker, P. R.. Wise, S. A., Koster, B. J., Zeisler, R., Alaskan Marine Mammal Tissue Archival Project: Revised Collection Protocol, NISTIR 4529, National Institute of Standards and Technology, Gaithersburg, USA, 1991.
17.  Holm, J. *Sci. Total Environ.* **1993,** 139/140, 237-249.
18.  Moody, J. R.; Lindstrom, R. M. *Anal. Chem.* **1977,** 49, 2264-67.
19.  Nordic Environmental Specimen Banking - Methods in Use in ESB: Manual for Nordic Countries. Nordic Council of Ministers, Copenhagen, Denmark, 1995.
20.  Zeisler, R., Greenberg, R.R., and Stone, S.F. *J. Radioanal. Nucl. Chem.* **1988,** 124, 47-63.

# Chapter 22

# Ethical and Legal Aspects of Human Tissue Banking

E.-H. W. Kluge[1] and K. S. Subramanian[2]

[1]University of Victoria, Victoria, British Columbia V8W 3P4, Canada
[2]Environmental Health Directorate, Postal Locator 0800B3, Health Canada, Tunney's Pasture, Ottawa, Ontario K1A 0L2, Canada

The collection and banking of human fluids and tissues is an integral part of biomonitoring on a global scale. However, this presents unique problems on the interface between ethics and law that are unlike those which are encountered in other contexts. They include issues of collection, disposition, ownership, access and use. Moreover, different jurisdictions have different laws that deal with these issues. This makes coordination and standardization difficult to achieve on an international scale. This chapter identifies the major ethical and legal issues involved and proposes a model for dealing with the problems based on analogous developments in medical informatics, but independent of juridical and legal variations.

Environmental degradation has become a matter of international concern, and concerted efforts are being made to ameliorate the situation. However, while protection of the environment is acknowledged to be of significance - especially when its degradation has a deleterious effect on its ability to support sustainable commercial usage and exploitation - it is the impact that environmental degradation has on human health that carries social significance. Biomonitoring of animal and plant models has proved itself to be a useful tool in this regard, and different countries have undertaken, or are currently undertaking, indicator studies with varying degrees of intensity and success.

The assumption that underlies these monitoring efforts has been that the environmental effects that are evident in animal and plant models sooner or later translate into analogous effects on human beings. As a working hypothesis this is fine. However, it is difficult to persuade policy makers to allocate the funds required for conducting biomonitoring programs unless it can be shown that environmental degradation does in fact have negative effects on human populations. This, in turn, can only be achieved by establishing, on the basis of scientifically sound data, that the indicators which can be found in animal and plant models are also present in human populations; and, further, that the effects that can be documented for animals and plants are indicative of effects that can be documented for human populations. To this end, the collection and banking of human fluids and tissues is a *sine qua non*.

Several jurisdictions have, or plan to have, human tissue banks which facilitate the long-term collection and storage of human tissues and fluids (*1*). However, without exception, these banks are rather specialized, concentrating on only a few types of tissues or fluids. Further, there is a growing realization that environmental degradation is not a limited phenomenon that respects national boundaries. The total global environment, in which humanity as whole is embedded, is implicated. Consequently the ultimate aim of biomonitoring, even when it is driven by national concerns, can be achieved only by biomonitoring on a global scale. This suggests that human tissue banks should be established on a national as well as international basis.

Tissue banks also are useful tools for the planning of national health care agendas. Current approaches to such planning typically involve identifying patterns in previous health care expenditures in order to determine what form future health care expenditures should take. Such an approach is liable to miss trends at the cellular level that have not as yet expressed themselves in overt disease processes but will ultimately manifest themselves in this fashion. As opposed to this, the materials collected by tissue banks would provide baseline snap-shots of the health status of the relevant populations at the organic, fluid and cellular levels. When compared with tissues and fluids that are collected at regular intervals and under controlled conditions, this would permit the scientific identification and evaluation of *whether* something has an impact on human health, *what sort* of impact it has, its *extent*, and *under what* conditions.

In summary, specimen banks for human tissues and fluids, therefore, are appropriate and cost-effective tools. Thus, they allow health care planners to avoid the pitfalls inherent in a purely expenditure-based approach; they satisfy the scientific needs of the environmental health community; they provide appropriate standardized materials for bio-scientific research and development; and they allow the development of strategies to deal with environmental changes that impact on human beings from the local to the global level.

### Current Rules and Regulations

The collection, storage and use of human tissues has generally proceeded independently of legal guidelines, except for the national equivalents of the Human Tissue Gift Acts that exist in Canada and the United States. Although these national acts (and their equivalents) are good initial attempts at providing a juridical framework, they fail to deal with many of the issues that have emerged since the drafting of the relevant laws themselves. The problematic that is here involved includes, but is not exhausted by, issues of collection, disposition, ownership, privacy, access, use etc. These are issues whose full import was not appreciated at the time the original legislation was passed because the scientific, social and political considerations that are now important did not obtain at that time. Therefore the acts and their equivalents are insufficient in nature and scope even at the national level for dealing with the issues that have since emerged. The fact that these rudimentary guidelines are national in scope also limits their usefulness in an international setting.

In addition to the legal problems, there is the ethical dilemma which centres on the relationship between the individual person and the human species, and of the rights of the human species as a whole. Implicated here are such issues as: (I) does the individual person have a property right in the tissue and in the components of her or his tissues, fluids and cells?; (ii) can ownership be claimed for the tissues and fluids themselves, but not to their structural features and to the structural features of their components - specifically, for the structural features of such components as chromosomal and mitochondrial DNA?; (iii) does the fact that all human beings have a shared biological heritage entail that all human beings the same right to claim ownership to the relevant structures?; and (iv) does the fact of a common heritage entail that the dignity of the human species as a whole is implicated in alterations and/or modifications of shared features?

These and similar questions are independent of any appeal to the rights of the individual person as these may be legally enshrined in a framework of laws that is unique to a particular jurisdiction. In other words, these problems cannot be resolved by the mere enactment of national laws. Instead, they concern the ethical status and rights of individual persons as members of the human species, as well as the ethical claims (if any) that humanity as a species may raise in its own regard. Arguably, therefore, the time has come to realize that the establishment of tissue banks and the development of collection processes demand more than merely the passage of enabling laws and the inception of standardized procedures within particular jurisdictions. Some efforts must be made to develop an international and integrated approach that is independent of variations in legal traditions, and whose results will be ethically acceptable irrespective of individual traditions.

## A Canadian Model

In this connection, it may be useful to consider the reasoning that underlies a current Canadian effort to establish a national bank. A basic premise underlying the Canadian approach is that the issues which arise in this connection are essentially ethical in nature, and that therefore a juridical approach, based as it would be on purely legal considerations, would fail to address the problems in an appropriate fashion.

The approach that has been adopted involves several stages: (I) identification of the relevant ethical parameters; (ii) examination of their nature and implications; (iii) consideration of how the matter could best be dealt within from a purely ethical perspective; and (iv) and integration of the results of these considerations into a process leading to appropriate legislation. The issues that have so far been identified include the following: collection and storage; ownership and disposition; use; access; and privacy. The list is not all-inclusive, but may give some indication of the complexity of the ethical paradigm.  Each of the issues is discussed below.

**Issues of Collection and Storage.**  Under this rubric, the following major issues have emerged as ethically significant: (I) disclosure of the existence, location, nature and purpose of the bank in which the tissues and/or fluids will be stored; (ii) informed consent by the biological originator of the relevant biota (or their duly empowered proxy decision-makers) to the collection of the relevant fluids and tissues; (iii) informed consent to the ultimate disposition of the banked fluids and/or tissues collected; and (iv) the establishment of nationally enforced controls over any aspect of collection and storage.

**Issues of Ownership and Disposition.**  The issues that have been identified under the rubric of ownership and disposition centre in questions such as the following: (I) is the current law prohibiting the sale of human tissue ethically defensible? In light of recent developments in biotechnology, should the law be changed so as to give the immediate biological originator of human tissue property rights over the tissues and fluids?; (ii) if property rights are recognized, should these rights be of the same nature and extent as those recognized for items that are otherwise produced by an individual in a non-biological fashion, inclusive of the whole panoply of rights to commercial transaction?; (iii) if rights to commercial transaction are recognized, should tissue that is functionally structured into organs be treated differently from tissue of essentially cellular nature?; (iv) if property rights are recognized for tissues at the cellular or organ level, should these rights extend merely to the material components of the tissues or should they extend to the structural features of the tissues and their various components, inclusive of nuclear and mitochondrial DNA?; (iv) if property rights are not recognized, should there be a recognition of the biological originator's right of disposition, inclusive of the right of usufructuary benefit deriving from use, modification etc. of the relevant tissues and their component structures?; and (v) should the benefits, if any, that derive or are derivable from the use, modification, or manipulation of human tissue be shared by society and any agent or agency that might have been involved   causally and materially in bringing about these interests or considerations?

**Issues of Use.**  The rubric of what constitutes ethically allowable use includes the following considerations: (I) to what use may human tissue ethically be put in terms of experimentation and research, commercial venture, health care or any other purposes?; (ii) what are the ethically allowable limits, if any, concerning the manipulation, modification, use, alteration, etc., of human tissues?; and (iii) how are such limits to be determined, and by whom?

  It may also be of interest to note that the evolving perspective in these matters holds that because of its unique nature, and for any manner of use: (I) human tissue should be accorded a dignity-status that is qualitatively distinct from, and greater than, the status that is accorded non-human tissue; (ii) the interests of relevant third-party who might be affected by the use of the tissue should be explored prior to any use; and (iii) these third-party interests include information-associated interests, and that therefore privacy and confidentiality may play an important role in this regard.

**Issues of Access.** Issues of use are distinct from issues of access. Therefore the following have emerged as considerations worthy of separate mention: (I) should there be special categories of access to the tissues and fluids banked in a tissue bank?; (ii) should access be controlled by nationally promulgated and enforced guidelines, or should other means of control of access be instituted?; (iii) should access to tissue be on the basis of prior and ethically validated protocols detailing use or should this matter be left open?; and (iv) in case prior ethically validated protocols are required, should these protocols be vetted (and adherence to them be supervised) by an appropriate national body having ethical, legal as well as scientific expertise, or should some other mechanism be instituted?

**Issues of Privacy.** In the Canadian context, as in many other jurisdictions, the individual person is protected by a series of concentric and increasingly stringent legal hurdles that safeguard the individual from unwanted and untoward external scrutiny. The existence of these hurdles is premised on the general presumption that everyone has a fundamental right to security of the person which may be overruled only for reasons that are demonstrably necessary to safeguard the equal and competing rights of others. As recent developments in forensic DNA analysis and similar undertakings have shown, this sphere of privacy can easily be breached on the basis of information that is derivable from the tissues and fluids originating in the body of the individual person. Consequently, the concern has arisen that the establishment of tissue banks may open the way to an otherwise unwarranted invasion of individual privacy.

These consideration have been focused in proposals like the following: (I) the collection, storage, access to and use of human tissue should be subject to appropriately structured laws, guidelines and regulations that meet all otherwise legitimate privacy concerns of the individual person; and (ii) the laws, guidelines and regulations governing such privacy concerns should be enforceable in a uniform and standard manner.

## General Considerations

In exploring these and other ethically based issues, Canada is proceeding on the assumption that human tissue has a special ethical status which is distinct from that of other species and consequently that it would be inappropriate to deal with human tissue on a property model. Canada inclines to the position that considerations which centre in the dignity of the human species must find appropriate reflection in any relevant legislation. Canada is also proceeding on the general assumption that there are certain basis ethical principles that govern the behaviour of human societies, and that these principles mandate the ascription of rights and demand the imposition of duties even if neither these rights nor these duties find explicit expression in legal tradition or in the laws as they are currently structured. This perspective finds support in the Nuremberg Code and the International Declaration of Human Rights, as well as in the condemnation of terrorism, oppressive regimes, etc.

If this Canadian perspective is correct, then it is arguable that there lies an universal obligation to ensure that the establishment of tissue banks will be governed by appropriately structured but ethically grounded rules and regulations. However, the mere adherence to this requirement would be insufficient. Different jurisdictions have different legal traditions and perspectives. To allow regulation to progress independently for each nation thereby run the risk of resulting in an international patchwork of regulations that are mutually incompatible and impossible to enforce. Therefore what is needed is some international mechanisms that reflects the above-mentioned ethical concerns. Several mechanisms suggest themselves. Three in particular deserve mention:

**Conventions Under the Auspices of the UN.** This mechanism has already been explored in UN Conventions concerning human rights. The advantage of such conventions is that they provide a global framework. Their disadvantage is that individual countries frequently do not consider them binding; and that in any case, they are not readily enforceable.

**Professional Codes of Ethics.** The various professional groups that are involved in the collection, storage and use of human tissue could incorporate the relevant ethical clauses into their respective codes of ethics. Although not specific to the issue at hand, the Code of Ethics of the World Medical Association could be seen as a model in this regard, as likewise the international professional agreements restricting the use of recombinant DNA techniques. The advantage of this approach is that adoption of the relevant clauses would be independent of national laws and would govern scientific conduct, which in turn would be enforceable through the adoption of ethically based policies that restrict publication in scientific journals. The ethical guidelines adopted by journals such as *Nature* and *The New England Journal of Medicine* clearly demonstrate that such a combined mechanism may work quite well. On the other hand, the disadvantage of such an approach is that it would be ineffective in cases of commercial exploitation where publication and professional behaviour is not an issue.

**International Treaties that Incorporate the Ethically-based Concerns Indicated Above.** The most promising model is a treaty model. The treaties that bind the various members of the European Economic Community provide a usable parallel. Particularly implicated here are the treaties that regulate the collection, use, storage and disposition of medical records. The relevant clauses anent this matter are grounded in the recognition that every person has a fundamental right to privacy, that this an ethically grounded right and not the creature of legislative accident, and that it should therefore be recognized irrespective of the juridical traditions of the respective jurisdictions. The clauses that detail the treatment of medical records are enforceable by the same mechanisms that enforce adherence to the other treaties that bind the European Economic Community. Arguably, similar provisions regarding tissue banks could be made within the framework of the international trade agreements that are currently being developed for the Asia-Pacific Rim region. Analogous measures could also be incorporated into international trade agreements on a global scale.

## Conclusions

The preceding discussion has explored some of the ethical implications that surround the establishment of human tissue banks. No doubt, there are sound social, scientific and pragmatic reasons for establishing such banks. However, care should be taken to address the ethical concerns that arise in this connection. A preliminary effort has been made here to identify some of these ethical concerns.

## Literature Cited

(1)     Subramanian, K. S. *Sci. Total Environ.* **1993,** *139/140,* 109-121.

Chapter 23

# Establishing Baseline Levels of Elements in Marine Mammals Through Analysis of Banked Liver Tissues

Paul R. Becker[1], Elizabeth A. Mackey[2], Rabia Demiralp[2], Barbara J. Koster[2], and Stephen A. Wise[2]

[1]National Institute of Standards and Technology, Analytical Chemistry Division, 217 Fort Johnson Road, Charleston, SC 29412
[2]National Institute of Standards and Technology, Analytical Chemistry Division, Gaithersburg, MD 20899

Tissues from marine mammals of the United States are routinely banked in the National Biomonitoring Specimen Bank through collaboration between the National Institute of Standards and Technology, the National Marine Fisheries Service, the National Biological Service, and numerous other agencies and organizations. A major part of this specimen banking involves collections from marine mammals taken in Alaska Native subsistence hunts. Subsamples of liver tissues banked from these collections are routinely analyzed for 39 elements, using instrumental neutron activation analysis, supplemented with other analytical techniques. Excluding the potentially toxic trace elements (e.g., Hg and Cd) and a few essential elements (e.g., Cu, Se, and Zn), the published database on element concentrations in marine mammals has been limited. The program described here offers an opportunity for establishing baseline levels for a wide range of major, minor, and trace elements in marine mammals of the western Arctic and North Pacific oceans.

In 1979 the National Institute of Standards and Technology (NIST) initiated the development of a formal bank for the long-term storage of well documented and preserved environmental specimens. Through sponsorship of several agencies (Table I), this formal bank, the National Biomonitoring Specimen Bank (NBSB), maintains collections of many different types of environmental specimens including: human liver tissue, human diet specimens, sediments, mussels, oysters, fish liver and muscle, and marine mammal blubber, liver, kidney, and muscle (1). In most cases these archived specimens are subsets of samples collected by ongoing or past contaminant monitoring programs; however, in other cases the specimens were collected solely for archival.

**Table I.  Materials Archived in the National Biomonitoring Specimen Bank**

| Specimen | Program | Sponsor[1] |
|---|---|---|
| Human liver | Environmental Specimen Bank Project | USEPA |
| Human food | Nutrients in Human Diet Project | IAEA, FDA,USDA |
| Mussels | Exxon Valdez Oil Spill Damage Assessment | NOAA/NOS USEPA |
| Mussels & oysters | Mussel Watch Project, NS&T[2] | NOAA/NOS |
| Sediments | Mussel Watch Project, NS&T<br>Benthic Surveillance Project, NS&T<br>Exxon Valdez Oil Spill Damage Assessment | NOAA/NOS<br>NOAA/NOS<br>NOAA/NOS |
| Fish liver & muscle | Benthic Surveillance Project, NS&T<br>Exxon Valdez Oil Spill Damage Assessment | NOAA/NOS<br>NOAA/NOS |
| Marine mammal tissues[3] | Marine Mammal Health & Stranding Response Program<br>Alaska Marine Mammal Tissue Archival Project | NOAA/NMFS<br>MMS, NBS |

[1]USEPA = U.S. Environmental Protection Agency; IAEA = International Atomic Energy Agency; FDA = Food and Drug Administration; USDA = U.S. Department of Agriculture; NOAA = National Oceanic and Atmospheric Administration; NOS = National Ocean Service; NMFS = National Marine Fisheries Service; MMS = Minerals Management Service; NBS = National Biological Service.
[2]NS&T = National Status and Trends Program
[3]blubber, liver, kidney, and muscle

Specimen banking is now recognized as an important part of real-time environmental monitoring. Collecting and archiving a subset of samples during ongoing monitoring programs provide a valuable resource for retrospective analysis and verification of data as analytical techniques improve, or for future measurement of yet to be identified contaminants of concern. Environmental specimen banking programs similar to the NBSB have been established in Germany, Sweden, Canada, and Japan, and other programs are being planned (2).

## Marine Mammal Tissue Bank Programs

The marine mammal tissues archived at the NBSB are the result of two ongoing collaborative programs. The Alaska Marine Mammal Tissue Archival Project (AMMTAP) began in 1987 as a joint venture between NIST, the National Oceanic and Atmospheric Administration's Arctic Environmental Assessment Center, and the U.S. Department of the Interior's Minerals Management Service (3). The project is presently sponsored by the U.S. Department of the Interior's National Biological Service. Under the AMMTAP, tissue samples are collected from animals taken by Alaska Natives for subsistence. This requires coordination and cooperation with numerous local hunters from coastal villages and with a number of Federal, State and local agencies, and private organizations.

The Marine Mammal Health and Stranding Response Program is conducted by the National Oceanic and Atmospheric Administration's National Marine Fisheries Service (4). As part of the program's research on the health status of marine mammals, NIST manages the Quality Assurance - Quality Control (QA/QC) Program for the chemical analysis of marine mammal tissues and maintains the National Marine Mammal Tissue Bank as a component of the NBSB. Tissues for the National Marine Mammal Tissue Bank are usually collected by the program during the necropsy of animals that have stranded or have been taken incidently during commercial fishing operations. In addition to specimens for banking, samples are also collected by the program for biotoxin assays, viral screening, and environmental contaminant analysis.

Marine mammal specimens archived in the NBSB are available for research; however, a specimen access policy must be followed in order to obtain samples (1). Requests for specimens must justify the use of banked samples rather than contemporary samples, demonstrate that good QA procedures are in place, and agree to provide all analytical results for incorporation into the NBSB database.

NIST analyzes selected specimens from the NBSB in order to: (a) provide organic and inorganic data for evaluating the stability of analytes and sample degradation during storage; (b) compare with results from samples collected in the future for long-term monitoring; and (c) compare with analytical results from other laboratories on samples collected at the same time for monitoring purposes. The environmental monitoring programs with which the NBSB is associated usually address specific contaminant analytes (e.g., Cd, Hg, Pb) or analytes affecting the action of contaminants (e.g., Se). The NIST analysis includes a relatively large number (39 in all) of major, minor, and trace elements not routinely measured in the contaminant monitoring programs. Therefore, the inorganic analysis conducted by NIST as part of its NBSB procedures provides an opportunity for establishing a database on the concentration of both essential and nonessential elements in marine mammal tissues.

## Sampling Protocol for Marine Mammals

Standard protocols have been developed by NIST for the collection of tissue specimens from various species of marine mammals and for archiving specimens. These protocols, described elsewhere *(4-6)*, were specifically designed to: (a) provide sufficient material (~300 g) for multiple analyses for many different kinds of analytes; (b) minimize the possibility of sample loss by storing duplicate portions (subsamples A and B) in separate freezers; (c) control collection, processing, and storage procedures and equipment so as to minimize inadvertent contamination during sample handling and insure sample integrity; (d) provide cryogenic storage conditions (-150 °C) so as to insure sample stability over relatively long periods of time (i.e., years); and (e) maintain a sample tracking system with all data resulting from sample analysis, sample collection history, and other data on the individual animals (e.g., necropsy reports), in order to maintain a complete database on the species sampled.

Of the two subsamples of each tissue collected, material selected for analysis is taken from subsample B. This subsample, ~150 g (wet weight), is homogenized using a grinding procedure specifically designed to maintain cryogenic conditions during the operation *(7)*. This procedure reduces potential losses of volatile compounds and avoids degradation of the sample due to thawing and refreezing. The procedure provides a homogeneous frozen fresh (not dried) powder with greater than 90% of the particles less than 0.46 mm in diameter and with subsampling errors due to inhomogeneity estimated at less than 2% *(7)*.

## Determination of Trace and Major Elements in Selected Tissues

Although trace element concentrations have been periodically determined for liver, kidney and muscle, liver is the tissue which is routinely analyzed because it is the organ which reflects best the trace element status of the animal for the largest number of elements. The NBSB has relied on several analytical procedures involving collaboration between several investigators and organizations.

The principal approach uses instrumental neutron activation analysis (INAA), a multi element analytical technique that provides data on a large number of trace elements using only a limited amount of a sample. INAA is routinely used to measure 37 elements in the NBSB specimens (Na, Mg, Al, Cl, K, Ca, Sc, V, Mn, Fe, Co, Cu, Zn, As Se, Br, Rb, Sr, Mo, Ag, Cd, Sn, Sb, I, Cs, Ba, La, Ce, Sm, Eu, Tb, Hf, Ta, Au, Hg, Th, and U). This method consists of exposing samples and standards to a neutron field to produce radioactivity and measuring the energy and amount of the resulting radiation. In the neutron field, many of the stable nuclides of elements comprising the sample undergo neutron capture which, for many elements, results in the formation of radioactive product nuclides. The gamma ray emissions from the resulting nuclides are collected using a germanium detector. The energy of the gamma ray indicates from which element the product nuclide was formed and the amount of radiation emitted is proportional to the concentration of that element. The INAA approach used by the program has been previously described in detail *(8-10*; Mackey, E. A., et al., *Arch. Environ. Contam. Toxicol.*, in press). For each analysis, aliquots of powdered Standard Reference Materials

(SRMs) were packaged in the same way as the tissue samples and were included in the analysis scheme for the purpose of quality control. Analyses of SRM 1577a Bovine Liver and, beginning in 1991, a QA pilot whale liver tissue homogenate (*11*) were included with all multi element INAA measurements. Analyses of SRM 2710 Montana Soil and SRM 1571 Orchard Leaves were included with all Hg measurements.

In addition to INAA, other analytical techniques have been used to provide data on elements that are not routinely measured by this technique (e.g., Pb and Ni) and to provide quality control data for selected elements (Co, Cu, Zn, Cd, and Hg) by comparing data from two different analytical techniques. Ni and Pb were determined at the Institute of Applied Physical Chemistry, Research Center of Jülich, Germany, by differential pulse and square wave voltammetry using previously published procedures (*12*) after high pressure ashing digestion with nitric acid (*13*). Mercury concentrations were also determined at Jülich using cold vapor atomic absorption spectrometry (CVAAS) as described elsewhere (*14,15*).

## Element Concentration Levels in Marine Mammal Liver Specimens from the NBSB

Element concentrations were measured in liver tissues from 52 individual marine mammals representing eight species. These included four pinniped species: northern fur seal *(Callorhinus ursinus)*, ringed seal (*Phoca hispida*), spotted seal (*P. largha*), and bearded seal (*Erignathus barbatus*); and four cetacean species: bowhead whale (*Balaena mysticetus*), beluga whale (*Delphinapterus leucas*), pilot whale (*Globicephala melas*), and harbor porpoise *(Phocoena phocoena)*. All of the pinnipeds and two of the cetaceans, bowhead and beluga whales, represent North Pacific/Western Arctic animals, while the pilot whales and harbor porpoise were from North Atlantic populations.

Hepatic concentrations of many trace elements may vary widely among different vertebrate species occupying different ecosystems and feeding within different food webs. For example, vertebrates of marine food webs may be exposed to relatively high levels of trace elements, such as Cd, Hg, Se, Cu, V, and As, through their prey. Any or several of these could be anthropogenically enriched in the ecosystem, thus, resulting in some difficulties in separating natural concentration levels from levels that may have been enhanced due to human induced pollution. In the case of marine mammals, this problem is compounded by the fact that, although the food habits of many species are fairly well known, this is certainly not the case for all species (particularly those occurring in the open ocean or moving over large geographic areas during the year). Marine mammals are inherently more difficult to study than terrestrial mammals; therefore, even for those species in which the database is best, some seasonal feeding information may be lacking.

The range of liver concentration values for 23 elements measured by the NBSB for the eight species of marine mammals are shown in Figure 1 and are compared with maximum and minimum values reported previously for human liver samples archived in the specimen bank (*16*); this represents an update of results reported earlier for animals from the NBSB (*15,17*). All values in this paper are presented as µg·g⁻¹, wet weight, unless noted otherwise. Levels for the following elements were below detection limits: Sc, Sr, Mo, Sn, I, Br, La, Ce, Sm, Eu, Tb, Hf, Ta, Au, Th, and U. The range of INAA detection limits for these elements in marine mammal liver has been presented elsewhere (8,10). Each value in Figure 1 represents the average concentration measured in two aliquots. In general, there has been good agreement between values from duplicate portions for all elements. Also, more recent results have been published on specific elements (9; Mackey, E. A., et al., *Arch. Environ. Contam. Toxicol.*, in press) and selected species (10). Both the marine mammal and human liver databases are comparable since collection protocols and analytical procedures were similar.

The concentrations of K, Cl, Na, Mg, Ca, Zn, Mn, Br, Al, and Rb, varied little between individual animals and species (Figure 1). Copper varied more widely, the highest concentrations occurring in the North Pacific/Western Arctic pinnipeds (6.5 to 45 $\mu g \cdot g^{-1}$), beluga whales (7 to 54 $\mu g \cdot g^{-1}$), and harbor porpoises (3.5 to 10 $\mu g \cdot g^{-1}$), while bowhead whales and pilot whales had relatively low levels (1 to 4 $\mu g \cdot g^{-1}$ for both species). The apparent similarity in hepatic Cu concentration in these two very different species is surprising since they occur in very widely separate geographic regions (bowhead in the Western Arctic and pilot whale in the North Atlantic) and feed at very different trophic levels (bowhead feeding on copepods and euphausiids and pilot whale feeding on squid and large fish).

The elements having the widest variation in liver concentration among marine mammals (between species and among individuals of the same species) were Hg, Se, Ag, Cd, As, and V (Figure 1). The wide concentration variation in hepatic selenium, even though it is an essential element, may be due to its close association with hepatic Hg, a potentially toxic trace element. Many species of marine mammals accumulate hepatic Hg to relatively high levels, and the metal may be incorporated into a selenium-mercury complex (9). This Hg accumulation continues as the animal ages, with little of the element being excreted (particularly in male animals). A strong correlation between hepatic Hg and Se, between Se and Ag, and between these three trace elements and age has been reported for the NBSB marine mammals (8, 9). Hepatic Hg levels were generally an order of magnitude higher in marine mammals than those reported for humans (Figure 1). The exception was bowhead whales (0.1 to 0.3 $\mu g \cdot g^{-1}$), the range of concentrations being very similar to that reported for human livers (16). With the exception of the bowhead whales, Hg concentration ranges were higher in the cetaceans than in the pinnipeds. This may be due to several factors, including their diet and physiological and age differences in the animals, since the cetaceans were generally older than any of the pinnipeds.

Hepatic Ag concentrations were at least an order of magnitude higher in marine mammals than in humans (<0.005 to 0.018 $\mu g \cdot g^{-1}$). The highest Ag concentrations were found in liver of beluga whales, which had levels one to two orders of magnitude higher than any of the other species (6 to 107 $\mu g \cdot g^{-1}$) and were comparable to the magnitude of both Hg and Se in these animals (9). Although not yet demonstrated for marine mammals, physiological mechanisms involving the interaction of Ag and Se with the anti oxidative functions of glutathione peroxidase and vitamin E have been shown for some species of mammals (18). The possible significance of this in beluga whales has been discussed elsewhere (9).

In mammals, the highest Cd concentrations occur in kidney followed by liver, with about 50% of the body burden in humans found in these two organs (19). The NBSB

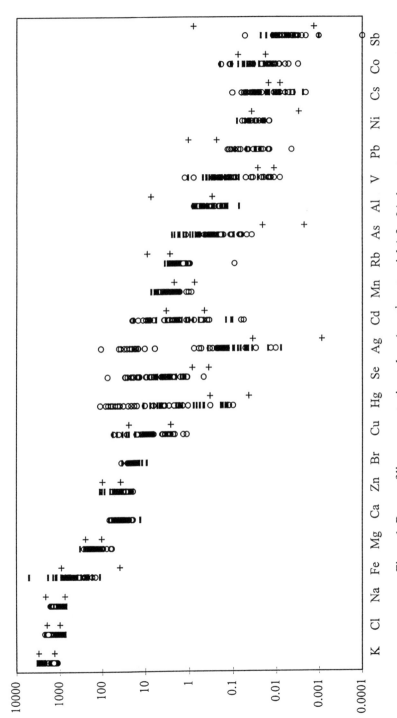

Figure 1. Range of liver concentration values ($\mu g \cdot g^{-1}$, wet weight) for 24 elements in cetaceans (open circles) and pinnipeds (bars) as compared to human maximum-minimum values (crosses).

banks both tissue types from marine mammals; however, since the liver is routinely analyzed by NIST for all trace elements, the data presented here are for liver only. The concentrations of Cd, a potentially toxic trace element, appeared to vary more widely in marine mammal liver than in human liver; the human values lie in the mid-range of the marine mammal concentrations (Figure 1). Cadmium, like Hg, Se, and Ag, accumulates with age. The highest hepatic Cd concentrations were found in the northern fur seal and bowhead whale (up to 19 and 20 $\mu g \cdot g^{-1}$, respectively). Similarly elevated levels in these two species have been reported by other investigators (20,21). Cadmium uptake and accumulation is probably determined by many factors, with the diet of the animals appearing to play a major role (10).

Marine organisms generally have higher concentrations of As than terrestrial or freshwater organisms and the majority of As is in organic form (22). Tivalent As is known to be toxic at elevated levels; however, in many marine animals (fish, crustaceans, and molluscs), this element occurs as the non-toxic pentavalent organic compound, arsenobetaine. As shown in Figure 1, the hepatic As levels in the marine mammals were orders of magnitude higher than that measured in the human livers (0.005 to 0.02 $\mu g \cdot g^{-1}$). The highest levels were in the ringed seals from Norton Sound, Bering Sea (0.16 to 2.4 $\mu g \cdot g^{-1}$). Since we did not analyze for arsenobetaine, it is not known whether the As is in this non-toxic form.

Vanadium concentrations in marine mammal liver varied more widely than those reported for the human liver, and were generally greater (Figure 1). Most other investigators do not routinely measure V in marine mammals. However it has been reported at relatively high levels in some invertebrates (23,24), many of which provide food for some species of marine mammals. Vanadium levels were present at or below detection limits in liver tissues of the North Atlantic marine mammals (pilot whales and harbor porpoises), but consistently higher in animals from the Western Arctic (particularly in ringed and bearded seal, bowhead and beluga whales). A discussion of the bioaccumulation of V in liver of Western Arctic cetaceans and pinnipeds and potential sources of this trace element has been presented elsewhere (Mackey, E. A., et al., *Arch. Environ. Contam. Toxicol.*, in press).

Elements that appear to vary more widely or occur at higher concentrations in human than in marine mammal liver are: Rb, Al, Pb, Ni, and Sb (Figure 1). Higher Pb levels would be expected in animals closest to anthropogenic sources; therefore, one would expect Pb levels to be somewhat higher in humans than in wide ranging oceanic animals. The same might be said for both Ni, Al and S; however, there is little information to even develop a hypothesis for the differences in Rb.

**Conclusion**

Tissues from marine mammals banked in the National Biomonitoring Specimen Bank provide a sufficient amount of cryogenically stored material to allow for a multitude of analyses for a variety of purposes. Through use of INAA supplemented by other analytical techniques, a database on levels of a wide range of major, minor, and trace elements is being developed for marine mammals that frequent the U.S. coastal waters. This includes animals of the Western Arctic and North Pacific oceans. Such a database should be useful beyond just defining the baseline levels in these species of the few contaminant related heavy metals.  The database has the potential for aiding the investigation of food web relationships and providing a baseline for understanding basic nutrient and health-related issues among populations of marine mammals.  Our understanding of the role of elements not commonly measured elements in maintaining the health balance of marine mammal species is very limited.  As new information becomes available, the specimens maintained by the NBSB will provide a valuable resource for further analysis and investigation.

## Acknowledgments

The success of the NBSB is due to many organizations and individuals, including the sponsors listed in this paper, as well as collaborating organizations.  For collections from marine mammals these include: the North Slope Borough Department of Wildlife Management, Kawerak Inc., New England Aquarium, California Marine Mammal Center, Cook Inlet Marine Mammal Council, Alaska Beluga Whale Committee, the Alaska Eskimo Whaling Commission, the U.S. Department of Fish and Wildlife, the Eskimo Walrus Commission, the residents of St. Paul Island and the TDX Corporation, the National Marine Fisheries Service offices in Anchorage and Juneau, and the National Marine Mammal Laboratory.  For tissue analysis these include: the Research Center of Jülich, Germany; University of Ulm, Germany; Department of Fisheries and Oceans Canada; and the Northwest Fisheries Science Center, National Marine Fisheries Service.

## Literature Cited

1. Wise, S. A.; Koster, B. J. *Envron. Health Persp.* **1995**, *103*, 61-67.
2. *Progress in Environmental Specimen Banking*; Wise, S. A.; Zeisler, R.; Goldstein, G.M., Eds.; NBS Spec. Publ. 740; National Bureau of Standards: Gaithersburg, MD, 1988.
3. Becker, P. R.; Koster, B. J.; Wise, S. A.; Zeisler, R.  *Sci. Total Environ.* **1993**, *139/140*, 69-95.
4. Becker, P.R; Wilkinson, D.; Lillestolen, T. I. *Marine Mammal Health and Stranding Response Program: Program Development Plan*; NOAA Techn. Memo. NMFS-OPR-94-2; National Marine Fisheries Service, Silver Spring, MD, 1994.
5. Becker, P. R.; Wise, S. A.; Koster, B. J.; Zeisler, R. *Alaskan Marine Mammal Tissue Archival Project: A Project Description Including Collection Protocols*; NBSIR 88-3750; National Bureau of Standards: Gaithersburg, MD, 1988.
6. Becker, P. R.; Wise, S. A.; Koster, B. J.; Zeisler, R. *Alaska Marine Mammal Tissue Archival Project: Revised Collection Protocol*; NISTIR 4529; National Institute of Standards and Technology: Gaithersburg, MD, 1991.
7. Zeisler, R.; Langland, J. K.; Harrison, S. H. *Anal. Chem.* **1983**, *55*, 2431-2434.
8. Becker, P. R.; Mackey, E. A.; Schantz, M. M.; Demiralp, R.; Greenberg, R. R.; Koster, B. J.; Wise, S. A.; Muir, D. C. G. *Concentrations of Chlorinated Hydrocarbons, Heavy Metals and Other Elements in Tissues Banked by the Alaska Marine Mammal Tissue Archival Project*; NISTIR 5620; National Institute of Standards and Technology: Gaithersburg, MD, 1995.

9. Becker, P. R.; Mackey, E. A.; Demiralp, R; Suydam, R; Early, G.; Koster, B. J.; Wise, S. A. *Mar. Poll. Bull.* **1995**, *30*, 262-271.

10. Mackey, E. A.; Demiralp, R.; Becker, P. R.; Greenberg, R. R.; Koster, B. J.; Wise, S. A. *Sci. Total Environ.* **1995**, *175*, 25-41.

11. Wise, S. A.; Schantz, M. M.; Koster, B. J.; Demiralp, R.; Mackey, E. A.; Greenberg, R. R.; Burow, M.; Ostapczuk, P; Lillestolen, T. I. *Fresenius Z Anal. Chem.* **1993**, *345*, 270-277.

12. Ostapczuk, P.; Valenta, P; Nurnberg, H. W. *J. Electroanal. Chem.* **1986**, *214*, 51-64.

13. Würfels, M.; Jackwerth, E.; Stoeppler, M. *Anal. Chim. Acta.* **1989**, *226*, 1-16.

14. May, K.; Stoeppler, M. *Fresenius Z. Anal. Chem.* **1984**, *317*, 248-251.

15. Zeisler, R.; Demiralp, R.; Koster, B. J; Becker, P. R.; Burow, M.; Ostapczuk, P.; Wise, S. A. *Sci. Total Environ.* **1993**, *139/140*, 365-386.

16. Zeisler, R; Harrison, S. H.; Wise, S. A. *Biol. Trace Elem. Res.* **1984**, *6*, 31-49.

17. Becker, P. R.; Koster, B. J.; Wise, S. A.; Zeisler, R. *Biol. Trace Elem. Res.* **1990**, *26/27*, 329-334.

18. Ridlington, J. W.; Whanger, P.D. *Fundam. Appl. Toxicol.* **1981**, *1*, 368-375.

19. Hammond, P. B.; Beliles, R. P. In *Casarett and Doull's Toxicology*; Doull, J.; Klaassen, C. D.; Amdur, M. A., Eds.; Macmillan Publishing Co., Inc.: New York, NY, 1980; Second Edition, pp 409-467.

20. Bratton, G. R.; Spainhour, C. B.; Flory, W.; Reed, M.; Jayko, J. In: *The Bowhead Whale*; Burns, J. J.; Montague, J. J.; Cowles, C. J., Eds; Special Publication No. 2; The Society for Marine Mammalogy; Lawrence, KS, 1993; pp 701-744.

21. Goldblatt, C. J.; Anthony, R. G. *J. Environ. Qual.* **1983**, *12*, 478-482.

22. Edmonds, J. S.; Francesconi, K. A. *Mar. Poll. Bull.* **1993**, *26*, 665-674.

23. Boehm, P; Steinhauer, M.; Crecelius, E.; Neff, J.; Tuckfield, C. *Beaufort Sea Monitoring Program: Analysis of Trace Metals and Hydrocarbons from Outer Continental Shelf (OCS) Activities*; MMS 87-0072; Batelle Ocean Sciences: Duxbury, MA, 1987.

24. Micibata, H.; Sakuri, H. In *Vanadium in Biological Systems*; Chasteen, N. D., Ed.; Kluwer Academic Press: Dordrecht, The Netherlands, 1990; pp 152-171.

Chapter 24

# A Pilot Biological Environmental Specimen Bank in China

P. Q. Zhang, C. F. Chai, X. L. Lu, Q. F. Qian, and W. Y. Feng

Institute of High Energy Physics, Academia Sinica, P.O. Box 2732, 100080 Beijing, China

A pilot Chinese Biological Environmental Bank (CBESB) has been in operation in the Laboratory of Nuclear Analysis Techniques, Academia Sinica, since 1994. The main purpose of CBESB is to collect and store specimens from the natural environment systems for real-time analysis and retrospective studies. The general guideline of CBESB was clearly defined. The main facilities available in CBESB laboratory are introduced, the target districts and types for the collection of first set of specimens are listed, some of the operating protocols and preliminary running practise on CBESB are briefly described and the future plans are mentioned in the present paper.

Biological environmental specimen banks (BESBs) have been established in more and more countries all over the world since the middle of 1970's, and are also emerging in developing countries in recent years. As its great significance is environmental monitoring and pollution control, scientists and governmental authorities are paying much attention to BESB activities. The Chinese economy is developing at rapid pace due to the implementation of reform and open policies. Many new economic zones, collective and private enterprises have begun to flourish. At the same time various environmental pollution problems are becoming increasingly serious. The Chinese government realizes the situation and does make a lot of efforts to protect ecosystems in which human beings live. It is obvious that the establishment of BESB in China is not only necessary, but also a pressing matter at the moment.

## PURPOSE AND GENERAL GUIDELINE

The establishment of the first pilot Chinese Biological Environmental Specimen Bank (CBESB) has been funded by National Natural Science Foundation of China (NSFC) and also supported by the Laboratory of Nuclear Analysis Techniques (LNAT), since 1994. It is housed in LNAT in the Institute of High Energy Physics, Academia Sinica.

The major objectives in establishing CBESB are :
1) Providing historical specimens and baseline levels for future environmental studies;.
2) Functioning as a component part of the currently envisaged environmental monitoring programmes;
3) Providing reliable recordings and methods for the long-term trend monitoring of environmental situation;
4) Performing retrospective replicate analysis of well-preserved samples for the determination of inorganic and organic constituents already examined when new environmental problems, new environmental contaminants or new (or modified) analytical methods emerge.
5) Providing reference materials for promoting the development of new analytical methods or the improvement of existing methods.

We envision CBESB to be harmonized with the international and domestic environmental monitoring programmes such that its operation will be along the guidelines of international standard operating procedures from the outset. The bank capacity will be extended gradually.

## CBESB LABORATORY FACILITIES

A special laboratory (about 40 m$^2$ ) with a class 100 clean room (about 10 m$^2$) as the main part has been built for sample processing and storing. It is equipped with some necessary facilities including a micro-dismembrator for sample homogenizing, an ultra-low temperature (-85°C) compressor freezer and two liquid nitrogen freezers for sample storage. Freeze-dryer, low-temperature asher etc. are also available in our laboratory. Besides, some important nuclear analysis facilities including γ-ray spectrometers with HPGe detectors for neutron activation analysis (NAA), proton induced X-ray emission (PIXE) equipment combined with a van de graaf accelerator, total-reflection X-ray fluorescence (TXRF) spectrometer and synchrotron radiation x-ray fluorescence (SRXRF) spectrometer are also available in our institute. Recently some facilities for chemical speciation studies such as high speed centrifuger, ultraviolet spectrophotometer, gel filtration and permeation chromatographer have also been installed in our laboratory .

## SELECTION OF DISTRICTS AND TYPES FOR SPECIMEN COLLECTION

The target districts for the collection of specimens were selected and shown in fig.
• Coastal developing economic zones such as the Great Bohai Sea area (1), Pudong and Yangtze River Delta (2);
• Important bays and estuaries such as Jiaozhou Bay (3);
• High incidence areas of endemic disease such as Goitre, Cretinism, Arsenism (4) ;
• Energy source base fields such as the North China Oil Field (5)and the Three Valleys area in the Yangtze River (6);
• Serious environmental pollution areas such as acid rain polluted districts in South China (7) and mercury polluted area in the Second Song-Hua-Jiang district (8);
• Unique Chinese environment, e.g. Shanxi loess area (9); and
• Metropolitan cities, such as Beijing (10) and Tianjing (11).

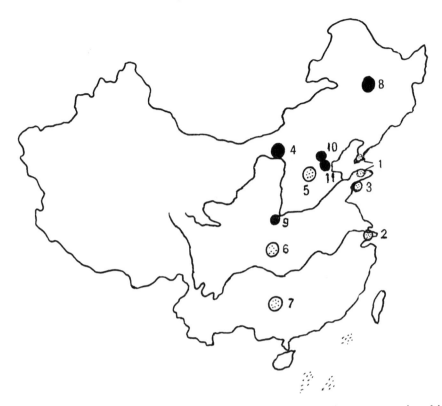

Fig. Sampling areas for CBESB. Dots and number refer to sampling areas mentioned in text . (●) areas already studied; (◎) proposed areas.

The species we selected for the first set of specimens collected were:
Human livers from the big cities, human scalp hair from the high incidence endemic disease areas, mussels from some typical bays and estuaries, lichens from some new developing economic zones, forest plants from some acid rain polluted areas, Chinese loess and so on.

SPECIMEN BANKING ACTIVITIES

(1) Specimen collections
● Eight human liver samples from normal people (violent death) living in the Tianjing area were collected for the long-term trend evaluation of environmental quality in that area. The human liver samples about 300g each, were collected within 48 hours of death and processed under the guidelines of international standard operating procedures (1) to minimize contamination of the sample and /or loss of constituents and immediately transported to the laboratory with ice box and then stored at -40°C in a compressor freezer. Detailed information about the donor and specimen were also obtained.
● Thirty three human scalp hair samples from normal males in the 20~40 age range in the Beijing area were collected for the trend monitoring of environment in this area. Only occipital hair less than 4 cm long from scalp were cut using stainless steel scissors for elemental analysis and storage.

- One hundred and thirty scalp hair samples from juveniles in some villages in Inner Mongolia, where endemic arsenism disease is prevalent. Higher arsenic content ( 0.12-0.95 mg/l ) than normal ( <0.05 mg/l ) in drinking water from some of wells was found in these villages .

- Twenty-nine pairs of scalp hair samples from pregnant women and their newborns at the Second Song-Hua-Jiang district, a mercury polluted area, and seventeen pairs of maternal and infant scalp hair samples from an unpolluted area were collected for observing mercury pollution situation and studying the transfer of mercury between mother's body and fetus.

- Twenty-six Chinese loess samples from Shaanxi Province for the studies of catastrophic and atmospheric environment in Quaternary.

Besides, the collections of pine needles from the forests seriously eroded by acid rain, marine organisms (e.g. mussels) from some bays and lichens from some new developing economic zones will be starting this year.

(2) Determination of elemental contents

The determination of elemental contents is carried out almost exclusively using nuclear analysis techniques such as NAA and PIXE, which are main analytical means in our laboratory. We also use other techniques such as atomic absorption spectroscopy (AAS) and inductively coupled plasma mass spectrometry (ICP-MS).

A) The contents of 15 trace elements in 33 scalp hair samples from normal people in the Beijing area were determined by INAA. The results were (in μg/g):

As  0.066±0.026,   Au 0.009±0.006,   Ba 4.0±1.8,   Br 2.0 ± 1.4,   Ca 1100 ± 600,
Co 0.024±0.008,   Cr  0.44 ± 0.26,   Fe 20±5.6,   Hg 0.53±0.30,   Mo 0.073±0.029,
La  0.020±0.013,   Sb 0.039±0.023,   Se  0.64±0.29,   Sm 0.005±0.003,   Zn 151±27.

Significant differences were found for the contents of some elements, such as Ca, Cr, Fe and Zn in the Beijing samples in comparison with those in the shanghai ones, and also the contents of Cr, Fe and Se in comparison with those in the Tianjing ones (2). The results provide the baseline level of the trace elemental contents in scalp hair of normal people in the Beijing area and information about the trend of the environmental quality in this area too.

B) The contents of 13 trace elements in 66 scalp hair samples (from 31 boys and 35 girls) from the endemic arsenism disease area in Inner Mongolia were determined by INAA . In comparison with the data obtained from the subjects at same age-bracket (19 boys and 21 girls) in Beijing area, it was found that besides the significantly high content of hair arsenic in the disease area, other elemental contents, such as Au, Co, Cr, Fe, Zn, La and Sm etc., were also significantly different in the two areas. The results were shown in table 1 and 2 for male and female, respectively. The correlation between the trace elemental contents of scalp hair and endemic arsenism resulting from the abnormal arsenic content in well (drinking) water is being further studied.

Table 1 Comparison of the elemental contents of boys' scalp hair
in the endemic area and the Beijing area

| Element | Arsenism area (µg/g) | Beijing area (µg/g) | P- value |
|---------|----------------------|---------------------|----------|
| As | 0.46 ± 0.32 | 0.11 ±0.06 | <0.01 |
| Au | 0.0010 ± 0.0008 | 0.0022 ± 0.0014 | <0.01 |
| Ba | 3.7 ± 1.7 | 1.8 ± 0.89 | <0.01 |
| Ca | 450 ± 290 | 499 ± 194 | >0.05 |
| Co | 0.036 ± 0.016 | 0.012 ± 0.003 | <0.01 |
| Cr | 0.92 ± 0.84 | 0.25 ± 0.09 | <0.01 |
| Fe | 45 ± 18 | 16 ± 3 | <0.01 |
| La | 0.053 ± 0.026 | 0.017 ± 0.005 | <0.01 |
| Sb | 0.059 ± 0.027 | 0.043 ± 0.018 | <0.05 |
| Se | 0.40 ±0.06 | | |
| Sm | 0.011 ± 0.005 | 0.0052 ± 0.0037 | <0.01 |
| Zn | 107 ± 25 | 136 ± 27 | <0.01 |

Table 2 Comparison of the elemental contents of girls' scalp hair
in the endemic area and the Beijing area

| Element | Arsenism area (µg/g) | Beijing area (µg/g) | P-value |
|---------|----------------------|---------------------|---------|
| As | 0.63 ± 0.58 | 0.16 ± 0.09 | <0.01 |
| Au | 0.0011 ± 0.0007 | 0.0026 ± 0.0027 | <0.01 |
| Ba | 6.5 ± 5.1 | 4.8 ± 2.3 | >0.05 |
| Ca | 880 ± 850 | 952 ± 545 | >0.05 |
| Co | 0.036 ± 0.014 | 0.018 ± 0.008 | <0.01 |
| Cr | 0.85 ± 0.52 | 0.40 ± 0.19 | <0.01 |
| Fe | 45 ± 14 | 21 ± 8.7 | <0.01 |
| La | 0.057 ± 0.023 | 0.019 ± 0.008 | <0.01 |
| Sb | 0.061 ± 0.019 | 0.056 ± 0.040 | >0.05 |
| Se | 0.38 ± 0.06 | | |
| Sm | 0.011 ± 0.004 | 0.0083 ± 0.0046 | <0.05 |
| Zn | 105 ± 29 | 143 ± 33 | <0.01 |

C) The total mercury (T-Hg) and methyl mercury (Me-Hg) contents in scalp hair samples from 29 pairs of pregnant women and their newborns living at the Second Song-Hua-Jiang district were determined by INAA and gas chromatography with electron capture (GC-EC), along with 17 pairs of maternal and infant hair samples taken from an unpolluted area as a control. Our results showed that the average contents of T-Hg in the mother and baby hair samples from the Second Song-Hua-Jiang district were 0.79±0.41 µg/g and 0.78±0.56µg/g, respectively, whereas from the unpolluted area the values were 0.38±0.23 and 0.51±0.18 µg/g, respectively. Thus, mercury pollution still remains at the Second Song-Hua-Jiang district, although the pollution level has been substantially alleviated since the seventies. The correlation between the hair Hg contents and fish consumption amounts, the hair T-Hg and Me-Hg contents of mothers and babies, and Hg and Se contents in the hair samples had been discussed elsewhere (3) .

D) The abundance of iridium and other elements for 26 Chinese loess across the Neogene/Quaternary boundary from Shaanxi Province were determined using radiochemical neutron activation analysis (RNAA) and (INAA). The results showed the content of iridium near the boundary to be 30~60 pg/g, about 10 times higher than the background values. From the distribution of Ir in different size fractions along with other experimental evidences, e.g. paleomagnetism, carbon and oxygen isotopic excursion etc., it was suggested that this weak Ir anomaly likely resulted from extraterrestrial impact, which might have caused environmental giant disturbance (4).

## FUTURE PLANS

• It is important to justify the rationale and significance through the practise of our pilot CBESB to ensure a continual commitment of support from Chinese authorities to meet financial and personnel requirements of continuously running the CBESB.
• Considerable work is required to establish the optimum conditions of collection, transportation, long term storage, preservation and processing procedures for various specimens.
• Improviing the analytical conditions and strengthening analytical quality control to continuously raise data reliability and sensitivity.
• Developing the methods for the analysis of inorganic elemental species and organic constituents of specimens.

## ACKNOWLEDGEMENTS

The authors are very grateful to Dr. Ma Peixue for supplying the loess samples and the analytical results concerned; and wish to thank Dr. Zhang Geyou and his colleagues for collecting the scalp hair samples in the endemic arsenism area. We would also like to thank Dr. K. S. Subramanian for his encouragement and valuable revision on our manuscript.

## LITERATURE CITED

1.   Zeisler. R.; Harrisson, S.H.; Wise, S.A. Bio. Trace Elem. Res. 1984, 6, 31-49
2.   Sha Yin et al. Nuclear Techniques 1987, 10, No.5, 42-43
3.   Feng W.Y.; Qian.Q.F.; Zhang P. Q.; Chai C.F.  J. Radioanal. Nucl. Chem. Articles 1995, 195, 67-73
4.   Ma PeiXue; Mao Xueying ; Hou Quanlin; Chai zhifang; Xu Heling; Ou Yanghong Science in China, 1996, 26 D, 74-79

INDEXES

# Author Index

Arashidani, Keiichi, 178
Barrios, Carlos A., 151
Becerra, José, 151
Becker, Paul R., 261
Bierenga, S., 83
Bruhn, Carlos G., 151
Carson, J., 83
Chai, C. F., 271
Dale, L. S., 49
Demiralp, Rabia, 261
Feng, W. Y., 271
Florence, T. M., 49
Frank, A., 57
Furimsky, E., 30
Galgan, V., 57
Gras, Nuri T., 151
Indulski, J. A., 195,206
Iyengar, G. V., 1,115,220
Jain, Jinesh, 30
Jaramillo, Victor H., 151
Jones, L., 49
Kawamoto, Toshihiro, 178,190
Kluge, E.-H. W., 254
Kodama, Yasushi, 178,190
Koster, Barbara J., 261
Lu, X. L., 271
Lund, Eiliv, 135
Lutz, W., 195, 206

Mackey, Elizabeth A., 261
Markert, Bernd, 18
Martin, R. R., 30
Matsuno, Koji, 178
Nieboer, E., 135,184
Nünez, Ernesto, 151
Odland, J. O., 135
Oehlmann, Jörg, 18
Qi, Min, 83
Qian, Q. F., 271
Que Hee, Shane S., 77
Raman, S., 96
Reyes, Olga C., 151
Rodriguez, Aldo A., 151
Romanova, N., 135
Roth, Mechthild, 18
Salbu, B., 135
Sand, G., 135
Skinner, W. M., 30
Stauber, J. L., 49
Stone, Susan F., 42
Subramanian, K. S., 1,96,115,220,254
Thomassen, Y., 135
Valentine, Jane L., 105
Wise, Stephen A., 261
Yano, Yoshiko, 65
Zeisler, Rolf, 246
Zhang, P. Q., 271

# Affiliation Index

Academia Sinica, 271
Agricultural University of Norway, 135
Biomineral Sciences International Inc.,
   1,115,220
CSIRO Division of Coal and Energy
   Technology, 49
Centre for Environmental & Health
   Science Pty. Limited, 49
Centre for Metal Biology, 57
Comisión Chilena de Energia Nuclear, 151
Grand Valley State University, 83

Health Canada, 1,96,115,220,254
International Atomic Energy Agency,
   42,246
International Graduate School Zittau, 18
McMaster University, 135,184
National Institute of Occupational
   Health, 135
National Institute of Standards and
   Technology, 261
National Veterinary Institute, 57
Natural Resources, Canada, 30

Nofer Institute of Occupational Medicine, 195,206
Secretaria Regional Ministerial de Salud Octava Región, 151
Technical University of Dresden, 18
Tokyo National College of Technology, 65
Universidad de Concepción, 151
University of California—Los Angeles, 77,105
University of Occupational and Environmental Health, 178,190
University of Ottawa School of Medicine, 96
University of Saskatchewan, 30
University of South Australia, 30
University of Tromso, 135
University of Victoria, 254
University of Western Ontario, 30

# Subject Index

## A

N-Acetyltransferase, genetic susceptibility to cancer, 216t,217
Adduct formation
biomarker of exposure, 197–199
environmental carcinogens with DNA and proteins, 207–209
Alces alces L., use as monitor of environmental changes, 57–62
Aldehyde concentration, saliva-leached components of chewing tobacco and simulated urine using bioluminescent bacteria, 77–82
Aldehyde dehydrogenase, effect on toluene metabolism, genetic polymorphism, 190–194
Aliphatic compounds, environmental pollution, 4
Aluminum
baseline levels in marine mammals, 265f,268
environmental pollution, 2–3
gastrointestinal absorption, 185
Alzeheimer's disease, gastrointestinal absorption of aluminum, 185
Ames test, description, 78
Antimony, baseline levels in marine mammals, 265f,268
Arctic areas of Russia and Norway, element concentrations in blood, urine, and placenta from mothers and newborns, 135–148

Arctic Monitoring and Assessment Programme, description, 135
Aromatic amine compounds, pollution, 4
Arsenic
baseline levels in marine mammals, 265–268
biomonitoring, 12
body burden concentrations in humans following low environmental exposure, 105–113
environmental pollution, 2–3
use of placenta as biomarker for environmental exposure assessment, 115–131
AS52 cells, nickel-induced mutations, 186–187
Australia, silver content of Pyura stolonifera as indicator of sewage pollution, 49–55

## B

Bacteria, studies on saliva-leached components of chewing tobacco and simulated urine, 77–82
Banking
human tissues, 254–260
See also Environmental specimen banking
Bear Lake, fish contamination, 83–84
Bioanalytical approaches to chemicals in environment, 4–7
Bioassay, 23

Bioconcentration, definition, 20
Bioenvironmental surveillance, 7–8
Bioindication, instruments, 23
Biological environmental specimen banks
  establishment, 271
  *See* Chinese Biological Environmental
    Specimen Bank
Biological markers, definition, 8
Bioluminescent bacteria, studies on
  saliva-leached components of chewing
  tobacco and simulated urine, 77–82
Biomagnification, definition, 20
Biomarker(s)
  adduct formation, 197–199
  carcinogens measured, 196–197
  cytogenetic aberrations, 200
  description, 23,195–196
  hypoxanthine phosphoribosyltransferase
    mutations, 200
  importance, 196,203
  inherited and acquired susceptibility
    to toxic substances, 184–188
  oncogene expression determination,
    200–201
  silver content of *Pyura Stolonifera*
    as indicator of sewage pollution, 49–55
  tumor markers, 201–203
  use of placenta, 115–131
Biomonitor(s)
  advantages, 254
  definitions, 19–20
  in International Atomic Energy Agency
    programs on health-related
    environmental studies
  methylmercury and human hair, 43–46
  quality assurance for biomonitoring
    programs, 47
  trace elements in air particulate
    matter and lichen, 46–47
  integrative function, 25
  quality assurance, 47
  use of human specimens, 10–12
Biosensors, 23
Biotest, description, 23
Bivalves, use of manganese concentration
  as biomonitor of water pollution, 65–75

Blood
  element concentrations from mothers
    and newborns, 135–148
  mercury and methylmercury levels
    of pregnant women, 151–173
Body burden concentrations in humans
  following low environmental exposure
  to trace elements
  arsenic body burden vs. low water
    arsenic, 108–113
  arsenic example, 106
  experimental description, 106–107
  selenium body burden vs. low water
    selenium, 107–109,111–112
  selenium example, 105–106


C

Cadmium
  baseline levels in marine mammals,
    265–268
  biomonitoring, 12
  concentrations in blood, urine, and
    placenta from mothers and newborns,
    135–148
  environmental pollution, 2–3
  use of placenta as biomarker for
    environmental exposure assessment,
    115–131
Canada
  environmental specimen banking,
    228–229
  ethical and legal aspects of human
    tissue banking, 255–259
Cancer, role of environmental chemical
  agents, 206
Cancer risk assessment
  biomarker use in areas of high potential
    carcinogenic hazard, 195–203
  use of molecular epidemiology, 206–217
Carboxylic acids, saliva-leached
  components of chewing tobacco and
  simulated urine using bioluminescent
  bacteria, 77–82
Carcinoembrional antigen, biomarker for
  cancer risk assessment, 201–203

Cesium-137, concentrations in blood, urine, and placenta from mothers and newborns, 135–148

Chemical carcinogens, role in cancer, 206

Chemical contaminants in St. Lawrence River, *See* Environmental chemical contaminants in St. Lawrence River

Chemical waste dumping sites, biomarkers for cancer risk assessment, 195–203

Chewing tobacco, saliva-leached components, 77–82

Chile, mercury and methylmercury levels in scalp hair and blood of pregnant women, 151–173

Chinese Biological Environmental Specimen Bank
  activities, 273–276
  guideline, 272
  laboratory facilities, 272
  selection, 272–273

Chromium concentration, moose as monitor of environmental changes, 61–62

Cold vapor atomic absorption spectrometry, description, 156–157

Communal dumping sites, biomarkers for cancer risk assessment, 195–203

Cooking stove, role in personal exposure to indoor nitrogen dioxide, 180,181*f*

Copper
  baseline levels in marine mammals, 265–268
  environmental pollution, 2–3
  moose as monitor of environmental changes, 61–62
  use of *Pyura Stolonifera* as biomonitor of sewage pollution, 52,54*t*

*Corbicula Japonica*, *See* Bivalves

Cunjevoi, 49

Cytogenetic aberrations, biomarker of exposure, 200,206–217

**D**

DDT, biomonitoring, 11

Dendrochemistry, function, 30

DNA, changes induced by chemical carcinogens, 206–207

Dumping sites, biomarkers for cancer risk assessment, 195–203

**E**

Element concentrations
  in blood, urine, and placenta from mothers and newborns
  cadmium, 140–143,145
  cesium-137, 141*t*,143,147–148
  experimental description, 138–140
  geographic sites, 137
  lead, 140,141*t*,142*f*,145
  mercury, 140–141,143,145–146
  nickel, 141,143–144,146–147
  in marine mammals through banked liver tissue analysis, 261–269

Environmental analytical chemistry, role of environmental specimen banking in quality management, 251–252

Environmental chemical carcinogens, identification of genetically related susceptibility to cancer, 215–217

Environmental chemical contaminants in St. Lawrence River
  environmental factors
  drinking water, 100
  fish consumption, 101
  epidemiologic strategies, 102
  experimental description, 97
  health effect, 97–98
  history of contaminants, 97
  host factors, 101–102
  role as agent, 98–100

Environmental exposure assessment, use of placenta as biomarker, 115–131

Environmental specimen banking
  advantages, 248,261,263
  analytical aspects, 234–242
  applications, 12–13
  development, 220–221
  establishment of banks
    access to banked specimens, 226
    biostatistical considerations, 224
    infrastructure, 222
    medical, legal and ethical issues, 225–226

Environmental specimen banking—
*Continued*
establishment of banks—*Continued*
multidisciplinary needs, 223
scope and size, 222–223
existing specimen banks
Canada, 228–229
China, 271–276
Germany, 229–231
Japan, 231
Sweden, 231
United Kingdom, 232
United States of America, 232–234
feasibility, 248
funding and management, 24
guidelines and protocols to assure
quality, 250–251
history, 220–221
objectives, 221,248
role in quality management in
environmental analytical chemistry,
251–252
technical concepts for improved quality
National Institute of Standards and
Technology Specimen Bank, 248–249
sample collection, transport, and
storage, 249
sample preparation and analysis,
249–250
Epidemiologic strategies, environmental
chemical contaminants in St. Lawrence
River, 102
Epidemiology in cancer risk assessment,
*See* Molecular epidemiology in cancer
risk assessment
Ethical and legal aspects of human tissue
banking
advantages, 255
Canadian model, 256–259
conventions under auspices of United
Nations, 259
current rules and regulations, 255–256
international treaties that incorporate
ethically based concerns, 259–260
professional codes of ethics, 259
use for planning national health care
agendas, 255

Exposure assessment
chemicals in environment, 8–9
methods employed in molecular
epidemiology, 207
*See also* Personal exposure to indoor
nitrogen dioxide

F

Fishing villages, mercury and
methylmercury levels in scalp hair
and blood of pregnant women, 151–173
Fluids, ethical and legal aspects of
banking, 254–260
Fluorescence methods, use for cancer risk
assessment, 210

G

Gastrointestinal absorption, aluminum, 185
GC with electron-capture detector
to analyze Hg and $CH_3Hg$ in hair,
157–158
Gene structure and expression, changes,
213–214
Genetic monitoring, function, 184
Genetic polymorphism of low $K_m$ aldehyde
dehydrogenase, effect on toluene
metabolism, 190–193
Genetically related susceptibility,
identification, 215–217
Genotoxicity test, requirements, 78
Geochemistry, role in levels of chemicals
in environment, 4
$S$-Glutathione transferase, genetic
susceptibility to cancer, 216t,217

H

Hair, use for methylmercury monitoring,
43–46
Halogenated aromatic compounds,
environmental pollution, 4
Halogenated hydrocarbons, occurrence in
food chain, 2

Health-related environmental studies,
    biomonitors, 42–47
Heaters, emissions, 178–179
Heating time, role in personal exposure
    to indoor nitrogen dioxide, 180,181*f*
Heavy metal monitoring
    definitions, 19–22
    development, 19
    instruments of bioindication, 23
    limitations, 25–28
    occurrence in food chain, 2
    possibilities, 24–25
    *Pyura Stolonifera* as biomonitor
        of sewage pollution, 52,54*t*
    quality assurance, 24
Host factors, environmental chemical
    contaminants in St. Lawrence River,
    101–102
Human(s), body burden concentrations
    following low environmental exposure
    to trace elements, 105–113
Human specimens for pollutant
    biomonitoring, 10–12
Human tissue banking, ethical and legal
    aspects, 254–260
8-Hydroxydeoxyguanosine, use for cancer
    risk assessment, 211
Hydroxyproline level, role in personal
    exposure to indoor nitrogen dioxide,
    180,182
Hypoxanthine phosphoribosyltransferase
    mutations, biomarker of exposure, 200


I

Immunoassay methods, use for cancer risk
    assessment, 210
Imposex phenomenon, description, 28
Indoor heater, source of nitrogen dioxide
    exposure, 178–179
Indoor nitrogen dioxide, personal
    exposure, 178–182
Inductively coupled plasma MS
    applications, 30–31
    trace metal distribution analysis of
        annual growth rings of trees, 30–40

Inherited susceptibility to toxic
    substances, biomarkers, 184–188
Inorganic constituents
    determination, 239–240
    use for pollutant biomonitoring, 12
Instrumental neutron activation analysis
    of total Hg in scalp hair, 157
International Atomic Energy Agency,
    biomonitors in health-related
    environmental studies, 42–47

J

Japanese ambient air quality standard,
    comparison to measured nitrogen
    dioxide levels, 179–180
Japanese brackish lakes, monitoring
    of water pollution using manganese
    concentration in bivalves, 65–75

K

Ketones, saliva-leached components
    of chewing tobacco and simulated urine
    using bioluminescent bacteria, 77–82

L

Lead
    baseline levels in marine mammals,
        265*f*,268
    biomonitoring, 12
    concentrations in blood, urine, and
        placenta from mothers and newborns,
        135–148
    environmental pollution, 2–3
    role of high altitude in environmental
        levels, 4
    use of placenta as biomarker for
        environmental exposure assessment,
        115–131
Legal aspects of human tissue banking,
    *See* Ethical and legal aspects of human
    tissue banking
Lichens, monitoring of trace elements
    in air particulate matter, 46–47

Liver tissues, baseline levels of elements
in marine mammals, 261–269
Long-term monitoring of pollutants, 9–10

**M**

Manganese, use of placenta as biomarker
for environmental exposure assessment,
115–131
Manganese concentration in bivalves as
biomonitor of water pollution
experimental description, 66–68
trace element concentrations, 68–75
Marine mammals, baseline levels
of elements through banked liver tissue
analysis, 261–269
Mercury
baseline levels in marine mammals,
265–268
concentrations in blood, urine, and
placenta from mothers and newborns,
135–148
concern of environmental exposure,
151–152
environmental pollution, 2–3
levels in scalp hair and blood of pregnant
women
experimental description, 154–163
total mercury levels
age, 166–168
fish and seafood consumption,
165–166,167*t*
fish meals per week, 164–165,166*t*
fishing village, 164,165*t*
high risk group, 168–173
location, 161,164,165*t*
residence period, 166–168
use of placenta as biomarker for
environmental exposure assessment,
115–131
Metal(s)
availability to plants and animals,
role of pH, 58–59
human toxicity, 115
physiological regulation mechanisms, 27

Metalloorganic compounds, environmental
pollution, 4
Metallothionein, induction in peripheral-
blood mononuclear leukocytes
of nonsmokers and smokers, 185–186
Methylmercury
biomonitoring, 12
levels in scalp hair and blood of
pregnant women, *See* Mercury
levels in scalp hair and blood
of pregnant women
monitoring using hair analysis, 43–46
toxicity, 43
Microtox test, saliva-leached components
of chewing tobacco and simulated urine
using bioluminescent bacteria, 77–82
Molecular epidemiology in cancer risk
assessment
adducts formed by environmental
carcinogens with DNA and proteins,
207–209
advances in DNA methods, 210–211
changes in gene structure and
expression, 213–214
cytogenetic changes, 215
identification of genetically related
susceptibility to cancer, 215–217
use of proteins, 211–213
Molybdenum concentration, moose as
monitor of environmental changes,
61–62
Moose as monitor of environmental
changes cadmium burden vs.
acidification, 59–60,62*f*
chromium concentration, 61–62
copper concentration, 61–62
experimental description, 58–59
molybdenum concentration, 61–62
Mutagenesis, saliva-leached components
of chewing tobacco and simulated
urine using bioluminescent bacteria,
77–82
Mutatox test, saliva-leached components
of chewing tobacco and simulated
urine using bioluminescent bacteria,
77–82

N

National Biomonitoring Specimen Bank, types of environmental specimens, 261,262t
Nickel
baseline levels in marine mammals, 265f,268
concentrations in blood, urine, and placenta from mothers and newborns, 135–148
Nickel-induced mutations, AS52 cells, 186–187
Nitrogen dioxide, personal exposure, 178–182
North Sea Task Force Monitoring Master Plans, description, 24
Norway, element concentrations in blood, urine, and placenta from mothers and newborns, 135–148
Norwegian–Russian border regions, industrial pollution studies, 135
Nutritional and Health-Related Environmental Studies Section of International Atomic Energy Agency, biomonitor studies, 42–43

O

Organic constituents
determination, 237–239
occurrence in environment, 4
use for pollutant biomonitoring, 10–12
Organohalogens, use for pollutant biomonitoring, 11–12
N-Oxidation, genetic susceptibility to cancer, 216t,217

P

Phosphorus-32 postlabeling test methods, use for cancer risk assessment, 210
Peripheral-blood mononuclear leukocytes of nonsmokers and smokers, metallothionein induction, 185–186

Personal exposure to indoor nitrogen dioxide
comparison to Japanese Ambient Air Quality Standard, 179–180
cooking stove, 180,181f
experimental procedure, 179
heating time, 180,181f
smoking, 179
subjects, 179
urinary hydroxyproline level, 180,182
pH effect
in metal availability to plants and animals, 58–59
in moose as monitor of environmental changes, 57–62
Photobacterium phosphoreum, use for studies on saliva-leached components of chewing tobacco and simulated urine, 77–82
Placenta
diameter and weight vs. developmental stages, 116,117t
element concentrations from mothers and newborns, 135–148
minor and trace elements, 118t–120t
physicochemical characteristics, 116,117t
placental transfer of toxic metals, 123–125
role in assessment of fetal effects of toxic metals, 115
sampling and sample preparation, 119,121–122
transportation of substances from maternal blood, 116,119
use as biomarker of pollutants, 125–131
variations in placental composition, 122
Plants, heavy metal monitoring, 19–28
Pollutant(s)
long-term monitoring, 9–10
monitoring program requirements, 2
monitoring using placenta as biomarker, 115–131
real-time monitoring, 9–10
use of human specimens for monitoring, 10–12

Polychlorinated biphenyl congener
  distribution in Bear Lake sediment
  concentration
    vs. depth, 88,90–91*t*
    vs. site, 88,92–93
  concentration ratio vs. depth, 88,89*f*,92
  experimental description, 84–87
  GC of sediment extract, 88,89*f*
  total volatile solids, 88
Polycyclic aromatic hydrocarbons,
  occurrence in food chain, 2
Polymorphism, genetic, *See* Genetic
  polymorphism
Proteins, use in cancer risk assessment,
  211–213
Proton-induced X-ray emission,
  application, 30
Pseudohermaphroditism phenomenon, 28
*Pyura stolonifera* concentration of
  heavy metals from seawater, 49


Q

Quality assurance
  biomonitoring programs, 47
  of heavy metal monitoring by plants
    and animals, 24
Quality management in environmental
  analytical chemistry, role of
  environmental specimen banking,
  251–252


R

Reaction indicator, definition, 20,22*f*
Real-time monitoring of pollutants,
  9–10
Renal disease, gastrointestinal absorption
  of aluminum, 185
Rubidium, baseline levels in marine
  mammals, 265*f*,268
Russia, element concentrations in blood,
  urine, and placenta from mothers and
  newborns, 135–148

S

Saliva-leached components of chewing
  tobacco and simulated urine using
  bioluminescent bacteria
  acute toxicity experiments, 78–81
  genotoxicity of nicotine and cotinine,
    81–82
  ketone concentration, 79–80
  saliva testing, 81
  test media, 78–79
*Salmonella typhimurium* reverse mutation
  assay, description, 78
Scalp hair of pregnant women, mercury
  and methylmercury levels, 151–173
Scanning electron microscopy, trace
  metal distribution analysis of annual
  growth rings of trees, 30–40
Sea squirt, *See Pyura stolonifera*
Secondary ion MS to analyze trace metal
  distribution of annual growth rings
  of trees
  experimental description, 31–33
  scanning electron microscopy
    copper–zinc rich feature, 34,38*f*
    sodium–potassium rich feature, 34,38*f*
  secondary ion MS
    analysis of paired samples subjected
      to inductively coupled plasma MS
      analysis, 39–40
    depth profiles, 34,36*t*
    ion counts for early vs. late wood, 34,39*t*
    ion images for 125 μm, 34,37*f*
    typical spectrum, 33,35*f*
Selenium
  baseline levels in marine mammals,
    265–268
  body burden concentrations in humans
    following low environmental
    exposure, 105–113
  environmental pollution, 2–3
  role of high altitude in environmental
    levels, 4
  use of placenta as biomarker for
    environmental exposure assessment,
    115–131

Sewage pollution, use of *Pyura stolonifera* as biomonitor, 49–55
Silicon, environmental pollution, 2–3
Silver content of *Pyura stolonifera* as indicator of sewage pollution
  copper concentrations, 52,54*t*
  experimental procedure, 50,51*f*,52*t*
  heavy metal concentrations, 52,54*t*
  reasons for use of *Pyura*, 50
  silver concentrations, 50,52,53*f*
  silver–uranium ratio vs. size of animals, 54–55
Smoking effect
  in metallothionein induction in peripheral-blood mononuclear leukocytes, 185–186
  in personal exposure to indoor nitrogen dioxide, 179
Specimen banking, *See* Environmental specimen banking
St. Lawrence River, environmental chemical contaminants, 96–102
Subacute toxicity, definition, 78
Subchronic toxicity, definition, 78
Sydney, Australia, silver content of *Pyura stolonifera* as indicator of sewage pollution, 49–55

T

Tissue(s), ethical and legal aspects of banking, 254–260
Tissue polypeptide antigen, biomarker for cancer risk assessment, 201–203
Titanium, environmental pollution, 2–3
Tobacco, saliva-leached components, 77–82
Toluene
  biotransformation to hippuric acid via low $K_m$ aldehyde dehydrogenase, 190–194
  exposure, 190
Total mercury levels, scalp hair and blood of pregnant women, 151–173
Toxic substances, inherited and acquired susceptibility biomarkers, 184–188
Toxicity to humans, types and tests to determine, 78

Trace elements
  body burden concentrations in humans following low environmental exposure, 105–113
  occurrence in food chain, 2
  role of surroundings on content, 65
Trace metal distribution analysis of annual growth rings of trees, secondary ion MS, 30–40
Tracers, function, 77–78
Tree(s), secondary ion MS for trace metal distribution analysis of annual growth rings, 30–40

U

Urinary hippuric acid, estimation of toluene exposure, 190
Urinary hydroxyproline level, role in personal exposure to indoor nitrogen dioxide, 180,182
Urine
  components, 77–82
  element concentrations from mothers and newborns, 135–148

V

Vanadium
  baseline levels in marine mammals, 265–268
  environmental pollution, 2–3
*Vibrio fischeri*, use for studies on saliva-leached components of chewing tobacco and simulated urine, 77–82

W

Water pollution, manganese concentration in bivalves as biomonitor, 65–75

Y

Yamatoshijimi, *See* Bivalves